ANALYTICAL CHEMISTRY OF THE ELEMENTS

ANALYTICAL CHEMISTRY OF ALUMINUM

Analytical Chemistry of the Elements

Series prepared by the Vernadskii Institute of Geochemistry and Analytical Chemistry
USSR Academy of Sciences
A. P. VINOGRADOV, *Editor*
English translation by ISRAEL PROGRAM FOR SCIENTIFIC TRANSLATIONS, JERUSALEM

ANALYTICAL CHEMISTRY OF ALUMINUM
ANALYTICAL CHEMISTRY OF BERYLLIUM
ANALYTICAL CHEMISTRY OF BORON
ANALYTICAL CHEMISTRY OF COBALT
ANALYTICAL CHEMISTRY OF FLUORINE
ANALYTICAL CHEMISTRY OF GALLIUM
ANALYTICAL CHEMISTRY OF MOLYBDENUM
ANALYTICAL CHEMISTRY OF NEPTUNIUM
ANALYTICAL CHEMISTRY OF NICKEL
ANALYTICAL CHEMISTRY OF NIOBIUM AND TANTALUM
ANALYTICAL CHEMISTRY OF PLUTONIUM
ANALYTICAL CHEMISTRY OF POTASSIUM
ANALYTICAL CHEMISTRY OF PROTACTINIUM
ANALYTICAL CHEMISTRY OF RUTHENIUM
ANALYTICAL CHEMISTRY OF SELENIUM AND TELLURIUM
ANALYTICAL CHEMISTRY OF TECHNETIUM, PROMETHIUM, ASTATINE
  AND FRANCIUM
ANALYTICAL CHEMISTRY OF THALLIUM
ANALYTICAL CHEMISTRY OF THORIUM
ANALYTICAL CHEMISTRY OF URANIUM
ANALYTICAL CHEMISTRY OF YTTRIUM AND THE LANDTHANIDE
  ELEMENTS
ANALYTICAL CHEMISTRY OF ZIRCONIUM AND HAFNIUM

ANALYTICAL CHEMISTRY OF

ALUMINUM

V. N. TIKHONOV

Translated by J. SCHMORAK

A HALSTED PRESS BOOK

JOHN WILEY & SONS
New York · Toronto

ISRAEL PROGRAM FOR SCIENTIFIC TRANSLATIONS
Jerusalem · London

546.6736
T568

Sole distributors for the Western Hemisphere and Japan

HALSTED PRESS, a division of
JOHN WILEY & SONS, INC., NEW YORK

Distributors for the U.K., Europe, Africa and
the Middle East

JOHN WILEY & SONS, LTD., CHICHESTER

Distributed in the rest of the world by

KETER PUBLISHING HOUSE, JERUSALEM

Library of Congress Catalog Card Number 72 4102
ISBN 0 7065 1223 5 IPST, Jerusalem
ISBN 0 470 86787 6 Halsted/Wiley, N.Y.
IPST cat. no. 22038

This book is a translation from Russian of
ANALITICHESKAYA KHIMIYA ALYUMINIYA
Izdatel'stvo "Nauka"
Moskva 1971

Printed in Israel

FOREWORD

The Vernadskii Institute of Geochemistry and Analytical Chemistry of the USSR Academy of Sciences has begun the publication of a series of monographs on the analytical chemistry of individual elements. This series, *Analytical Chemistry of the Elements,* will include about fifty volumes, and it is planned to complete the publication during the coming five years. The necessity for such a project has been felt for a long time. We also have at our disposal the accumulated experience of numerous laboratories which now can (and should) be summarized. In this way the present series originated, the first of its kind. The analytical chemistry of any element with its different compounds is at present extremely varied; this is due to the complexity of the modern materials investigated and the wide range of concentrations determined, as well as to the variety of the methods employed.

With this in mind, a general plan has been worked out for the present series, both with respect to the contents and to the presentation of the material.

The monographs contain general information on the properties of the elements and their compounds, followed by a discussion of the chemical reactions which are the basis of the analytical work. The physical, physicochemical, and chemical methods for the quantitative determination of the element are given in the following order: first, the analysis of raw materials, next the analysis of the typical semimanufactured products, and last, that of the finished products — metals or alloys, oxides, salts, and other compounds and materials. The underlying principles are always explained; whenever necessary, the exhaustive description of the entire analytical procedure is given. Due attention is paid to rapid analytical methods. A separate place is allotted to techniques for the determination of impurities in high purity materials.

Particular stress is placed on the accuracy and the sensitivity of the various methods, in view of the general tendency toward increased sensitivity of determination of traces of impurity elements.

The monographs contain an exhaustive and up-to-date bibliography. They are intended for a wide circle of chemists: in the first place, for the analysts of research institutes and industrial laboratories of various branches of the national economy, and also for teachers and students of chemistry in institutions of higher learning.

The most prominent Soviet experts participated in the preparation of the monographs, drawing upon their own extensive experience in the analytical chemistry of the element in question.

Each volume of the series will be published separately as soon as it has been made ready for publication. [A list of currently available volumes will be found facing the title page. Further titles are in preparation.]

We invite all our readers to send us their comments and criticisms on the monographs.

Editorial Board

TABLE OF CONTENTS

PREFACE

Aluminum is determined during the study of different natural and industrial products, in production control, soil analysis, etc. The analytical problems involved are becoming increasingly difficult, since the amounts of aluminum which must be determined in pure metals and other materials are decreasing continually, so that reliable, highly sensitive methods must be employed. If the purpose of the determination is industrial process control, it should be rapid, and therefore interfering elements must be effectively masked or rapidly separated.

Considerable advances have been made in the analytical chemistry of aluminum during the past 10–15 years. The determination of aluminum by the modern volumetric technique of complexometry is certainly the most important development in this field. Moreover, new, highly sensitive organic reagents have been proposed for the photometric determination of aluminum, and various methods for the separation of aluminum from interfering elements have been developed. The number of publications on the determination of aluminum is now several thousand. But in spite of this large number, there is only one monograph on the analytical chemistry of aluminum, which was compiled by Fischer and forms part of the handbook by Fresenius and Jander [733]. This monograph first appeared in 1942, and is now largely obsolete. The monographs by Přibil "Complexometric and Chemical Analysis" [347], and Sandell "Colorimetric Determination of Traces of Metals" [360], contain descriptions of complexometric and photometric methods for determining aluminum, but the many highly important methods published during the past 8–10 years are not mentioned

The present monograph is a systematic presentation of all the known methods for the determination of aluminum, taken from the published literature and from our own experience. It describes chemical, physicochemical and physical methods for the determination of aluminum, but the main stress is placed on rapid methods, i.e., those which involve the least number of separations. Unfortunately, not all of these can be practiced in the laboratory because the reagents required are not yet commercially available, but probably this difficulty will be overcome in time.

For reasons of space, the chapter on physical methods for the determination of aluminum had to be considerably abbreviated, even though a special monograph could be written on this subject.

The literature has been reviewed up to October 1968; some of the more recent literature has also been included. Publications which are no longer of interest, and those which deal with the applications of known methods to new substrates, have been omitted.

The author wishes to express his gratitude to Professor A. I. Busev, Dr. Chem. Sci., for his assistance in the compilation of the present monograph, and to the reviewers P. Ya. Yakovlev, Dr. Chem. Sci., and S. B. Savvin, Dr. Chem. Sci. for a number of comments, which led to improvements in the monograph.

The Author

CHAPTER I

Physicochemical and Analytical Description of Aluminum and its Compounds

OCCURRENCE IN NATURE

Aluminum is the third most abundant element in the Earth's crust. According to A. P. Vinogradov, its content in the lithosphere is 8.80%. It occurs in nature in the combined state only, and forms part of 270 different minerals. The most widely occurring aluminum minerals are double silicates (felspars, micas, etc.), and clays, the products of erosion of these minerals. The more important double silicates include potassium felspar or orthoclase $K[AlSi_3O_8]$, sodium felspar or albite $Na[AlSi_3O_8]$, calcium felspar or anorthite $Ca[Al_2Si_2O_8]$, plagioclase (isomorphic mixtures of calcium and sodium felspars: oligoclase, andesite, labradorite); micas: biotite, muscovite, zinnwaldite and lepidolite. Nepheline $Na[AlSiO_4]$ and leucite $K[AlSi_2O_6]$ have a composition close to that of felspars. Among the known calcium and aluminum double silicates are zoisite, epidote and vesuvianite, and also cordierite, a double sulfate of magnesium and aluminum. Aluminum silicate Al_2SiO_5 is found in a number of minerals: kyanite, sillimanite and andalusite. Topaze, $Al_2(OH,F)_2[SiO_4]$, is one of the fluorine-containing alumosilicates.

Aluminum oxide occurs as corundum and emery. The most important source of aluminum, bauxite, consists of the minerals boehmite and diaspore, $AlOOH$, and hydrargillite (gibbsite), $Al(OH)_3$. Another important aluminum mineral is cryolite, Na_3AlF_6.

1

Table 1. Ore-forming aluminum minerals [505]

Mineral	Formula	Al_2O_3, %	Accompanying minerals	Industrial importance
Hydrargillite	$Al(OH)_3$; $\leqslant 2\%$ Fe_2O_3 and $\leqslant 0.006\%$ Ga_2O_3	65.4	Diaspore, boehmite, iron oxides	Major aluminum ore
Boehmite	$AlOOH$, and SiO_2, Fe_2O_3 and Ga_2O_3 as impurities	84.97	Hydrargillite, diaspore, iron oxides, zeolites	"
Diaspore	$AlOOH$; $\leqslant 7\%$ Fe_2O_3; $\leqslant 5\%$ Mn_2O_3, Cr_2O_3, $\leqslant 4\%$ SiO_2, $\leqslant n \cdot 10^{-2}\%$ Ga_2O_3	84.97	In bauxites with hydrargillite and boehmite; in metamorphic rocks with corundum, disthene, hematite, rutile, etc.	"
Alunite	$KAl_3(SO_4)_2(OH)_6$	37.0	Hydrargillite, kaolin, alkali felspars	Of limited value
Kyanite (disthene)	Al_2SiO_5	63.1	Andalusite, staurolite, corundum, turmaline, rutile, etc.	Important raw material
Andalusite	Al_2SiO_5	63.1	Corundum, muscovite, diaspore, pyrophyllite, etc.	"
Sillimanite	Al_2SiO_5	63.1	Andalusite, corundum, spinel, cordierite, etc.	"
Nepheline	$Na(AlSiO_4)$	35.0	Aegirite, albite, microcline, cancrinite, sodalite, zeolites, sphene, ilmenite, etc.	"
Leucite	$K(AlSi_2O_6)$	23.5	Alkali pyroxenes, nepheline, orthoclase, sericite	Of limited value
Kaolinite	$Al_4(Si_4O_{10})(OH)_8$	39.5	Usually forms monomineral masses	Potential raw material for the manufacture of aluminum

Table 1 lists the minerals that form part of ores used for the production of metallic aluminum; bauxites are the most important. Nepheline syenites and their varieties (urtites, sillimanite and kyanite shales and alunite rocks) are also employed industrially.

The first step in the manufacture of metallic aluminum is the extraction of alumina, Al_2O_3, from the ores. The alumina is then dissolved in molten cryolite (6–8% Al_2O_3 and 92–94% Na_3AlF_6) and electrolyzed.

Metallic aluminum was first obtained by Oersted in 1825.

Aluminum is highly important in the national economy. Aluminum-base alloys are an important material for the construction of airplanes. Aluminum alloys are also employed in other branches of industry. Aluminum as an alloying element forms part of many important alloys based on copper, magnesium, titanium, nickel, zinc, and iron. Aluminum is used to deoxidize steel, and in the aluminothermal production of certain metals. At the present time the annual world production of aluminum outside the Soviet Union is about 5 million tons.

PHYSICAL AND CHEMICAL PROPERTIES OF ALUMINUM

Aluminum is a silvery-white, light metal, and a good conductor of heat and electricity; it is plastic and can be mechanically worked. It has a face-centered cubic lattice with $a = 4.0494$ Å. The most important physical constants of aluminum are given below [206, 366]:

Density at 20°C, g/cm³	2.6989
M.p., °C	660
B.p., °C	2500
Heat of fusion, cal/g-atom	2520
Atomic heat capacity at 0°C, cal/g-at · °C	5.99
Thermal conductivity at 20°C, cal/cm · sec · °C	0.520
Electrical conductivity, ohm⁻¹ · cm⁻¹	$3.82 \cdot 10^{-3}$
Thermal neutron capture, cross section, barn	0.215
Atomic radius, Å	1.43
Ionic radius of Al^{3+} ion, Å	0.57
Ionization potentials, eV:	
$Al^\circ \rightarrow Al^+$	5.984
$Al^+ \rightarrow Al^{2+}$	18.82
$Al^{2+} \rightarrow Al^{3+}$	28.44
$Al^\circ \rightarrow Al^{3+}$ ionization energy, kcal/g-atom	83
Electron affinity, kcal/g-atom	12

Aluminum consists of the stable isotope Al^{27} (100%). Isotopes with atomic masses of 24, 25, 26, 28 and 29 have been artificially obtained (Table 2).

Table 2. Radioactive isotopes of aluminum [361, 374]

Isotope	Type of radiation	Half-life	Radiation energy of particles, MeV	Reaction of formation
Al^{24}	β^+; α ($\sim 10^{-2}$%); γ	2.1 sec	~ 8.5 (β^+); 2 (α)	Mg (p, n)
Al^{25}	β^+	7.6 sec	3.24	Mg (p, n); Mg (p, γ)
Al^{26} *	β^+	6.7 sec	3.20	Na (α, n); Mg (p, γ); Mg (p, n); Mg (d, n); Al (γ, n)
Al^{26}	β^+ (84%); EC (16%); γ	$8 \cdot 10^5$ years	1.17	Na (α, n); Mg (p, γ); Al (p, pn); Si (d, α)
Al^{28}	β^-; γ	2.3 min	2.865	O (N, $2p$); Mg (α, p); Mg $\to \beta^-$; Al (d, p); Al (n, γ); Si (n, p); Si (γ, p); P (n, α)
Al^{29}	β^-; γ	6.6 min	2.7 (~ 70%); 1.4 (~ 30%)	Mg (α, p); Al ($\alpha, 2p$); Al (He^3, p); Si (n, p); Si (γ, p); P ($\gamma, 2p$)

* Nucleus in excited isomeric state.

Aluminum is located in the main subgroup of Group III in the Periodic Table. Its atomic number is 13, and its atomic mass 26.9815. The electron configuration of the aluminum atom in the ground state is $1s^2 2s^2 2p^6 3s^2 3p^1$. The three s- and p-electrons of the outermost shell are the valence electrons, so that the maximum valence displayed by aluminum is +3.

At high temperatures aluminum can assume the valences of 1, or more rarely, 2. These valences are unstable, and aluminum is trivalent in all its stable compounds.

Aluminum has a strong affinity for oxygen, and is therefore always covered by an oxide film, which protects it from further oxidation. This oxide film renders aluminum stable to water even at high temperatures. In the potential series aluminum is close to the most active metals (alkali and alkaline-earth metals); its normal electrode potential is –1.66 V in acid solution and –2.35 V in alkaline solution. Nevertheless, owing to the presence of the protective oxide film, aluminum is practically insoluble in very dilute acids and in conc. HNO_3, and dissolves very slowly in dilute and concentrated sulfuric acid. It gradually dissolves in nitric and sulfuric acid solutions of medium concentration. Pure aluminum is very slowly dissolved in HCl, but aluminum with other metals as impurities dissolves quite readily. Aluminum is inert in acetic and phosphoric acids.

Aluminum readily dissolves in solutions of alkalis with the formation of aluminates and the liberation of hydrogen. The metal dissolves as a result of the removal of the protective oxide film from the surface by hydroxyl ions:

$$Al_2O_3 + 2OH^- + 3H_2O = 2\,[Al(OH)_4]^-,$$

and then the metal without the film dissolves as follows:

$$2Al + 6H^+ \text{ (from water)} = 2Al^{3+} + 3H_2,$$

$$2Al^{3+} + 8OH^- = 2\,[Al(OH)_4]^-.$$

Finely pulverized aluminum burns in the air with the formation of Al_2O_3. Because of its high affinity to oxygen, aluminum displaces many metals from their oxides; this reaction is the principle of aluminothermy, by which a number of metals are obtained. Aluminum reacts with chlorine with the evolution of large amounts of heat; it reacts with liquid bromine, but with iodine only when heated. At room temperature aluminum does not react with fluorine owing to the formation of a film of AlF_3, but the reaction proceeds vigorously at high temperatures. Aluminum reacts with sulfur at red heat with the formation of the sulfide Al_2S_3. Above $800°C$ it reacts with nitrogen to form the nitride AlN, and at $1400°C$ it reacts vigorously with carbon to form the carbide Al_4C_3. When heated, aluminum reacts explosively with selenium and tellurium, with the formation of Al_2Se_3 and Al_2Te_3. Powdered aluminum burns in CO_2 and CO at high temperatures to form Al_2O_3 and carbon. When a mixture of aluminum and phosphorus powders is heated, the phosphide AlP is formed. Aluminum does not react with hydrogen directly; the hydride $(AlH_3)_x$ is prepared by indirect methods.

COMPOUNDS OF ALUMINUM

Aluminum oxide

Aluminum oxide plays an important part in the analytical chemistry of aluminum, since this is the gravimetric form in analysis. It is the product of the ignition of aluminum hydroxide and other aluminum compounds.   It exists in three polymorphic forms: α-Al_2O_3, β-Al_2O_3 and γ-Al_2O_3 [761, 1107].

α-Alumina forms white crystals with a hexagonal rhombohedral lattice, $a=5.13\,\text{Å}$, $\alpha=55.16°$, specific gravity 3.96, m.p. 2050°C, b.p. above 3000°C. It can be prepared by igniting aluminum hydroxide or an aluminum salt at 900–1200°C or above. It occurs in nature as ruby, corundum and sapphire.

β-Alumina forms hexagonal crystals with a specific gravity of 3.30. This modification is stable at very high temperatures. It is formed when molten α-alumina is slowly cooled. At 1500–1800°C α-alumina is completely converted to β-alumina.

γ-Alumina is a cubic modification with a specific gravity of 3.40. It is formed when natural or synthetic aluminum hydroxide or aluminum salts are heated to 600–900°C. Above 1000°C γ-alumina is converted to α-alumina; this conversion is complete at 1200°C.

A few unstable modifications of alumina have also been described.

The solubility of the oxides in acids depends on the manner of their preparation. α-Alumina which has been obtained by igniting the hydroxide above 1000°C is practically insoluble in acids and alkalis. γ-Alumina is distinguished from the other forms by its higher degree of dispersion and hygroscopicity, and its higher solubility in acids. The solubility of alumina in water at 20°C, found by the conductometric method, is 1.04 mg Al_2O_3/ liter or $1.02 \cdot 10^{-5}$ moles per liter [1108], while the determination by potentiometric titration gave the value of $0.96 \cdot 10^{-5}$ mole/liter [602]. Alumina prepared at higher temperatures is less hygroscopic.

Aluminum hydroxide

Aluminum hydroxide may be prepared in amorphous and crystalline forms.

When precipitated from acid solutions in the cold by a small excess of ammonia, the α-hydroxide $Al(OH)_3$ separates out as a white, flocculent, amorphous gel. This hydroxide is highly reactive and unstable. It is readily soluble in dilute solutions of HCl and NaOH, and is highly adsorbent. When stored under water, it is spontaneously converted to other forms. This process may take two simultaneous (parallel) courses: at 70°C the β-hydroxide is formed rapidly and directly, while at 20°C it is formed slowly through the intermediate α/β form.

The β-form of the hydroxide is a polyhydroxide, i.e., the product of the association of a number of molecules accompanied by the liberation of water. The β-hydroxide is almost inert, and its absorptive capacity is insignificant. The γ-hydroxide is obtained by dispersing the crystalline trihydrate bayerite produced in the β-hydroxide.

The γ-hydroxide is also chemically inert. The β-hydroxide is unstable, and the aging of the precipitated hydroxide is accompanied by a gradual conversion to the crystalline boehmite AlOOH (γ-metahydroxide). If the precipitation is effected from hot solutions, boehmite with admixed amorphous hydroxide is formed immediately. The isomer of boehmite, diaspore (α-metahydroxide), occurs in nature. Both isomers form rhombic crystals. α-AlOOH is converted to α-Al$_2$O$_3$ at 350–420°C, while γ-AlOOH is converted to α-Al$_2$O$_3$ above 600°. If the solution is made more alkaline, the crystalline trihydrate, bayerite (Al$_2$O$_3 \cdot$ 3H$_2$O, monoclinic crystals), is formed. The isomer of bayerite, hydrargillite, is found in nature.

If a large excess of ammonia is employed in the precipitation, the product consists not only of the α-hydroxide, but also of the polyhydroxide; a few hydroxide molecules combine to form a chain, with the liberation of water. The formation of the simplest polyhydroxide in the series, the dihydrate, can be represented as follows:

$$2Al(OH)_3 + H_2O \rightarrow Al(OH)_2 \cdot O \cdot Al(OH)_2 \text{ (or } Al_2O_3 \cdot 2H_2O).$$

These compounds are unstable, and when kept under water or in an ammoniacal solution they are converted to the stable trihydrate (bayerite):

$$Al(OH)_2 \cdot O \cdot Al(OH)_2 + H_2O = 2Al(OH)_3.$$

The crystalline monohydrate (boehmite) and the trihydrate (bayerite, hydrargillite) are the final products of the slow aging of the gel. The transformations of the α-hydroxide can be represented by the following diagram:

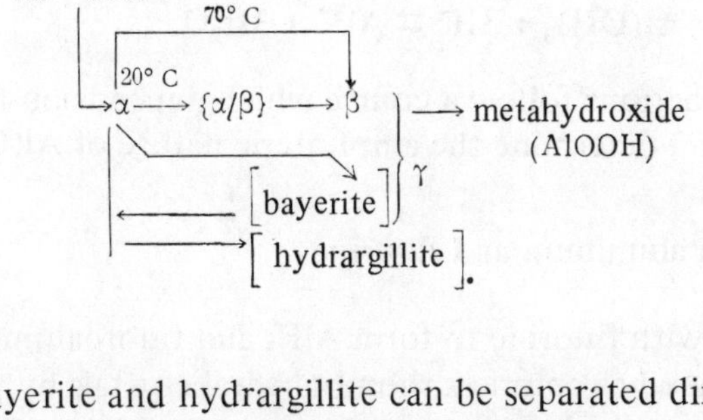

Crystalline bayerite and hydrargillite can be separated directly by passing CO$_2$ through a solution of an alkali aluminate (if the gas is passed at a rapid rate, bayerite is formed; if at a slow rate, hydrargillite is obtained):

$$2[Al(OH)_4]^- + CO_2 = 2Al(OH)_3 + CO_3^{2-} + H_2O.$$

Boehmite and bayerite are very stable chemically. Boehmite is practically insoluble in concentrated alkali solutions and is only sparingly soluble in HCl. Bayerite which has been separated from an aluminate solution is sparingly soluble in cold HCl, HNO_3 and H_2SO_4, and is solubilized only on heating; the hydroxide is soluble in hot 30% NaOH.

The following literature data are available on the solubility product of $HAlO_2$: $4 \cdot 10^{-13}$ at 15°C [889], $3.7 \cdot 10^{-15}$ at 25°C [1172], and $1.3 \cdot 10^{-14}$ at 18°C [734]. The solubility product of $Al(OH)_3$ has been reported as $1.06 \cdot 10^{-33}$ [805] and as $5.1 \cdot 10^{-33}$ [376].

Aluminum hydroxide is an amphoteric compound and may react with both acids and bases with the formation of salts. It is therefore commonly considered to be both acidic and basic. The dissociation constant of the so-called monobasic aluminic acid has been calculated as $K = 6.3 \cdot 10^{-13}$ at 25°C [1279]. However, according to Remy [1107], the formation of salts between aluminum hydroxide and a base is not due to the formation of H^+-ions by dissociation, but to the continued addition of OH^- ions to $Al(OH)_3$. According to Remy, the reaction

$$Al(OH)_3 = [AlO(OH)_2]^- + H^+$$

cannot take place.

Thus, in alkaline solutions $Al(OH)_3$ behaves as an acid anhydride rather than as an acid, and forms hydroxo salts such as potassium hydroxoaluminate $K[Al(OH)_4]$.

The H^+-ions are able to bind the OH^--ions in $[Al(OH)_4]^-$-ions. The equilibria which are involved are:

$$[Al(OH)_4]^- + H^+ \rightleftarrows Al(OH)_3 + H_2O;$$

$$Al(OH)_3 + 3H^+ \rightleftarrows Al^{3+} + 3H_2O.$$

These equilibrium reactions follow a course which depends on the concentration of H^+-ions, and determine the amphoteric nature of $Al(OH)_3$.

Compounds between aluminum and fluorine

Aluminum reacts with fluorine to form AlF_3 and fluoroaluminates. The fluoride can be prepared as colorless rhombohedral crystals by passing hydrogen fluoride through aluminum or alumina at red heat. When heated, it sublimes without melting. It is sparingly soluble in water, 0.559 g AlF_3 per 100 g of water at 25°C [620]. Aluminum trifluoride is chemically very stable: it is resistant to boiling KOH and NaOH solutions, and to cold and hot acids, and decomposes when fused with Na_2CO_3 + NaOH. It forms a number of hydrates: $AlF_3 \cdot 9H_2O$, $AlF_3 \cdot 3H_2O$, $AlF_3 \cdot 2H_2O$, etc. The trihydrate is the most stable at room temperature; the other hydrates are converted to the trihydrate when brought into contact with water.

Aluminum trifluoride reacts with alkali fluorides to form complex fluoroaluminates of the type $MAlF_4$, M_2AlF_5 and M_3AlF_6, where M is an alkali metal. Fluoroaluminates are white, crystalline powders, which are sparingly soluble in water. The most important fluoroaluminate is cryolite, Na_3AlF_6. The formation of cryolite is the principle of one of the more important methods for the gravimetric determination of aluminum. Cryolite occurs in nature, and can be synthesized. It is dimorphous (monoclinic and octahedral); m.p. 1000°C.

The solubility of artificially prepared cryolite in water is 0.041 and 0.061 g per 100 g of water at 12 and 15°C, respectively [761].

Fluoride complexes of aluminum are very stable; their instability constants are tabulated below [67]:

Ion	K_{inst}	Ion	K_{inst}
$[AlF_6]^{3-}$	$1.44 \cdot 10^{-20}$	$[AlF_3]$	$1.0 \cdot 10^{-15}$
$[AlF_5]^{2+}$	$64.3 \cdot 10^{-20}$	$[AlF_2]^+$	$7.1 \cdot 10^{-12}$
$[AlF_4]^-$	$1.8 \cdot 10^{-18}$	$[AlF]^{2+}$	$7.4 \cdot 10^{-7}$

Aluminum chloride. Anhydrous aluminum chloride, $AlCl_3$, forms colorless monoclinic crystals. Even at ordinary temperatures the compound is markedly volatile, and sublimes at 183°C without melting. The melting point (192.6°C) can only be attained at elevated pressure. Near the sublimation temperature the vapor of aluminum chloride is dimeric, and its formula is Al_2Cl_6; near 800°C the dimer is completely decomposed into single molecules. $AlCl_3$ is soluble in many organic solvents. It fumes in the air owing to hydrolysis. It is readily soluble in water; at 25°C the solubility is 44.38 g $AlCl_3$ in 100 g of water. Anhydrous $AlCl_3$ hydrolyzes in water, and therefore its aqueous solution is strongly acid. A 0.1 N solution is 2% hydrolyzed, and a 0.001 N solution is 4.5% hydrolyzed [1107].

When aqueous solutions of $AlCl_3$ are evaporated, the hexahydrate is formed as colorless, deliquescent crystals. The hexahydrate is readily soluble in water. The table below [1222] gives its solubility at different temperatures:

Temperature, °C	Solubility, wt %	Temperature, °C	Solubility, wt %
0	30.48	45	32.17
5	30.82	65	32.32
10	31.66	98	33.23
30	31.96		

Due to hydrolysis, aqueous solutions of $AlCl_3 \cdot 6H_2O$ are acid to litmus. The pH of these solutions varies between 2.5 and 3.7 as the concentration of the hexahydrate varies from 0.01 M to 0.5 M. The hydrolysis is accompanied by the formation of oxychlorides. The addition of neutral salts, such as KCl, suppresses the hydrolysis.

When HCl is added, the solubility of aluminum chloride in water decreases. The solubility at 25°C is as follows [761]:

Conc. of HCl, %	Solubility of $AlCl_3$, wt %
5.09	27.98
19.43	10.11
40.98	0.98

The solubility is even lower in hydrochloric acid saturated with ether. According to Gooch and Havens [766], 100 g of an ethereal solution of HCl dissolve 4 mg of $AlCl_3 \cdot 6H_2O$. This forms the principle of the separation of aluminum.

Aluminum oxychlorides with composition Al_2OHCl_5, $Al_2(OH)_2Cl_4$, $Al_2(OH)_3Cl_3$, $Al_2(OH)_4Cl_2$, $Al_2(OH)_5Cl$ are known. During the peptization of $Al(OH)_3$ gel washed free of ammonium ions with small amounts of HCl, the oxychlorides $Al(OH)Cl_2$, $Al(OH)_2Cl$ (or AlOCl) are formed.

Aluminum nitrate

When aqueous solutions of aluminum nitrate are evaporated, the compound is obtained as the nonahydrate $Al(NO_3)_3 \cdot 9H_2O$. Hydrates with 8.6 and 4 molecules of water are also known. The nonahydrate forms colorless orthorhombic crystals which deliquesce in the air. At 73.5°C the hydrate is converted to $Al(NO_3)_6 \cdot 6H_2O$, which forms a basic nitrate at 140°C, and is converted to Al_2O_3 at 200°C.

The nonahydrate is readily soluble in water and alcohol; at 25°C 100 g of water dissolve 63.7 g of nitrate, calculated as anhydrous salt. Aluminum nitrate hydrolyzes in water, and therefore its solutions have a pH varying between 2.5 and 3.7 in the concentration range between 0.01 and 0.5 M, respectively. When solutions of nitrates are heated, the basic nitrates $Al(OH)(NO_3)_2$, $Al(OH)_2NO_3$, $Al_2(OH)_3(NO_3)_3$, etc. are obtained.

Aluminum sulfate

Anhydrous aluminum sulfate is a white powder; when heated it decomposes with liberation of SO_2 or SO_3. The temperature of the complete decomposition of sulfates has been variously given as 960, 800, 770, 760 and 750°C. Hydrates with 27, 18, 16, 10 and 6 molecules of water are known. At ordinary temperatures, octahedra of $Al_2(SO_4)_3 \cdot 18H_2O$ crystallize from aqueous solutions. This hydrate is readily soluble in water: 100 g of water dissolve 107.35 g of the hydrate (i.e., 36.15 g of the anhydrous salt) at 20°C [761]. When the hydrate is heated, it loses the water of crystallization without melting. According to the data of different authors, the complete dehydration of this hydrate occurs between 200 and 450°C. Owing to hydrolysis, solutions of aluminum sulfate have an acid reaction, and the pH varies between 2.8 and 3.7 in the concentration range of 0.01– 0.5 M. The hydrolysis proceeds according to the reaction

$$Al_2(SO_4)_3 + 2H_2O \rightleftarrows Al_2(OH)_2(SO_4)_2 + H_2SO_4.$$

In more dilute solutions the compound is hydrolyzed to the second stage:

$$Al_2(SO_4)_3 + 4H_2O \rightleftarrows Al_2(OH)_4SO_4 + 2H_2SO_4.$$

The products of hydrolysis can be isolated in crystalline form; a number of such basic sulfates are known.

The hydrolysis of aluminum sulfate must be taken into account when aluminum hydroxide is precipitated from solutions of sulfates.

Aluminum sulfate tends to form double salts with alkali metal sulfates and ammonium sulfate, the so-called alums. Their general formula is $M^I Al(SO_4)_2 \cdot 12H_2O$ (M^I = K, Na, NH_4^+, etc.). Alums are readily soluble in water and form octahedral crystals.

Aluminum perchlorate

The anhydrous salt has not been isolated; hydrates with 15, 12, 9 and 6 molecules of water are known. The hydrates are readily soluble in water. At 14°C, 100 g of water dissolve 564 g of $Al(ClO_4)_3 \cdot 9H_2O$.

Aluminum phosphate

Aluminum phosphate separates out as a sparingly soluble, flocculent precipitate when aluminum salts react with soluble phosphates. If the pH of the precipitation is not higher than 4.5, the composition of the resulting precipitate corresponds to the formula $AlPO_4 \cdot xH_2O$ [971], while basic salts are formed at higher pH values. When aluminum phosphates are heated they lose water, and at 1200–1300°C the water is completely lost. The minimum solubility of aluminum phosphate occurs at pH 4.07–6.93. The solubility product is $3.87 \cdot 10^{-11}$ [399] or $1.64 \cdot 10^{-20}$ [170]. Aluminum phosphate is readily soluble in HCl and HNO_3, and is sparingly soluble in CH_3COOH.

The hydrates $AlPO_4 \cdot 3.5H_2O$ and $AlPO_4 \cdot 2H_2O$ have been prepared. The following acid aluminum phosphates are known: $Al(H_2PO_4)_3$, $Al_2(HPO_4)_3 \cdot 8H_2O$, $Al_2H_3(PO_4)_2 \cdot 2.5H_2O$ [sic]. When aluminum is precipitated from a solution of chloride by $Na_4P_2O_7$, aluminum pyrophosphate is formed as a white, gelatinous precipitate which, after drying at 110°C, has the composition $Al_4(P_2O_7)_3 \cdot 10H_2O$, and dissolves in excess precipitant with the formation of a complex compound.

Sodium aluminate

Sodium aluminate, $NaAlO_2$, is a white, enamel-like mass, which decomposes in the air and is readily soluble in water. It is formed when Al_2O_3 is fused with NaOH or Na_2CO_3. Aluminum hydroxide which has been precipitated by NaOH forms the hydroxyaluminate with excess precipitant:

$$Al(OH)_3 + NaOH = Na[Al(OH)_4].$$

When solutions of aluminates are diluted, a hydrolytic decomposition occurs, with the separation of the crystalline $Al(OH)_3$ (hydrargillite).

Compounds with organic acids

Compounds of aluminum with certain organic acids (benzoic, succinic, cinnamic, diphenic) are employed in the gravimetric determinations of aluminum. Many other organic acids (see below) form complex compounds with aluminum and interfere with its quantitative determination, and the analyst must take this fact into account.

The list of organic acids which follows is arranged in the order of increasing complexity of the acid structure.

Aluminum acetates. Solid aluminum acetate, $Al(CH_3COO)_3$, cannot be obtained from aqueous solutions, since it is strongly hydrolyzed. It may be obtained in the anhydrous state by moderately heating $AlCl_3$ with acetic anhydride. The hydrolytic precipitation of aluminum by acetates always yields basic acetates, such as $Al(OH)(CH_3COO)_2$, $Al_2O(CH_3COO)_4$, $Al(CH_3COO)_3 \cdot 2AlO(CH_3COO)$, etc.

Solutions of aluminum acetates undergo hydrolytic decomposition with the liberation of free acetic acid at pH between 3.65 and 4.64, depending on the concentration. The basic acetate $Al(OH)(CH_3COO)_2$ is only sparingly soluble in water (2–3%), and is insoluble in alcohols, acetone, and ether; it is soluble in chloroform in an amount of about 0.03%. It decomposes at 250–300°C:

$$Al(OH)(CH_3COO)_2 = AlO(OH) + (CH_3CO)_2O.$$

Pure *ammonium oxalate* cannot be isolated, but the isolation of the hydrate $Al_2(C_2O_4)_3 \cdot nH_2O$ has been given. The formation of soluble aluminum oxalate complexes is very important in analytical chemistry. Aluminum forms several complexes with oxalate ions. The complex ion $[Al(C_2O_4)]^+$ is formed in 0.5 M HCl, while in less acid solutions complexes with a larger number of ligands are present. Thus, the complex anion $[Al(C_2O_4)_3]^{3-}$ appears at pH 5, while at pH 2 the anion $[Al(C_2O_4)_2]^-$ is formed. According to [30], the complexes $[Al(C_2O_4)]^+$ and $[Al(C_2O_4)_2]^-$ are formed to an insignificant extent only. The following K_{inst} values have been found for the various oxalate complexes of aluminum:

Complex	K_{inst}
$[Al(C_2O_4)]^+$	$5 \cdot 10^{-8}$ [30]
$[Al(C_2O_4)_2]^-$	$1 \cdot 10^{-13}$ [515]
$[Al(C_2O_4)_3]^{3-}$	$1.6 \cdot 10^{-17}$ [515]

Malonic acid, $HOOC-CH_2-COOH$, reacts with aluminum to give a fairly stable complex, but this is less stable than the oxalate complex. In the presence of malonic acid, $Al(OH)_3$ is precipitated by ammonia only on standing or on boiling. The complex aluminum malonate of composition $K_3[AlMal_3] \cdot 3H_2O$ can probably [693] be isolated from solution; its instability constant, which has been determined by the potentiometric method, is $1.47 \cdot 10^{-16}$ [694].

Succinic acid, $HOOC(CH_2)_2COOH$, reacts with aluminum in weakly acid medium with the formation of basic succinates which are sparingly soluble in water. These are employed in the gravimetric determination of aluminum; for details, see the section "Gravimetric Methods" in the following chapter.

Aluminum reacts with *glycolic acid* to give a 1 : 2 complex. The stability of this complex is low, and only 55—60% of the aluminum is bound in the complex [1030].

Aluminum reacts with the salts of *lactic acid,* $CH_3CHOHCOOH$, to give 1 : 2 and 1 : 3 complexes; the compound $H_3[Al(C_3H_4O_3)_3] \cdot 5H_2O$ has been isolated in the solid state [1159].

Aluminum gives a 1 : 1 complex with *malic acid* [981]. At pH 4, an ionic strength of 1.0, and at 27—30°C the instability constant is $4.84 \cdot 10^{-4}$ [981].

Aluminum reacts with *tartaric acid* and with tartrates to form a complex, in both acid and alkaline media. According to Cǎdariu et al. [606—608], a complex of the composition Al : tartrate = 2 : 1 is formed in acid solutions; the equilibrium constant of this reaction is about $7.2 \cdot 10^{-5}$. At a pH above 3.5—3.6, a 1 : 1 complex is formed, with an instability constant of the order of 10^4. During complex formation, two H^+ ions per aluminum atom are liberated [607]. The structure of the complex can be represented as follows:

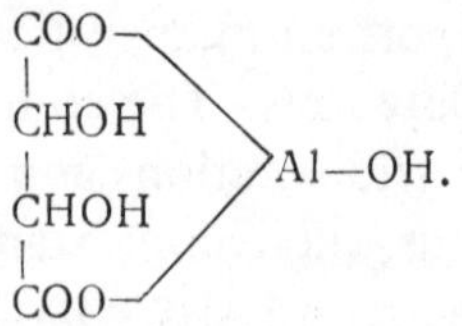

$$
\begin{array}{l}
\text{COO} \\
\quad | \\
\text{CHOH} \\
\quad | \\
\text{CHOH} \\
\quad | \\
\text{COO}
\end{array} \Big\rangle \text{Al—OH.}
$$

There are indications that complexes with an Al : tartrate ratio of 1 : 4 are formed in alkaline solutions [606].

In acid media, at pH below 5, aluminum gives a 1 : 1 complex with *citrates* [575, 606, 1050], while in alkaline solutions a 1 : 2 complex is formed [470]. The complex formed in acid medium has the composition

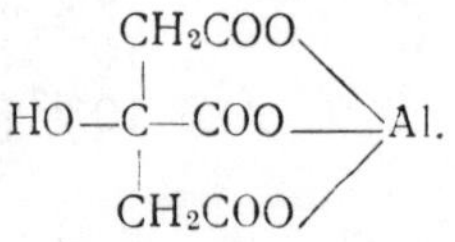

$$
\begin{array}{l}
\text{CH}_2\text{COO} \\
\quad | \\
\text{HO—C—COO} \\
\quad | \\
\text{CH}_2\text{COO}
\end{array} \Big\rangle \text{Al.}
$$

Three coordination links of aluminum are occupied by the citrate radical, and the remaining ones by water molecules.

The salts of *benzoic* and *cinnamic* (β-phenylacrylic) *acids* react with aluminum to form sparingly soluble precipitates, which can be used in the gravimetric determination of aluminum (see section "Gravimetric Methods" in the following chapter).

The formation of *phthalic acid* complexes of aluminum has been studied by heterometric titration [573]. It was found that a complex ion of structure $[Al(C_8H_4O_4)_2]^-$ was formed. When more phthalic acid was added, $K_3Al(C_8H_4O_4)_3$ precipitated out. If a solution of aluminum is added to this precipitate, the precipitate dissolves at an Al : phthalate ratio of 2 : 3. According to the authors, the undissociated covalent compound $Al_2(C_8H_4O_4)_3$ is formed.

Aluminum reacts with cresotic acid [765] to give a 1 : 3 complex.

In weakly acid medium (pH 4–5) aluminum reacts with salicylic acid to form a 1 : 1 soluble complex [605, 609, 674], with the formula

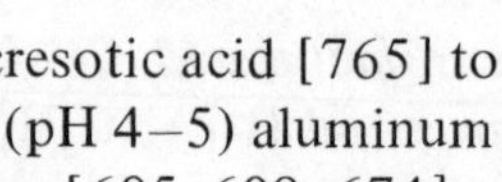

The table below lists the values of the formation constant of this complex obtained spectrophotometrically in solutions of different ionic strengths at 26–28°C [674]:

Ionic strength	K_{stab}
0.02	$(2.86 \pm 1.1) \cdot 10^4$
0.05	$(2.43 \pm 0.95) \cdot 10^4$
0.20	$(1.65 \pm 0.67) \cdot 10^4$

The value of K_{stab} extrapolated to an ionic strength of zero is $4.6 \cdot 10^4$. At $\lambda = 305$ mμ the molar extinction coefficient of the complex is $4.35 \cdot 10^3$ [674].

Other values for the constant of the 1 : 1 salicylate complex have also been reported: $K_{stab} = 5.42 \cdot 10^{11}$ [1032] and $K_{inst} = 8 \cdot 10^{-15}$ [36].

There are reasons for believing [1032] that other salicylate complexes of aluminum also exist: $AlSal_2^-$ with $K_{stab} = 4.09 \cdot 10^9$ and $AlSal_3^{3-}$ with $K_{stab} = 4.87 \cdot 10^7$.

Sulfosalicylic acid, $C_6H_3(OH)(COOH)SO_3H$, forms a 1 : 1 complex with aluminum [1005]. The following values of the instability constants were found by the spectrophotometric method at $28 \pm 0.5°C$ and $\mu = 0.2$:

pH	K_{inst}
2.4	$(2.48 \pm 0.05) \cdot 10^{-3}$
3.0	$(5.37 \pm 0.11) \cdot 10^{-4}$
4.0	$(5.94 \pm 0.14) \cdot 10^{-5}$

β-Resorcylic acid gives only the 1 : 1 complex when it reacts with aluminum in ratios from 1 : 1 to 1 : 3; the absorption maximum of the complex is at 300 mμ. The optimum pH of the reaction is 4.5–5. The structure of the complex can be represented as follows [1105]:

According to [1031, 1032], complexes with Al : reagent ratios of 1 : 2 and 1 : 3 also exist; their stability constants are $5.37 \cdot 10^6$ and $8.33 \cdot 10^3$, respectively. The stability constant of the 1 : 1 complex is $4.52 \cdot 10^9$ [1032], which is in contradiction with the data of [1031]; accordingly, these results do not appear to be reliable.

Aluminum reacts with *mandelic acid* C_6H_5–CHOH–COOH to form three complexes, with compositions 1 : 1, 1 : 2, and 1 : 3 [765, 1190] at pH values of < 5.8, 6.5, and 7.8, respectively. The following complex ions are formed:

and

Compounds formed with acids derived from carbohydrates

At pH < 5, aluminum reacts with d-gluconic acid, $C_5H_{11}O_5COOH$, to give the polynuclear aluminogluconic acid, $Al_2C_{18}O_{33}H_{34}$; at pH > 5, salts with the complex anion $[AlC_6H_8O_4]^-$ [452] are formed. The existence of an unstable complex of composition AlG_3 (where G is the gluconate ion), with overall $K_{stab} = 2.2 \cdot 10^{-6}$ [1242] has been reported.

Aluminum reacts with *galacturonic acid*, $C_4H_4(OH)_4CHOCOOH$, to form polynuclear compounds $AlC_{18}H_{27}O_{21}$ and $Al_2(OH)C_{18}H_{26}O_{21}$, and the mono-nuclear compound $AlC_6H_7O_7$, depending on the pH [88].

Saccharic acid $C_4H_4(OH)_4(COOH)_2$ reacts with aluminum to form a complex containing three Al atoms per two sugar molecules [324].

Complexes with polyaminopolyacetic acids (complexones). Complexes with this class of compounds are highly important in the analytical chemistry of aluminum. The formation of such complexes forms the principle of the complexometric methods for the determination of aluminum. The complexes with nitrilotriacetic acid NTA, ethylenediaminetetraacetic acid EDTA, 1,2-diaminocyclohexanetetraacetic acid DCTA, diethylenetriaminepenta-acetic acid DTPA, and hydroxyethylenediaminetriacetic acid HEDTA have been studied most. The following table lists the pK values of the stepwise dissociation of the four most important complexones, determined at $\mu = 0.1$ (KNO$_3$) and 20°C.

Complexone	pK_1	pK_2	pK_3	pK_4	pK_5
EDTA	2.02	2.66	6.21	10.31	
DTPA	1.80	2.55	4.33	8.60	10.58
DCTA	2.40	3.55	6.14	11.70	
HEDTA	2.51	5.31	9.86		

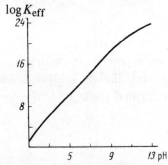

Figure 1. Dependence of the effective stability constant of the complex formed between aluminum and Complexone III on the pH [519].

All these complexones form 1 : 1 complexes with aluminum. The following values of the instability constants have been found for these complexes [985]:

Complexone	pK_{inst}	Complexone	pK_{inst}
EDTA	16.78 ± 0.03	DCTA	18.73 ± 0.05
DTPA	18.51 ± 0.01	HEDTA	12.54 ± 0.05

Similar results were obtained by other workers [519]. These data, together with the pK of formation of acidic and basic substituted aluminum complexonates, are given below.

Complexone	pK_{abs} AlY	pK_{form} AlYH	pK_{form} AlYOH
EDTA	16.5	3.4	8.0
HEDTA	14.4	2.4	9.3
DTPA	18.7	4.3	6.6
DCTA	18.9	3.4	6.3

Figure 1 shows the dependence of the effective stability constant of aluminum complexonate on the pH [519].

Compounds with phenols and phenol derivatives

Except for salicylal-o-aminophenol, phenols and their derivatives are not used in the quantitative determination of aluminum. The data given below on the compounds of aluminum with these compounds are intended to remind the analyst of the potential interference by such substances.

Pyrocatechol. In acid solution, even in the presence of a large excess of pyrocatechol, the 1 : 1 complex only is formed. This contains the ion

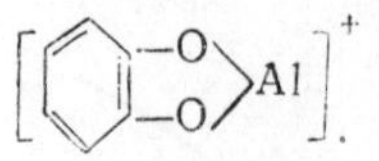

For this complex $\log K_{stab} = 16.56$ [691]; according to other data [500], $K_{inst} = 0.4 \cdot 10^{-23.4}$. In neutral and alkaline solutions, complex anions of composition 1 : 2 and 1 : 3 are formed [500, 691]:

The values of $\log K_{stab}$ found for the 1 : 2 and 1 : 3 complexes are 15.64 and 13.65, respectively [691].

Tiron (disodium salt of pyrocatecholdisulfonic acid) behaves similarly to catechol in many respects. It forms 1 : 1, 1 : 2 and 1 : 3 complexes with aluminum [690, 1008] in acid, neutral, and alkaline solutions, respectively. The complex ions which are then formed have the following structure:

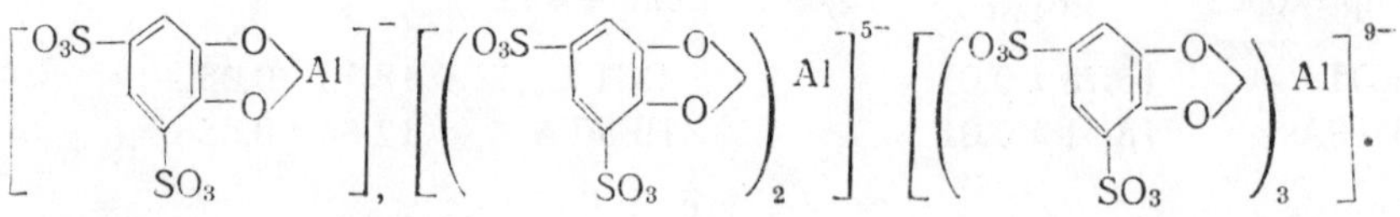

The logarithms of the stepwise stability constants of 1 : 1, 1 : 2, and 1 : 3 complexes are 17.02 ± 0.23, 16.49 ± 0,21, and 14.3 ± 0.07, respectively [690].

Chromotropic acid reacts with aluminum to form two complexes, 1 : 1 and 1 : 2, with values of $\log K_{stab}$ of 17.53 ± 0.13 and 16.78 ± 0.08, respectively [690]. The 1 : 1 complex has the following structure:

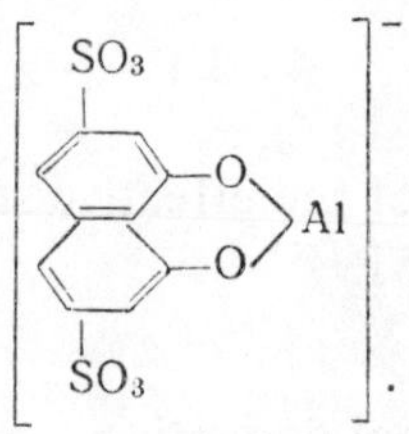

Juglone. A few complexes have been reported [327] between juglone and aluminum. However, definite data confirming the existence of complexes of any particular composition are lacking.

Salicylal-o-aminophenol reacts with aluminum to give a fluorescent compound, which is used in the fluorometric determination of aluminum. According to Tumanov and Efimychev [435], another, nonfluorescent compound is also formed and is suitable for the photometric determination of aluminum. A description of these complexes is given below [435].

	pH	Time required to establish chemical equilibrium, min	λ_{max}, $m\mu$	Al : reagent	K_{inst}	ϵ at λ_{max}
Fluorescent complex	5.5—5.8	10—15	405	1 : 2	$2.02 \cdot 10^{-7}$	9,200
Nonfluorescent complex	7.1—7.6	20—30	415	1 : 3	$2.02 \cdot 10^{-7}$	34,500

Similar values for the composition, λ_{max}, and optimum pH for the fluorescent complex were reported by Saylor and Ledbetter [1144]. A 1 : 1 complex has been reported [55].

Compounds with 8-hydroxyquinoline and its derivatives

8-Hydroxyquinoline (*o*-hydroxyquinoline, oxine). For aluminum hydroxyquinolate, see Chapter II.

8-Hydroxyquinaldine (2-methyl-8-hydroxyquinoline) gives a 1 : 1 complex with aluminum in a medium of absolute alcohol [1025].

The complex is fluorescent; the absorption spectrum has peaks at 315 and 358 $m\mu$. It is less stable than aluminum hydroxyquinolate; it decomposes rapidly when as little as 5% of water is added. Therefore, 8-hydroxyquinaldine cannot be used as a reagent for the determination of aluminum, but is very useful for the removal of elements interfering with the determination of this element.

7-Iodo-8-hydroxy-5-quinolinesulfonic acid reacts with aluminum to form a complex with $\lambda_{max} = 370$ $m\mu$, which is used for the photometric determination of aluminum.

Compounds containing two 8-hydroxyquinoline groups. Phillips et al. [1059] studied the formation of complexes between aluminum and compounds containing two 8-hydroxyquinoline groups. Of these, 8,8'-dihydroxy-5,5'-diquinolyl and 8,8'-dihydroxy-5,5'-diquinolylmethane give complexes with aluminum with the ratio Al : reagent = 1 : 3. These complexes resemble aluminum 8-hydroxyquinolate.

Compounds with triphenylmethane dyes

Triphenylmethane dyes are a very important group of reagents for the analytical chemistry of aluminum. Complexes of aluminum with a large number of triphenylmethane dyes have been investigated.

The conditions under which these complexes are formed will be discussed in detail in the section "Photometric Methods." Table 3 merely gives a few data on these complexes.

Table 3. Properties of complexes of aluminum with triphenylmethane dyes

Reagent	Al : reagent	λ_{max}, mμ, of		ϵ	pH_{max}
		reagent	complex		
Aluminon	1 : 1		535	12,400	3.8
Eriochrome Cyanine R	1 : 3	430	530–535	67,500	6.0
Chrome Azurol S	1 : 2*	430	545	59,300	5.8
Xylenol Orange	1 : 1**	435	555	21,100	3.3–3.5
	(pH < 3);				
	1 : 2				
	(pH > 4.5)				
Methylthymol Blue	1 : 1	435	585–590	19,000	3.5
Chromoxane Violet P	1 : 1	500	500	30,000	4.5
Sulfochrome	1 : 1†	520	560	38,000	3.8–4.1
Acid Chrome Pure Blue	1 : 3		540–550	20,000	6.0
Pyrocatechol Violet	1 : 2††	445–450	580	25,000–29,000	5.8
Pyrogallol Red			525		4.8–5.2
Alumocresone			500	13,500	3.3

* $K_{inst} = 10^{4.3}$; ** $K_{form} = 2.0 \cdot 10^{14}$; † $K_{inst} = (3.10 + 0.93) \cdot 10^{-5}$; †† $K_{form} = 10^{19.13}$ (for monometallic complex); $K_{form} = 10^{24.08}$ (total, for bimetallic complex).

Compounds with hydroxyanthraquinones

The formation of complexes between aluminum and *alizarin* (1,2-dihydroxyanthraquinone), *Alizarin S* (1,2-dihydroxyanthraquinone-3-sulfonic acid, sodium salt), *quinalizarin* (1, 2, 5, 8-tetrahydroxyanthraquinone), and *sodium quinalizarinsulfonate* has been studied most thoroughly. For details see section "Photometric Methods."

At pH 6 aluminum reacts with 2-phenoxyquinizarin-3,4-disulfonic acid to give a 1 : 1 complex, with $\lambda_{max} = 550-560$ mμ [1039].

Complexes with azo compounds

Aluminum reacts with *o,o'-dihydroxyazobenzene* (*o*-azophenol) to form a 1 : 1 complex; in the presence of excess reagent, the 1 : 2 complex is formed. For the 1 : 1 complex, $pK = 3.1$ [874].

Nazarenko et al. [293] studied the complexes between aluminum and a number of azo compounds. Table 4 shows some parameters determined for these complexes.

Table 4. Properties of complexes formed between aluminum and a number of azo compounds

Reagent	Composition of complex	K_{inst}	λ_{max} of reagent, $m\mu$	λ_{max} of complex, $m\mu$	pH
2,2',4'-Trihydroxy-azobenzene	1 : 1 (pH 5) 1 : 2 (pH 6)	$5.2 \cdot 10^{-13}$ $2.8 \cdot 10^{-30}$	400—420 450—465	480 500—515	4.5 7
5-Chloro-2-hydroxy-benzeneazo-2'-naphthol	1 : 2 (pH 4—5); 1 : 2 (pH 7—8)	$1.1 \cdot 10^{-42}$	510 540	526 544	4 7
2-Naphtholazo-2'-naphthol-4'-sulfonic acid (Eriochrome Blue-Black R)	1 : 2 (pH 4—5) 1 : 2 (pH 7—8)	$2.4 \cdot 10^{-41}$	520 530	530 590	5 7

These authors proposed the following structure for the trihydroxyazo-benzene complexes:

and a similar structure for the complex formed by aluminum with 5-chloro-2-hydroxybenzeneazo-2'-naphthol.

Aluminum reacts with *4-(2-pyridylazo)resorcinol* to form a 1 : 1 complex [1182]. This complex is not used for the photometric determination of aluminum since many cations interfere.

The formation of complexes between aluminum and many azo compounds based on chromotropic acid has been studied [14a, 475, 947]. Table 5 shows the properties of complexes of aluminum with two chromotropic acid derivatives.

Table 5. Properties of complexes of aluminum with two derivatives of chromotropic acid at pH 5 [947]

Reagent	Composition of complex	K_{inst}	λ_{max}, mμ	ϵ	Sensitivity, μg Al/cm^3 (after Sandell)
2-(Pyridyl-3-azo)-chromotropic acid, Na salt	1 : 3	10^{-13}	580	5946	0.00453
2-(2-Carboxypyridyl-3-azo)-chromotropic acid, Na salt	1 : 1	10^{-5}	590	9747	0.00276

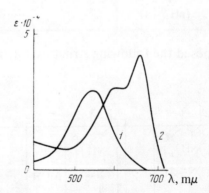

Figure 2. Absorption spectra of picraminazochrome (1) and its complex with aluminum (2) [475].

Below are some of the chromotropic acid derivatives studied by Cherkesov et al. [14a, 475, 477]: picraminazochrome (1,8-dihydroxynaphthalene-3,6-disulfonic acid-2,7-*bis*-[(azo-1)-2-hydroxy-3,5-dinitrobenzene]) and stilbazochrome (stilbene-2,2'-disulfonic acid-4,4'-bis-[(azo-1)-1,8-dihydroxy-naphthalene-3,6-disulfonic acid]). Picraminazochrome forms an intensely colored complex with aluminum with an absorption maximum at 650 mμ (the absorption maximum of the reagent is at 540 mμ) (Figure 2). The composition of the complex is 1 : 1, pH 5.5, pK_H = 5.5. The sensitivity of the reaction is 0.02 μg Al/ml. With stilbazochrome a 1 : 1 complex is formed, with pK = 5.6, optimum pH value 5, molar extinction coefficient 5.8 · 10^4.

Stilbazo and Arsenazo I have also been proposed for the photometric determination of aluminum. For details, see section "Photometric Determinations."

Complexes with dipyridyl and o-phenanthroline

Aluminum reacts with *2,2'-dipyridyl* (α, α'-dipyridyl) to form a complex with composition Al(Dipy)$_3$. The complex has been isolated in the form of green crystals; it is soluble in tetrahydrofuran, benzene and dioxane, and is decomposed by water.

The reaction with *1,10-phenanthroline* (o-phenanthroline) yields a complex with composition [Al(Phen)$_3$]$^{3+}$. This complex has been isolated as white crystals, which are soluble in water and in alkali solutions.

CHAPTER II

Chemical and Physicochemical Methods for the Determination of Aluminum

QUALITATIVE REACTIONS GIVEN BY ALUMINUM ION

The qualitative detection of aluminum is usually based on the formation of colored compounds, precipitation of sparingly soluble compounds, fluorescence reactions, and spot tests. Table 6 shows the most important methods for the detection of aluminum.

Aluminum is detected by the color reaction with alizarin [115, 762] and many other reagents, and by fluorescence reactions with morin [1048], Pontachrome Blue-Black R, and other reagents.

Aluminum may be precipitated as the hydroxide, which is then treated with a solution of a fluoride; this results in the formation of cryolite and in the liberation of OH^- ions, then detected by phenolphthalein [132]. This reaction may be carried out as spot test, when its sensitivity is 0.4 μg of aluminum at the limiting dilution of $1 : 7.3 \cdot 10^4$ [928]. The exchange reaction between Al^{3+} ion and CaF_2 yields cryolite with the liberation of Ca^{2+} ions, that can be detected by bis-(2-hydroxyanil)glyoxal; the sensitivity of the method is 0.5 μg Al at the limiting dilution of $1 : 10^5$ [852].

Aluminum reduces the rate of the catalytic oxidation of alizarin by perborate or hydrogen peroxide in the presence of cobalt. This effect forms the principle of a method for detecting aluminum with a sensitivity of 0.1 μg Al, at the limiting dilution of $1 : 10^7$ [918].

Table 6. Methods for the detection of aluminum

Reagent	Limiting dilution in detection test	Spot test		Color	Notes	Reference
		sensitivity, μg	limiting dilution			
Aluminon	$1 : 5 \cdot 10^5$			Red	Large amounts of Fe removed by extraction as thiocyanate. Effect of Cr, Th, Ga and Ti suppressed by treating their aluminon compounds with dilute H_2SO_3. Large amounts of phosphates, oxalates and fluorides reduce the sensitivity	[1014]
Eriochrome Cyanine R		0.03	$1 : 1.67 \cdot 10^6$	Violet	The following do not interfere: up to 0.1 g-ion/l Fe(III) bound by ascorbic acid, Cu(II) bound by thioglycolic acid, Co, Ni, Mn, Zn. Ti, Zr, Th separated by NaOH V(IV) and Be interfere	[745]
Alumocresone		0.007	$1 : 5 \cdot 10^6$	Pink	Fe(III), Be, Ga, U(VI) interfere	[218]

Table 6 (continued)

Reagent	Limiting dilution in detection test	Spot test		Color	Notes	Reference
		sensitivity, μg	limiting dilution			
2,3-Hydroxy-naphthoic acid*	$1:10^8$	0.0002	$1:10^7$	Blue fluorescence on green background	The following do not interfere: Ca, Sr, Ba, Mg, Zn, Cd, Hg, Pb, Sn, Ti, V, As, Sb, Bi, Cr, Mo, W, Mn, Co, Ni, Cu	[474]
5,8-Dichloroquinizarin	$1:2 \cdot 10^6$ (color reaction)** $1:10^7$ (fluorescence reaction)†	0.002	$1:10^6$	Bright-pink and yellow-green fluorescence (in UV light also orange-pink fluorescence)	Fe(III), Cu(II), Be, Th interfere	[219]
Salicylal-o-aminophenol	$2\,\mu$g/ml (in daylight) $0.05\,\mu$g/ml (in UV light)	0.01		Yellow-green fluorescence	On acidification, the fluorescence given by metals other than Al, Be and Zr disappears. The following do not interfere: Ag, Pb, Tl(I), Cd, Mn, La, Ce(III), Ca, Sr, Ba, Mg, Na, K, NH_4^+, Cl^-, SO_4^{2-}, BO_3^{3-}. Salts of Cu, Bi, Hg(II), Sb(II), Sn(II), Zn, Cr, Ti, Th, Zr, Ni, Co, Fe(III), Be, UO_2^{2+} and F^- reduce the fluorescence	[816]

* Sensitivity of detection test, 0.01 μg.   ** Sensitivity of detection test, 0.5 μg.   † Sensitivity of detection test, 0.1 μg.

Aluminum can be detected by the formation of characteristic crystals of $CsAl(SO_4)_2 \cdot 12H_2O$ [762], and by the formation of a blue bead when $Al(OH)_3$ precipitate is fused with a solution of a cobalt salt.

Aluminum can be semiquantitatively determined by the "in situ" method in magnesium [93, 405], copper [338], zinc [339], and nickel [93] alloys, using aluminon and Arsenazo I.

METHODS FOR THE DETERMINATION OF ALUMINUM

GRAVIMETRIC METHODS

Precipitation as hydroxyquinolate

8-Hydroxyquinoline reacts with aluminum to form a sparingly soluble chelate compound of composition $Al(C_9H_6ON)_3$. Its solubility product is $1 \cdot 10^{-19}$ [399] or $1.03 \cdot 10^{-29}$ [414, 415]. Table 7 shows that 8-hydroxyquinoline is not a specific reagent for aluminum. However, if the interfering elements are masked or separated, aluminum can be simply and quite rapidly determined in substances of rather complex and variable composition.

Table 7. The pH of the quantitative precipitation of metal hydroxyquinolates

Metal	pH	References	Metal	pH	References
Al	4.2–9.8	[819]	Ru(III)	Weakly acid medium	[360]
Sb(III)	>1.5	[360]	Bi	5.0–8.3	[578]
In	2.5–3.0	[360]	W	5.0–5.7	[722]
V(V)	2.7–6.1	[819]	Hg(I)	5.2–>8.2	[578]
Fe(III)	2.8–11.2	[819]	Cu(II)	5.3–14.6	[722]
Pd(II)	3.0–11.6	[578]	Cd	5.4–13.3	[819]
Pu(VI)	3.5–9.0	[360]	Ce	>5.6	[360]
Ga	3.6–11	[578]	U	5.7–9.8	[722]
Mo(VI)	3.6–7.3	[722]	Mn(II)	5.9–9.5	[819]
Ti	<3.7–8.7	[578]	Y	5.9–9.3	[999]
Cr(III)	3.7–12*	[360]	Ag	6.1–11.6	[578]
Co	4.3–14.5	[722]	Tl(III)	6.5–7.0	[982]
Ni	4.3–14.6	[722]	La	6.5–>10.3	[578]
Th	4.4–8.8	[819]	Sc	6.5–8.5	[1070]
Hf	4.5–11.3	[578]	Pb	8.4–12.3	[819]
Zn	4.6–13.4	[722]	Mg	9.4–12.7	[722]
Zr	4.7–12.5	[360]	Ca	9.2–12.7	[819]
Hg(II)	4.8–7.4	[578]			

* Precipitation not quantitative.

The hydroxyquinoline method is the most accurate of all the gravimetric methods for the determination of aluminum. This method was introduced by Berg [559] more than 40 years ago, and has been critically reviewed in the papers [627, 644].

Optimum pH for precipitation. Aluminum hydroxyquinolate can be precipitated in both weakly acid and alkaline media. The most popular technique is the precipitation from acetic acid solutions, even though precipitation from ammoniacal solutions may prove to be more selective in the presence of masking substances.

Gotô [770] recommended pH 4.15−9.80 for the complete precipitation of aluminum hydroxyquinolate. According to [631], accurate results are obtained at pH 4.80−5.14 in precipitation from weakly acid media, while at pH below 4.48 the results are low. According to Miller and Chalmers [970], correct results are obtained at pH 4.4−6.7. The discrepancy in the values reported by different workers at the optimum pH can be explained by the different experimental conditions of the precipitation. According to Classen and Bastings [644], the following amounts of aluminum remain in the filtrate when 5−30 mg Al are precipitated at 50°C by an excess (10 ml) of a 2.5% solution of hydroxyquinoline (total volume of solution 150 ml):

pH	Al, μg	pH	Al, μg
3.8	1000	4.8	30
4.0	75	4.9	28
4.3	50	5.0	26
4.5	41	5.3−7.0	22−20
4.7	31		

Thus, aluminum is precipitated most completely at pH 5.2 and above.

*Sequence of addition of reagents and
amount of precipitant*

The best procedure is to add the solution of hydroxyquinoline to the weakly acid solution, and then slowly introduce ammonium acetate. The crystals thus obtained are larger than when the reagents are added in the reverse sequence [644, 1208]. Views on the required excess of the precipitant vary. Classen and Bastings [644] and Miller and Chalmers [970] believe that if the amount of the aluminum being precipitated is 5−30 mg, an excess of 8−15 ml of a 2.5% solution of hydroxyquinoline should be added. According to [970], there is an approximately linear relationship between the amount of the precipitant in excess and the percentage error; if the excess of hydroxyquinoline is less than 250 mg, the error is negative, while if the excess is larger, the error is positive. This conclusion was subsequently verified by a statistical series of experiments [964].

Stumpf [1208] showed that the rate of addition of ammonium acetate is also important. If this reagent is added slowly, drop by drop, with constant stirring, a compact, purer precipitate is obtained and the results are more accurate. If the acetate solution is introduced in a jet, a voluminous, impure precipitate is formed.

Various suggestions have been made on the mode of heating the reaction mixture. Berg [559] recommended that all the reagents be added to the cold aluminum-containing solution, then the pH be adjusted to the desired value, and the mixture heated to complete coagulation. More recent publications recommend that the solution be heated to $50-60°C$ [891] or to $100°C$ [1177] before the addition of the hydroxyquinoline or the adjustment of the pH to the desired value [178, and others]. Chalmers and Basit [627] found that if the solution is heated only after all the reagents have been added and the pH has been adjusted to 5.2, the results are accurate whatever the excess hydroxyquinoline, whereas if ammonium acetate is added to the solution preheated to $90°C$, negative errors are obtained in the presence of a small excess of hydroxyquinoline and positive errors if the excess is large [964, 970].

The positive errors (high results) are explained by coprecipitation of the hydroxyquinoline. According to the above authors, the low results are due to losses of the precipitant by volatilization, as well as occlusion of polynuclear complexes with mixed ligands, containing more aluminum than the ordinary hydroxyquinolate $Al(C_9H_6ON)_3$.

Mass-spectrometric studies indicated the existence of complexes with compositions $Al_2(H_2O)_2(C_9H_6ON)_4^{2+}$ and $Al(H_2O)(C_9H_6ON)_2^+$. Thus, more accurate results are obtained if the solutions are heated after all the reagents have been added. •

Hydroxyquinoline can be employed as a solution in acetic or dilute hydrochloric acid. Ethanol solutions are not advisable, since in the presence of ethanol the solubility of aluminum hydroxyquinolate increases [51, 887, 891]. According to Classen and Bastings [644], it is best to work with a solution of hydroxyquinoline in dilute HCl (pH about 3). This reagent is not recommended as a solution in $1-2M$ acetic acid, since during evaporation the precipitate adheres to the vessel walls and does not dissolve in the reaction mixture. Moreover, under these conditions it is more difficult to adjust the mixture to the desired pH of 5.2–5.8. At 20–70°C and pH 5.7–7.0, 20 μg Al is not precipitated and remains in solution. Classen and Bastings [644] suggest that ammonium acetate be introduced into solutions which have been preheated to 70°C. However, the results of Chalmers and Basit [627] indicate that it is better to introduce ammonium acetate into cold solutions, and to coagulate the precipitate by heating. Despite Kampf's suggestion [860] that small amounts of tartrates be introduced to prevent the partial precipitation of aluminum as basic acetate, this is not advisable, as the solubility of aluminum hydroxyquinolate is thus increased.

Washing the precipitate. The precipitate of aluminum hydroxyquinolate can be washed with hot, warm, or cold water. According to Classen and Bastings [644], the amounts of aluminum which pass into the filtrate are insignificant:

Temperature of wash water, °C	μg Al/100 ml of wash water
100	27
80	15
60	7
25	6

As aluminum hydroxyquinolate is partially hydrolyzed, some of the hydroxyquinoline passes into the filtrate, so that washing the precipitate with hot or warm water results in material losses [1196, 1208]. If the precipitate is washed with a small quantity of cold water, these losses are insignificant.

It has been recommended [644] that the precipitates be washed with a neutral 0.02% solution of hydroxyquinoline. In this case, 100 ml of the washings will contain about 5 μg Al irrespective of the experimental temperature (between 20 and 100°C). Satisfactory results are obtained if the precipitate is washed with a 2% solution of ammonium acetate. Washing with dilute acetic acid, suggested by a number of authors, is not feasible, since aluminum hydroxyquinolate is partially soluble in this solvent [51, 556]. The precipitate can be washed with water and then with 20 ml of ether, and the time of drying at 100°C is reduced to 10 min [527].

Drying the precipitate. The optimum temperature for drying and decomposing aluminum hydroxyquinolate has frequently been discussed. Thermogravimetric studies showed that a constant weight of the precipitate is attained at 102°C [692], 135°C [583], and 160°C [867]. According to the data of different workers, the conversion of the hydroxyquinolate to the oxide occurs at 1000°C [692], 700°C [583], and 600°C [867]. We consider that a drying temperature of 140–150°C is the optimum for aluminum hydroxyquinolate.

Some workers ignite the hydroxyquinolate to alumina. However, in this case, the advantages of determining aluminum in its most accurate gravimetric form are lost. The conversion factor from the oxide is much higher than from aluminum hydroxyquinolate, and besides the oxide is hygroscopic, which introduces a new source of error. Accordingly, in the method of precipitation by hydroxyquinoline, it is better to weigh aluminum as the hydroxyquinolate or else determine it by titrimetry (bromatometric method).

The following procedure [644] is recommended for the gravimetric determination of aluminum by precipitation with hydroxyquinoline from acetate solution.

The sample solution, containing 2–20 mg Al, is diluted to 100–125 ml. To the cold solution, 2 M ammonia solution is added drop by drop, with constant stirring, until a small amount of $Al(OH)_3$ precipitate appears which does not redissolve on stirring. Hydrochloric acid (2 M) is then added until the precipitate has dissolved, and then 5–10 drops in excess. Then, 0.70 ml of a 2.5% solution of hydroxyquinoline (25 g of hydroxyquinoline are dissolved in 29 ml of 6 M HCl and the solution is diluted to one liter) is added for each mg of aluminum present, and then 10 ml in excess. Twenty ml of a 20% solution of CH_3COONH_4 at pH 6.5–7.0 are then slowly introduced, with stirring. The mixture is heated to boiling, placed on a steam bath for 30 min, and then cooled to 50°C. The pH of the solution after precipitation is 5.2–5.8.

The solution is decanted through a tared No.4 Schott crucible. The precipitate is washed with not more than 100 ml of warm (50–60°C) wash liquid* in small portions, and then with two 5-ml portions of cold water. During the washing, the precipitate should not be sucked dry or a compact mass will be formed, from which salts are washed out with difficulty.

The crucible with the precipitate is dried to constant weight at 140–150°C. The conversion factor to aluminum is 0.05873; the conversion factor to the oxide is 0.1110.

The procedure for precipitating aluminum hydroxyquinolate from an ammoniacal solution is given below [644].

To 100 ml of the sample solution, a fivefold excess of tartaric acid (with reference to aluminum) is added, then 1–2 g NH_4Cl and 8–10 drops of a 0.04% alcoholic solution of bromocresol purple. The solution is neutralized with 1 : 1 ammonia solution until the color of the indicator turns purple. The aluminum is precipitated by the dropwise addition of a 2–3% solution of hydroxyquinoline in acetic acid. The reaction mixture is then boiled and left to simmer for one minute. It is then cooled to 60°C, and the precipitate is filtered through a No.4 Schott crucible by gentle suction, and washed with 100 ml of cold water. The procedure is continued as in the precipitation from acetate solution.

* To prepare the wash liquid, 8 ml of a 2.5% solution of hydroxyquinoline are diluted with water to 500 ml, 3 drops of a 0.1% solution of bromocresol purple in 20% ethanol and 2 M ammonia are added until a purple color (pH 6), and the solution is made up to one liter.

During precipitation from an ammoniacal solution, tartaric acid is introduced to keep aluminum in solution. If the amounts of aluminum to be
determined are small (less than 5 mg per 50 ml of solution), the precipitates
should be filtered only after cooling to room temperature [51]. If the
ammonia is added too rapidly, the results may be too high [51].

Effect of cations and anions. The precipitation of aluminum hydroxyquinolate from acetate buffer solution is not selective; the only ions which
do not interfere when present in significant amounts are Mg, Be, alkali and
alkaline-earth metals. Many studies were carried out in the search for
suitable masking agents for the interfering elements. Studies on the determination of aluminum in the presence of iron are particularly numerous.
Attempts at the fractional precipitation of aluminum and iron at different
pH values did not give satisfactory results [747].

Berg [51, 561] separated aluminum from iron by means of oxalic, malonic,
tartaric and salicylic acids, which form complexes with aluminum in acetate
solutions; iron is precipitated as hydroxyquinolate. According to Berg
[51, 561], small amounts of iron are best separated from large amounts of
aluminum in the presence of tartaric acid, whereas large amounts of iron
should be separated from small amounts of aluminum in the presence of
malonic acid or a mixture of malonic and tartaric acids. With this method
it is possible to rapidly determine aluminum and iron present together.

In the presence of malonic or oxalic acid, titanium is quantitatively
precipitated as hydroxyquinolate; in this way titanium can be quantitatively separated from aluminum [51, 128, 417]. It is possible to separate Al,
Fe and Ti as follows: iron is precipitated from a strong acetic acid solution
in the presence of tartaric acid, and then titanium is precipitated after the
addition of malonic (or oxalic) acid. Aluminum hydroxyquinolate can be
precipitated from the filtrate after the addition of ammonia. The experimental procedures for the analysis of mixtures of Al and Fe, and of Al, Fe
and Ti, can be found in the monograph by Berg [51]. The determination
of aluminum in such mixtures is also described in [120, 121, 638, 995].

Pyatnitskii [348] studied the solubilities of aluminum and titanium
hydroxyquinolates in solutions of tartaric and oxalic acids, and reported
the optimum conditions for the precipitation of titanium in the presence
of aluminum. Since titanium hydroxyquinolate is fairly soluble in solutions
of such acids, a 10-ml excess of a 2% solution of hydroxyquinoline should
be added. Titanium is best precipitated at pH 6. According to Pyatnitskii,
tartaric acid is not necessary. Satisfactory results are obtained with oxalic
acid alone, and the author claims that the addition of tartaric acid does

more harm than good, since titanium hydroxyquinolate is more soluble in tartaric acid than in oxalic. The optimum concentration of the latter is 0.25 M.

Small amounts of aluminum can be separated from large amounts of magnesium by simple precipitation from acetate solution. If large amounts of aluminum are to be separated from small amounts of magnesium, magnesium hydroxyquinolate must first be separated from a tartrate-containing solution of sodium hydroxide, and then aluminum is determined as hydroxyquinolate in the filtrate [560].

Aluminum can also be separated from Zn, Cd and Cu similarly [51]. The separation of Cu from Al from nitrate [472], acetate, and ammoniacal [612] tartrate-containing solutions has been described. The precipitation of aluminum as hydroxyquinolate from an acetate solution is used to separate aluminum from beryllium [1013] and in the presence of high concentrations of ammonium carbonate from uranium also, as uranium forms a stable complex with ammonium carbonate [942]. Reprecipitation is necessary for quantitative separation from uranium.

Chlorides, sulfates, nitrates and perchlorates do not interfere with the determination of aluminum by the hydroxyquinolate method. Fluorides interfere even when present in small amounts [77, 644]. The addition of $1-5$ g H_3BO_3 reduces the error, but not more than 2 mg of F^- ion can be masked [644]. Beryllium in an amount equal to or greater than that of aluminum eliminates interference by fluorides, owing to the formation of a fluoride complex of beryllium [398]. Tartrates do not interfere, but the precipitate formed in their presence is finer grained. If citrates and oxalates are present, the precipitation is quantitative at pH above $7-8$ only.

In the presence of phosphate ions aluminum hydroxyquinolate should be precipitated from an ammoniacal solution containing tartrates [51, 851, 942, 1100], and the precipitate is then analyzed volumetrically. Wilson [1275] showed that this separation can be conducted from an acetate buffer solution if hydroxyquinoline is first added to a solution $1 N$ in HCl, and the pH is adjusted to about 5 by ammonia or ammonium acetate. Under these conditions 25 mg Al_2O_3 can be quite accurately determined in the presence of 50 mg P_2O_5. If the pH is adjusted to about 5 before the introduction of hydroxyquinoline, aluminum phosphate is precipitated and the results are too low.

Silicic acid does not interfere with the bromatometric determination of aluminum when this element is precipitated by hydroxyquinoline from acetate solutions; the gravimetric method cannot be employed under these conditions.

In weakly acid solutions the interfering cations are difficult to mask, whereas masking is easy in ammoniacal solutions, and cyanides are usually employed. Lang and Reifer [916] and, in particular, Heczko [793], were the first to discuss the theory of masking by cyanides. Heczko used cyanides to determine aluminum in the presence of Fe, Co, Ni, Mo, Cu and Cr. Lundell and Knowles [942] determined aluminum in the presence of Ti, Ta, Nb, V and Mo in an ammoniacal solution containing hydrogen peroxide. Kassner and Ozier [865] combined these two methods and developed a method for the determination of aluminum in the presence of many elements.

If Complexone III and cyanide are used, aluminum can be determined in zinc alloys without any preliminary separation [157]. Gassner [747] determined aluminum in the presence of Fe, Ca and H_3PO_4; iron was reduced by ascorbic acid and converted to a ferrocyanide complex, while calcium was converted to a citrate complex.

Methods for determining aluminum with cyanide masking have been developed for steels [585, 793, 1061, 1196], ferromolybdenum [857], and alloys of copper [698, 702, 1250] and zinc [157, 1109].

To concentrate small amounts of aluminum, aluminum hydroxyquinolate is precipitated on scavengers, such as calcium hydroxyquinolate. The method was employed by Marczenko [955], who isolated 0.001–0.01% of aluminum from rock salt.

Precipitation of aluminum hydroxyquinolate
from homogeneous solutions

In homogeneous precipitation, the precipitant is formed in a uniform manner throughout the bulk of the solution, or else the pH of the solution is slowly adjusted to the value at which precipitation takes place. Precipitation from homogeneous solutions has a number of advantages over conventional precipitation: the crystalline precipitate is coarse grained, and thus coprecipitation of foreign substances is reduced. Such precipitates are more rapidly filtered and are easier to wash. However, the methods are time-consuming, and large amounts of reagents, often in short supply, are needed. These methods are therefore not employed in routine analyses, but only when high accuracy is required, e.g., in the analysis of standard specimens.

The homogeneous precipitation of aluminum hydroxyquinolate is carried out by one of three techniques: the slow formation of the precipitate in solution; the gradual adjustment of the pH until precipitation occurs; homogeneous precipitation from a solvent mixture.

Slow formation of precipitate in solution. Aluminum is precipitated by a derivative of hydroxyquinoline and not by hydroxyquinoline. This derivative hydrolyzes to produce hydroxyquinoline. The best reagent is 8-acetoxyquinoline [827, 956], which hydrolyzes at pH 5 and 60°C when aluminum hydroxyquinolate is quantitatively produced. Satisfactory results are obtained by heating on a water bath at 60°C for 5 hours.

Gradual establishment of the pH of precipitation of the hydroxyquinolate. The required pH is produced in the homogeneous solution by adding acetamide, urotropin, or urea, which hydrolyze to yield ammonium acetate or ammonia. Urotropin and acetamide are not suitable for the homogeneous precipitation of aluminum, since the former is hydrolyzed too rapidly, and the latter too slowly. Urea is the most suitable. For the relevant procedure, see [1208]. Urea hydrolyzes at a satisfactory rate above 90°C only. At 95°C the precipitation of aluminum hydroxyquinolate is complete within 2−3 hours, while the final pH of the precipitation is 4.4.

The solution should contain not more than 25−50 mg Al and 1.25−2 ml HCl (sp.gr. 1.19), and is diluted to 150−200 ml. Five or six ml of a 10% solution of hydroxyquinoline in 20% acetic acid and 5 g of urea are added per 25 mg of aluminum present. The mixture is heated to just below the boiling point, and the beaker is covered by a watchglass and held at 95°C for 2−3 hours. The precipitation is considered to be complete if the supernatant solution above the precipitate is not greenish yellow, but orange yellow in color. When cool, the precipitate is filtered through a No.3 Schott crucible; a No.4 Schott crucible is used with small (about 10 mg) amounts of aluminum. The precipitate is washed in the usual way, dried at 140°C, and weighed [1208].

The relative error involved in the determination of Al_2O_3 in alumina (about 65%) is about 0.10%, while the conventional precipitation results in an error of 0.26% [1208].

Homogeneous precipitation from a mixture of solvents. These methods are based on the fact that in the presence of large amounts of organic solvents the formation of certain precipitates is hindered. After the organic solvent has been evaporated by heating, the optimum conditions for the separation of the precipitate are produced. This technique has certain advantages over those just described. The required amount of the precipitant is introduced into the solution at the beginning, and the desired medium is thus produced. The volatilization of the organic solvent results in the gradual formation of the precipitate, while the pH of the solution remains perfectly constant, which is important if aluminum is to be precipitated in the presence of other ions.

Howick and Jones [825, 826] precipitated aluminum hydroxyquinolate from a water-acetone solution.

To 2–10 ml of the sample solution are added 50 ml of water, 60 ml of acetone, 4 ml of a 5% solution of hydroxyquinoline, and 40 ml of 2 N ammonium acetate. The solution is heated on a water bath at 70–75°C for three hours and then cooled. The precipitate is filtered through a Schott crucible, washed three times with water, and dried at 135–140°C.

According to these workers, the error involved in the determination of 2–10 mg Al is 0.2–0.7%.

Large amounts of Mg and Ca do not interfere. In the presence of 420 mg Mg and 800 mg Ca, less than 0.1 mg Mg and Ca passes into the hydroxyquinolate precipitate. Up to 10 mg of cadmium do not interfere at pH 4.7– 4.9; outside this pH range cadmium interferes strongly. With the aid of radioactive isotopes it was shown that significant amounts of indium and about 0.5% of Y, Ce and Sc are coprecipitated [945]. Iron is coprecipitated over a wide range of acetate concentrations, while the precipitation of zinc is insignificant [542].

Precipitation as hydroxide

The precipitation of aluminum as hydroxide for its separation from other elements or its gravimetric determination is the oldest and the most popular method. At present, the gravimetric determination of the element as hydroxide is rarely used, since other, more accurate methods are available, but even now aluminum is frequently precipitated as the hydroxide to separate it from interfering elements. The precipitation of aluminum hydroxide begins at a pH somewhat above 4 [61, 591, 755] or even at pH 3.5–4 [9]. Hillebrand et al. [89] reported the pH of precipitation of the hydroxides of a large number of metals. We can supplement his list by quoting the pH of precipitation of gallium (3.4), indium (3.7) and scandium (4.7), as reported by Ostroumov [318]. Aluminum hydroxide can be precipitated by ammonia, weak organic bases, and by compounds which evolve ammonia when heated, or else hydrolytic precipitation with the aid of salts of inorganic acids can be carried out.

Precipitation by ammonia

Aluminum hydroxide is usually precipitated by ammonia. This is the most popular, but not the most accurate, method for the precipitation of $Al(OH)_3$.

The most important conditions necessary for accurate results are the presence of sufficient amounts of ammonium salts, and of ammonia in a very small excess, and precipitation at the boiling point. The addition of ammonium salts is necessary to keep magnesium, manganese, and other metals in solution. Moreover, the presence of ammonium salts prevents the pH of the solution from rising unduly, and favors coagulation of the precipitate. In some older textbooks [1231] it is suggested that excess ammonia be completely removed by boiling the mixture after precipitation, but in this case small amounts of $Al(OH)_3$ will remain in solution [841, 842]. It has been reported [1123] that the results of the two methods agree.

According to Jander and Ruperti [841], the solubility of $Al(OH)_3$ filtered immediately after boiling is 1.2 mg Al_2O_3 per liter of water. The solubility decreases on standing. Thus, the solubility of the hydroxide filtered at 12–15°C within 1–3 days is only 0.6 mg Al_2O_3 per liter [841]. Similar values were obtained by Remy and Kuhlman also [1108]. Blum [570] washed 0.1 g of Al_2O_3 with 75 ml of hot water, and found that the washings contained 0.5–2.9 mg Al_2O_3.

The solubility of $Al(OH)_3$ is a function of the pH. The solubility is a minimum at pH 6.0–7.8. Blum [570] recommended that the addition of ammonia be controlled by the color change of methyl red indicator (pH 4.4–6.2). Freru [732] prefers to precipitate in the presence of phenol red (pH 7.5); rosolic acid (pH 6.2–8.0) has also been recommended.

The precipitate of $Al(OH)_3$ should not be washed with water, since it easily passes into the colloidal form and is also partially solubilized. According to Murawleff and Krassnowski [998], the $Al(OH)_3$ precipitate should be washed with a hot 2% solution of ammonium nitrate with ammonia added until it is alkaline to methyl red; if ammonia is not added, some of the $Al(OH)_3$ will pass into solution.

The excess ammonia should be as small as possible, since aluminum hydroxide is markedly soluble at pH values above 9 [842], and even at pH 7–8 an insignificant proportion of the hydroxide is dissolved [662, 732].

The following table shows the solubilities of freshly precipitated hydroxide in solutions of ammonia of various concentrations at 20°C [761]:

NH_4OH, M	NH_4Cl, wt %	g $Al(OH)_3$/100 ml of saturated solution
0.50	5.0	0.0148
0.50	10.0	0.0010
1.00	5.0	0.0193
1.00	10.0	0.0085
1.00	30.0	0.0038

To reduce the solubility of $Al(OH)_3$, the precipitation should be carried out from volumes of solution as small as possible. According to Tananaev [404], if the precipitation is carried out from concentrated solutions, the amorphous precipitate is less bulky, and foreign ions are sorbed on it to a smaller extent. The precipitate should be formed in chloride and nitrate solutions; if sulfate solutions are employed, the precipitate will be strongly contaminated by sulfate ions, owing to the tendency of these ions to hydrolyze with the formation of basic sulfates. The sulfate ions can be removed only by prolonged ignition at high temperatures. If the sample solution contains large amounts of sulfates, the hydroxide should be reprecipitated [732].

The reader is referred to Chapter I for an account of the forms of $Al(OH)_3$ obtained by precipitation with ammonia.

Aluminum hydroxide should be ignited at temperatures of at least 1200°C. According to Milner and Gordon [978], and Willard and Tang [1273], the hydroxide, the basic sulfate, and the basic succinate of aluminum should be ignited at 1200°C. According to Karch [862], $Al(OH)_3$ should be ignited at 1300°C for one hour, but if Al_2O_3 and Fe_2O_3 are determined together, the ignition temperature should be 1200°C to avoid material losses due to the volatilization of Fe_2O_3.

A number of authors, including Duval [695], claim that the ignition temperature of aluminum hydroxide need not be above 1030°C, but this cannot be accepted. It was shown by Erdey and Paulik [701] that the recommendations made by Duval are applicable under certain conditions of precipitation only, and that the ignition temperature varies greatly with the experimental conditions. Other authors also came to similar conclusions [832, 923]. Imelik et al. [832] showed that constant weight can also be attained without total expulsion of water. Thus, if constant weight has been attained under these conditions, there is no indication that the ignition is complete.

It is well known that Al_2O_3 is hygroscopic. After ignition at 1400°C it becomes completely nonhygroscopic. Wiele [1266] recommended ignition of the precipitate for three hours at 1400°C or for one hour at 1500°C. Milner and Gordon [978] also showed that even after ignition at 1100–1200°C, the precipitates of aluminum oxide are still hygroscopic. According to Miehr et al. [968], the weight of aluminum oxide that has been ignited at 1300°C is practically constant. Fricke and Meyring [735] state that a nonhygroscopic precipitate is obtained at 1200–1300°C.

All these data indicate that aluminum hydroxide is preferably ignited at 1200°C, but must be weighed quickly. Ignition at 1400 or 1500°C may not be feasible, since such high temperatures cannot easily be produced in most laboratories.

A procedure for the gravimetric determination of aluminum precipitated by ammonia is given below [89].

The sample solution is diluted to 200 ml and 5 g NH_4Cl are introduced. This step may be omitted if the solution is strongly acid, since sufficient amounts of ammonium chloride will be formed during its neutralization with ammonia. A few drops of a 0.2% alcoholic solution of methyl red are introduced and the solution is heated to boiling. Ammonia is cautiously introduced drop by drop until the solution turns yellow. The solution is boiled for another 1–2 min, and after the precipitate has settled (which takes a few minutes), it is filtered through a white ribbon filter paper. The beaker is wiped with a piece of ashless paper, which is placed on the filter paper containing the precipitate.

The precipitate is thoroughly washed with a hot 2% solution of ammonium nitrate neutralized by ammonia to methyl red. If necessary, the precipitate is redissolved in hot 1 : 3 HCl, and reprecipitated. The precipitate is ignited at 1200°C for one hour. The conversion factor to aluminum is 0.5291.

Effect of other ions. Many other metals are precipitated by ammonia under the experimental conditions employed in the determination of aluminum. The only metals from which aluminum can be separated in this way are alkali and alkaline-earth metals, Mg, and small amounts of Mn and Ni. Even then, reprecipitation may be necessary, depending on the amounts of these ions. Aluminum can be quantitatively separated from Mn and Ni if the sample solution at pH 3 is slowly neutralized to a pH of not more than 6.6–6.7. Under these conditions aluminum cannot be separated from Cu, Zn and Co [570, 941]. Ishibashi et al. [835] found that when 50.4 mg Al were precipitated by ammonia at pH 7.1–10.5 in the presence of 20-fold amounts of copper, up to 5% of the copper is entrained together with the precipitate of aluminum hydroxide. After reprecipitation, the precipitate contained 8% of the copper entrained in the first precipitation. The amount of the coprecipitated copper is a function of the pH of the solution; it is maximum at pH 7.5–8.5, and decreases rapidly at higher and lower pH values [411]. Similar data were reported by Kovalenko [169].

Morachevskii and Bashun [270] studied the separation of aluminum from zinc by precipitation with ammonia in the presence of Zn^{65}, and found that the best results are obtained at pH 5.5–5.8. However, 7–8% of the zinc passed into the precipitate (initial amounts: 27 mg Al and 13.53 mg Zn). When the pH was raised to about 7, the amounts of coprecipitated zinc increased, and when the pH was raised further to 9.4, the amount of the coprecipitated zinc decreased. After reprecipitation at pH 5.5–5.8, the precipitate was free of zinc. At pH 5.9–6.0 and 9.1–9.4, the precipitate contained up to 1–2% Zn, while at pH 6.6 it contained 18% of the originally introduced amount of zinc. When the amounts of aluminum and zinc were increased to 80 and 40 mg, respectively, the precipitate contained 1–1.25% Zn even when the precipitation had been conducted under optimum conditions (pH 5.5–5.8 with reprecipitation).

It was found that after one precipitation under similar conditions 10–20% Zn is coprecipitated, while after reprecipitation the amount of zinc in the precipitate is negligible [821].

According to Bloch [569], if aluminum is precipitated from solutions containing large amounts of zinc, there is too much zinc in the precipitate for a gravimetric determination to be possible, even after four precipitations. Satisfactory results can be obtained after two precipitations if the sample solution is slowly introduced into a cold solution of ammonia containing ammonium nitrate [569]. Under these conditions the aluminum hydroxide precipitate contains not more than 0.1% Zn. Bloch used this technique to determine aluminum in zinc alloys. By this method aluminum can be more efficiently separated from Ni, Co and Mn also; in the presence of phosphates the pH must be maintained at 4.8–6.3 [905a].

Morachevskii and Bashun [270] studied the coprecipitation of cobalt with aluminum hydroxide by the method of labeled atoms. The maximum separation was at pH 5.5–6.0 in the presence of 5% NH_4Cl, with reprecipitation. Plotnikov [334] used Cd^{115}, and found that the maximum coprecipitation of cadmium with aluminum hydroxide occurs at pH 7–8.

Numerous studies were carried out to find masking techniques for the interfering elements. Mayr and Gebauer [961] precipitate aluminum in the presence of significant amounts of iron, which is reduced by sulfurous acid and held in solution by thioglycolic acid. Iron is masked best in hydrochloric acid solutions; nitrates interfere. Hummel and Sandell [828] verified the method of Mayr and Gebauer, and found it was highly efficient; up to 40 mg Al_2O_3 could be separated from equal amounts of iron with satisfactory accuracy in one precipitation, while in the presence of larger amounts of iron reprecipitation proved necessary. If the Mayr-Gebauer method is used, it is very important that aluminum hydroxide be filtered and washed as rapidly as possible; prolonged heating must be avoided, or thioglycolic acid may decompose, and FeS be precipitated. The wash water should contain thioglycolic acid and ammonium bisulfite (1 g NH_4Cl, 1 ml thioglycolic acid, and 2 ml of a 45% solution of NH_4HSO_3 in 100 ml water; the solution is neutralized to methyl red by ammonia).

In another technique, hydroxylamine hydrochloride is used instead of the sulfite [760]. Ferrous iron can also be masked by α, α'-dipyridyl, when obtained by reducing ferric iron with sulfurous acid [710]. The results obtained by precipitating $Al(OH)_3$ after the reduction of Fe(III) by hydro-xylamine are slightly less satisfactory [736]. When aluminum is determined in copper alloys, it is precipitated by ammonia from solutions containing ascorbic acid and Complexone III to keep Fe, Sn and Pb in solution [261]. Other authors mask iron and nickel by cyanides [637, 1246].

Aluminum is sometimes precipitated together with iron, and aluminum is found by difference after iron is determined in an aliquot. This method gives satisfactory results if the composition of the sample is known to be relatively simple. In the analysis of materials which contain a large number of components, the results must be corrected for titanium, phosphorus and some other elements, and this markedly lowers the accuracy of the determination.

When $Al(OH)_3$ is precipitated from solutions containing large concentrations of phosphates, phosphates of alkaline-earth metals may be coprecipitated. In these cases the precipitation of aluminum by benzoate at lower pH values is preferable [1275]. If the precipitation is carried out with ammonia containing ammonium carbonate, alkaline-earth carbonates may precipitate. If vanadium and arsenic are present in the sample solution, they may coprecipitate with $Al(OH)_3$ as sparingly soluble vanadates and arsenates. The vanadyl ion is coprecipitated with $Al(OH)_3$ even at pH 3.4–4.1 [31]. The coprecipitation of small amounts of La, Y, Ce and Eu begins at pH 5–6, and increases with increase in pH [273–275].

In the presence of silicates, phosphates, arsenates, molybdates and tungstates, the basic salts of the acids are formed. In the presence of fluoride ions, $Al(OH)_3$ is solubilized. If ammonia and fluoride are introduced at the same time, aluminum hydroxide is not formed, but basic aluminum fluoride is precipitated instead [622, 812]. In the presence of borates the precipitate contains the sparingly soluble aluminum borate and borates of alkaline-earth metals [643, 741]. Before the precipitation of aluminum, borates are removed by boiling the solution with hydrochloric acid and methanol [1168]. According to Krassnowski [902], up to 30% of borates do not interfere with the determination of aluminum present in amounts of 10% or less. It has been repeatedly suggested that chromium be oxidized to chromate by chlorine water, bromine water [1232], or by persulfate in the precipitation of aluminum from chromium-containing solutions. However, chromate ion is also coprecipitated in amounts depending on the conditions of precipitation. With increase in the pH and in the concentration of ammonium salts, the coprecipitation of the chromate ion decreases [309, 368]. The coprecipitated chromate ion is difficult to wash out. The maximum coprecipitation of the selenate ion with aluminum hydroxide occurs at pH 6–7, and decreases with increase in the pH [335].

Many organic compounds containing hydroxyl groups (citric and tartaric acids, starch, sucrose, glucose, lactose, isopropanol, butanol and isovaleric alcohol) suppress the formation of the $Al(OH)_3$ precipitate [666, 839]; in the presence of organic hydroxy acids the results are too low [501, 666].

Precipitation by ammonia may also be employed to deposit small amounts of aluminum on scavengers (iron hydroxide [758], zirconium hydroxide [954] or lanthanum hydroxide [953]).

*Precipitation of aluminum hydroxide by weak organic bases
and by compounds which evolve ammonia during hydrolysis*

The only organic bases of interest for the precipitation of aluminum are those which are weaker than ammonia. Bases stronger than ammonia (ethylamine, methylamine, piperazine) have been proposed as precipitants for aluminum, but should not be employed. Hydrazine is somewhat weaker than ammonia, but the aluminum hydroxide precipitated by this compound tends to sorb alkali metals and zinc strongly [1148].

Weak organic bases for the precipitation of aluminum include pyridine [270, 318–320], α-picoline [318], phenylhydrazine [804, 919, 976], acridine [1142], p-chloroaniline [1142], and o-phenetidine [628]. Aluminum hydroxide can also be precipitated by compounds which evolve ammonia when they hydrolyze, namely, urea, urotropin, mercury amidochlorides.

Precipitation by pyridine. Ostroumov [318–320] introduced pyridine as an analytical reagent in the separation of aluminum and other trivalent and tetravalent elements from divalent elements. Pyridine, C_5H_5N, is a weak organic base with $K_{diss} = 1.6 \cdot 10^{-9}$. Pyridine is a better precipitant for aluminum than ammonia. During precipitation by pyridine, the pH of the solution increases slowly; moreover, many divalent metals (Mn, Ni, Co, Zn, Cu, and Cd) form readily soluble complex compounds with pyridine, and are retained in solution without contaminating the $Al(OH)_3$ precipitate.

The precipitation of aluminum by pyridine takes place according to the reaction

$$AlCl_3 + 3 C_5H_5N + 3 H_2O = 3 C_5H_5N \cdot HCl + Al(OH)_3.$$

The pyridine salt formed during the reaction, together with free pyridine, forms a buffer mixture. If excess pyridine is introduced, the buffer action of the salt is suppressed, and optimum conditions for the precipitation of $Al(OH)_3$ are produced.

The acid solution, 100 ml in volume, is neutralized with ammonia to a faint turbidity, which is dissolved in 1–2 drops of HCl. Then 10 g of NH_4Cl are added, and the mixture is heated to boiling. To the hot solution, a 20% solution of pyridine is carefully added, drop by drop, with careful stirring, until methyl red indicator changes color, and then 10–15 ml are added in excess. The mixture is heated to boiling, and held on a water bath for 30 min, until the precipitate has coagulated and settled. The precipitate is filtered and washed with a 3% solution of ammonium chloride containing a few drops of pyridine. The determination is then continued as in the precipitation by ammonia.

During precipitation by the method just described, the pH of the reaction mixture is about 6.5. Under these experimental conditions, Fe, Cr, I, In, Ga, Ti, Zr, Th and Sc are precipitated together with $Al(OH)_3$; Mn, Ni, Co, Zn, Cu, Cd, Mg, alkali and alkaline-earth metals are not precipitated. Beryllium is precipitated, but not quantitatively. The precipitate of $Al(OH)_3$ obtained by means of pyridine settles readily and is easily filtered, without sorbing foreign ions. In the presence of ammonium salts and at high temperatures, denser precipitates are formed. If large amounts of ammonium salts are introduced, the sorption of divalent metals on the precipitate is greatly reduced.

Aluminum hydroxide precipitated by pyridine from chloride and nitrate solutions does not contain basic salts as impurity. In precipitation from sulfate solutions, the precipitate is contaminated by basic salts; the precipitation is not quantitative, the precipitate settles with difficulty and tends to pass through the filter. The harmful effect of up to 2 g of potassium sulfate, sodium sulfate, or ammonium sulfate can be eliminated by adding 10–20 g of ammonium chloride, but it is preferable not to introduce sulfates. One precipitation will usually separate aluminum from divalent metals. Reprecipitation is necessary in the presence of very high concentrations of Mn, Co, and Ni, as in the analysis of pyrolusite, certain cobalt ores, and cobalt and nickel salts.

When aluminum is separated from zinc, it should be borne in mind that a precipitate of the pyridine hydrochloride complex of zinc is formed in the cold, and is redissolved on heating. Therefore, if zinc is present, the precipitate should be filtered from hot solutions.

Morachevskii and Bashun [270] used the labeled atoms method to study the completeness of the separation of aluminum from zinc and cobalt by precipitation with pyridine. The separation of cobalt was quantitative after one precipitation at pH > 6, even in the presence of relatively small amounts of ammonium chloride (2 g per 100 ml). However, when the NH_4Cl concentration was less than 8 g per 100 ml, there was appreciable coprecipitation of zinc with $Al(OH)_3$. When the concentrations of aluminum and zinc were about equal (about 28 mg each), only about 95% of zinc could be separated in one precipitation, even in the presence of 10 g NH_4Cl. The separation was quantitative after reprecipitation.

Although it is more efficient than ammonia [320], pyridine is little used in analytical practice owing to its toxicity and repulsive small.

Precipitation by α-picoline. α-Picoline (methylpyridine) $CH_3C_5H_4N$, $K_{diss} = 1.7 \cdot 10^{-8}$, is a somewhat stronger base than pyridine. The reagent quantitatively precipitates all metals which are precipitated by pyridine, and also beryllium [318]. The pH of precipitation is 7.0. The reagent can be used to separate aluminum from Mn, Zn, Co and Ni.

Precipitation by urotropin. Urotropin (hexamethylenetetramine) decomposes in boiling solutions:

$$(CH_2)_6N_4 + 6 H_2O \rightleftarrows 6 HCHO + 4 NH_3.$$

The ammonia which is evolved in the reaction increases the pH of the solution, thus producing conditions favoring the quantitative precipitation of aluminum hydroxide.

According to Ray [1103], the precipitation of $Al(OH)_3$ from nitrate and chloride solutions is not quantitative; complete precipitation can be attained from sulfate solutions, owing to the strong hydrolysis of aluminum sulfate. Under these experimental conditions, Fe(III), Cr and Ti are precipitated as well. In the presence of ammonium salts, Mn, Zn, Co and Ni remain in solution. Aluminum can be separated from Mn, Mg and Ca by one precipitation; the separation from calcium is more satisfactory than it is when ammonia is employed [886]. Reprecipitation is required when aluminum is separated from zinc and cobalt. According to Shushich [502], aluminum can be separated from cobalt by one precipitation in the presence of $HgCl_2$, since the latter prevents the sorption of cobalt on the precipitated aluminum hydroxide. Quantitative separation from nickel cannot be attained even by reprecipitation.

Some workers have suggested that aluminum be separated from beryllium by precipitation in the cold, but in fact even then the precipitation of aluminum is unsatisfactory.

Precipitation by urea. When urea is boiled in a weakly acid solution, it decomposes into NH_3 and CO_2. The pH of the solution rises steadily and the optimum medium for the precipitation of $Al(OH)_3$ is produced. Urea is preferable to ammonia as a precipitant, since it yields dense precipitates, which are readily filtered and washed. Since the precipitation takes place from a homogeneous solution, foreign ions are absorbed less than during precipitation from ammonia [1270, 1273]. The disadvantage of the method is that the time required for the precipitation (1½–2 hours) is longer than in the ammonia method. The precipitation is preferably carried out from a sulfate solution, or else in the presence of salts of organic acids (succinates, benzoates, phthalates, formates, etc.). Precipitation from nitrates and chlorides is less satisfactory. According to Willard and Tang [1273], aluminum can be precipitated as basic sulfate from a gently boiling solution containing urea, ammonium chloride, and ammonium sulfate in 1–2 hours. The optimum pH value of 6.5–7.5 is produced if the solution contains 4 g of urea, 10–20 g of ammonium chloride, and 1 g of ammonium sulfate in 500 ml. The separation from Ca, Mg, Mn and Cd is satisfactory after one precipitation; for separation from Ni, Co and Zn reprecipitation is required.

Precipitation by other reagents. Before aluminum is precipitated by *phenylhydrazine* [804], iron must be reduced by ammonium sulfite. It has also been suggested that aluminum be precipitated by a mixture of *thiosulfate with phenylhydrazine* in the presence of sodium sulfite to reduce the iron [256]. Iron does not interfere even when present in large amounts; titanium and phosphorus only are precipitated together with aluminum. A serious disadvantage of phenylhydrazine is its toxicity, and for this reason the reagent cannot be recommended for the precipitation of aluminum.

Aluminum can also be precipitated as hydroxide by *acridine* [1142]. By this method aluminum can be separated from beryllium if the Al_2O_3 : BeO ratio is between 1 : 5 and 10 : 1. Aluminum can be separated from beryllium at Al_2O_3 : BeO ratios of between 1 : 16 and 1 : 3 by *p-chloroaniline* [1142]. The separation of aluminum from iron by *o-phenetidine* has been described [628]; the iron must first be reduced to ferrous iron by hydrogen sulfide.

Aluminum in the presence of ammonium chloride can also be precipitated as hydroxide by an aqueous suspension of *mercury amidochloride* $ClHgNH_2$ [1180] at room temperature; the precipitate is filtered on the following day. Aluminum can be separated from manganese in this way. A method for the separation of aluminum from divalent metals, and in the presence of hydroxylamine from iron as well, has been described [1181], with mercury amidochloride as precipitant. However, the method cannot be recommended for practical use; many elements interfere, the method is time-consuming, and mercury compounds are toxic. In the presence of ammonium chloride aluminum is quantitatively precipitated as $Al(OH)_3$ by monoethanolamine [358]. Korenman and Faiziev [191, 192] used this method to separate aluminum from magnesium and zinc. The separation from magnesium is quantitative, irrespective of whether the precipitation takes place at room temperature or from a hot solution. The separation of zinc is not satisfactory; accurate results are obtained after two reprecipitations only.

Aluminum can be precipitated by hexamethylenediamine [194]. A method for the precipitation of aluminum hydroxide by ethylene chlorohydrin has been described [438].

Hydrolytic precipitation of aluminum hydroxide
by salts of weak inorganic acids

These methods are based on the hydrolytic decomposition of aluminum salts of weak inorganic acids. The products formed are $Al(OH)_3$ and the free weak acid, which is easily expelled from solution by heating. These techniques can be considered as precipitation from homogeneous solutions. The $Al(OH)_3$ precipitates thus obtained are purer than those formed by precipitation with ammonia, and are denser and readily filtered.

Reagents proposed for the precipitation of aluminum include sodium thiosulfate [618, 629], potassium, sodium and ammonium nitrites [1147], iodine-iodate mixture [1200], ammonium carbonate [907], hydrazine carbonate [847], etc. However, these methods are not used much, since they are unsuitable for polycomponent samples (many metals which accompany aluminum interfere).

Precipitation by thiosulfate. In the thiosulfate method, the H^+ ions set free during the hydrolysis of aluminum salts are bound by thiosulfate ions to form unstable $H_2S_2O_3$, which rapidly decomposes into S, SO_2 and H_2O. The SO_2 is removed by boiling, while the sulfur burns away when the precipitate is ignited. Clennel [647] studied the various precipitants in this group, and found that the best results are obtained with thiosulfate. Thiosulfate reduces ferric iron to ferrous, but quantitative separation of aluminum from iron by thiosulfate alone has not been attained. Better results are obtained with mixtures of thiosulfate with sodium sulfite [367] or phenylhydrazine [919].

Precipitation by ammonium nitrite. In this method, nitrous acid, HNO_2, is formed as a result of hydrolysis, and decomposes into NO and NO_2. When aluminum is precipitated by the nitrite method, the sample solution is neutralized by ammonia to the appearance of the precipitate, which is dissolved in acid.

To a solution containing a sufficient amount of ammonium salts, 20 ml of a 6% solution of ammonium nitrite are added. The solution is diluted to 200 ml, and boiled for about 20 min until nitrogen oxides are no longer evolved, and is then heated for

15–30 min on a water bath. The precipitate is then filtered, washed, and ignited, as in the ammonia method [877a].

According to Congton and Carter [652], the nitrite method is highly accurate.

Precipitation by potassium iodide-potassium iodate mixture. The weakly acid sample solution is neutralized by ammonia to the first turbidity, which is then redissolved by acid, and a mixture of equal volumes of a 25% solution of KI and a 7% solution of KIO_3 is added. After 5 min the solution is decolorized by a 20% solution of $Na_2S_2O_3$. Small amounts of iodide-iodate mixture are then added; if iodine is liberated as a result, the solution is again decolorized by the addition of 1–2 ml of thiosulfate solution. It is then heated on a water bath for 30 min. When the precipitation is complete, the precipitate is filtered off, and washed and dried, as described in the ammonia method [877a].

Precipitation by carbonates. Majumdar and Sen [949] obtained satisfactory results when precipitating aluminum by ammonium carbonate in the presence of Fe(II) and Cr bound into stable complexes by α-picolinic(2-pyridinecarboxylic) acid. The Fe(II) complex is not decomposed by ammonium carbonate even at pH > 7.7. The complex with Cr(III) is formed if the solution is boiled for a few minutes with α-picolinic acid, and is stable up to pH 7.0. The optimum pH values for the separation from Fe and Cr are 5.8–6.85 and 6.25–7.05, respectively. At higher pH values the results are too high and at lower pH values too low. If aluminum is precipitated at pH 6.25–6.8, neither metal interferes. Smith and Cagle [1176] succeeded in separating 10 mg Al_2O_3 from 90 mg Fe_2O_3 by precipitating the aluminum by ammonium carbonate in the presence of dipyridyl.

It is difficult to achieve a quantitative precipitation of aluminum by ammonium carbonate, and it is only possible if the pH is maintained within very narrow limits [907, 989]. Precipitation by ammonium bicarbonate is of even less importance [882]; also, precipitation of aluminum hydroxide by barium or calcium carbonates is not practical [1231].

The precipitation by hydrazine carbonate gives a better separation of aluminum from divalent metals than by the carbonates just mentioned [847, 882]. Aluminum can be quantitatively separated from Mn, Zn, Co and Ni by double precipitation; if hydrazine is present, it can be separated from iron as well.

Precipitation by other reagents

Ammonium sulfide precipitates $Al(OH)_3$ from solutions of aluminum salts. The precipitation is not quantitative; moreover, aluminum can be separated from alkali metals only by this method, and therefore the method is of no interest. A dense precipitate is obtained [735] when $Al(OH)_3$ is precipitated from aluminate solution by carbon dioxide.

By precipitating $Al(OH)_3$ by a suspension of HgO, aluminum can be separated from a small number of metals only; Zn, Co, Ni, Be and Cr are partially coprecipitated. Separation of aluminum from manganese is possible, since manganese separates out on long standing only.

Precipitation of aluminum as basic salts of organic acids

Precipitation as benzoate. The gravimetric benzoate method is one of the most accurate methods for the determination of aluminum. Benzoic acid was first employed in chemical analysis by Kolthoff et al. [892] to separate titanium and trivalent metals from divalent metals. Precipitation with benzoate results in the formation of basic aluminum benzoate, which is easier to filter and to wash than the precipitate obtained with ammonia; furthermore, the sorption of foreign ions is much less.

Various published data are available on the optimum pH of precipitation of benzoate. It has been shown [846, 1175] that titanium and trivalent metals are quantitatively precipitated by benzoate at pH 3–4. The following pH values corresponding to quantitative precipitation of aluminum benzoate have been reported: 3.7 [834]; not less than 3.0 [1258]; not less than 3.5 [1175]; 3.5–4.0 [1033].

For quantitative precipitation of aluminum the pH should be at least 3.5 [1175, 1275].

Under conditions suitable for the separation of aluminum, the following elements are also quantitatively precipitated: Fe(III), Ti, Bi, Zr, Hf, Th, Sn(IV), Ce(IV), Cr(III)* [846, 892, 1258], V(IV) [344, 491]; the elements V(V), Co, Ni, Mn, Zn, Cd, Ba, Sr, Mg, Fe(II), Ca, Hg(II), Ce(III) and alkali metals are not precipitated [892]; U(VI), Ti(III), Cu(II), Pb and Sn are partially precipitated [892]. Beryllium is not precipitated at pH below 4.3, but is coprecipitated with aluminum benzoate [834]. Of the anions, phosphates precipitate out in considerable amounts.

According to Wilson [1275], the coprecipitation of Ca, Mg and Mn with aluminum benzoate increases with increase in pH; it is insignificant at pH 3, and after reprecipitation, even at pH 3.5. Thus, if Ca, Mg and Mn are present, aluminum benzoate should be precipitated at pH 3.5, while in the presence

* According to other data [344], the precipitation of chromium is not quantitative.

of large amounts of these metals, aluminum benzoate should be reprecipitated. According to Smalls [1175], up to 50 mg Co, Cr(VI), Ni and Zn do not interfere with the precipitation of aluminum at pH 3.7–3.9; if reprecipitation is effected, up to 100 mg of these metals may be present. At pH 5.5 and above, the aluminum precipitate is gelatinous, since $Al(OH)_3$ is partially precipitated and is difficult to filter; in addition, occlusion of foreign ions becomes more marked. Thus, during the precipitation of aluminum benzoate the pH should be maintained at not less than 3.5 and not more than 4.0.

When aluminum is determined by the benzoate method, iron is sometimes preliminarily separated by sodium hydroxide [230], but this is inconvenient, as reprecipitation is then required. A better procedure is to reduce the iron to ferrous iron by sulfite [117], hydrosulfite [344], thioglycolic acid [1277], or hydroxylamine, or else to complex the metal. According to Wilson [1277], reduction by sulfurous acid often results in the contamination of aluminum benzoate precipitate by iron; it is better to use thioglycolic acid and, according to this worker, up to 1 g of iron does not interfere with the determination of 20 mg of aluminum. Thioglycolic acid masks and reduces a number of other ions as well, and at the same time the number of interfering ions becomes much smaller. Fe, Cu, Zn, Pb, Mn, Sn, tungstates and molybdates do not interfere; Ti, Cr, Zr, V(IV) and Th are the only remaining interfering ions. Jurczyk [854] successfully employed thioglycolic acid to mask the interfering elements in the analyses of ores, slags, pig iron and alloyed steels. Aluminum in concentrations between 0.5 and 5% could be determined in the presence of 3% Mo, 3% Co, 6% Ni, and 8% Mn; in the presence of V, Cr and Cu the results were high.

Alfonsi and Bussi [521] determined aluminum in bronzes and brasses with preliminary precipitation of copper as thioglycolate. The resulting precipitate is voluminous, and therefore it is advisable to make up the solution together with the precipitate to the mark in a volumetric flask, and filter off only an aliquot of the solution for the determination of aluminum.

If aluminum is to be precipitated from chromium-containing solutions, the chromium is oxidized to Cr(VI) by hydrogen peroxide in an alkaline medium [117]. In the presence of phosphates the precipitate of aluminum benzoate contains some aluminum phosphate. In such cases it may be better [1275] to dissolve the benzoate and determine aluminum gravimetrically by the hydroxyquinoline method. Other workers, too, employed the benzoate method for preliminary separation of aluminum only. For example, aluminum is determined in magnesium, zinc and copper alloys by the complexometric method after preliminary separation as benzoate [976].

The benzoate method for the separation and determination of aluminum is applicable to copper [520, 521, 676, 1015], magnesium [362, 976, 1199] and zinc [976] alloys, titanium concentrates [209] and phosphate rocks [1275]. Benzoic acid can be used as a qualitative reagent in the presence of phosphates [537].

The following procedure [1277], given in a slightly modified form, may be employed in the gravimetric determination of aluminum by the benzoate method.

To 100 ml of a weakly acid sample solution, preheated to 60–80°C, are added 2 g NH_4Cl, 2 g CH_3COONH_4, and 1 ml of 90% thioglycolic acid. When the iron has been reduced, 20 ml of a 10% solution of ammonium benzoate are introduced, and then Bromophenol Blue indicator and 1 : 1 ammonia, which is slowly added with constant stirring until the indicator has changed color. The mixture is brought to the boil, and gently boiled for another 5 min. The solution is placed on a boiling water bath for 15 min.

The precipitate is immediately filtered through a white ribbon paper, and is washed with five portions of a hot washing solution containing 1% of benzoate and 2% of acetic acid. The precipitate is transferred back to the beaker used for the precipitation by water, and hot 5 N HCl is poured onto the filter; the filtrate is collected in the same beaker. The filter paper is then thoroughly washed with hot water. The precipitation is repeated, and the precipitate filtered off and washed as above. The filter together with the precipitate is then ignited at 1200°C for two hours. The conversion factor to aluminum is 0.5291.

The benzoate can be precipitated from homogeneous solutions.

To 350 ml of benzoate buffer solution (48 g NH_4Cl and 18 g benzoic acid in 2 liters of water) are added 3 g of urea and 50 ml of a weakly acid sample solution, which should contain 0.01–0.05 g of aluminum. The mixture is rapidly brought to the boil, and held for 1½–2 hr on a steam bath. The precipitate is filtered through medium porous filter paper and washed with a 1% solution of ammonium nitrate. It is then ignited at 1250°C to constant weight in a platinum crucible, and weighed as Al_2O_3 [834].

The dependence of the completeness of separation of aluminum on the duration of the precipitation process is illustrated by the following table [834]:

Time, min	pH	mg Al_2O_3/400 ml of filtrate	Time, min	pH	mg Al_2O_3/400 ml of filtrate
30	3.15	6	60	3.80	0.01
50	3.63	0.2	90	4.00	0.01
55	3.70	0.01	120	4.08	0.01

As usual in homogeneous precipitation, the precipitates are purer than in conventional precipitation. Homogeneous precipitation as benzoate is preferable to homogeneous precipitation as succinate; the benzoate precipitate is less dense than the succinate, but is more easily filtered and washed.

Precipitation as succinate. This is an old and undeservedly neglected method for the separation and gravimetric determination of aluminum. It has several advantages over the ammonia method, the main being that aluminum can be separated from a larger number of elements than by the ammonia method. In accuracy and convenience it is only slightly inferior to the benzoate method.

Aluminum is precipitated from boiling solutions containing succinic acid and urea. The pH of the solution at the moment of the first appearance of the aluminum precipitate is pH 4.0−4.1, while its final pH is 4.4−4.5 [1273]. Fe(III) begins to precipitate at pH 2.0−2.1, while Fe(II) is only precipitated at pH 5.5. Thus, interference by iron may be eliminated if the iron is reduced to the ferrous state. The procedure recommended by Willard and Tang [1273] is time-consuming, as it involves two precipitations, each requiring two hours of boiling. Palmer [1042] succeeded in considerably shortening the duration of the determination and adapting the procedure to routine analysis.

Willard and Tang [1273] reduce iron by ammonium metabisulfite and phenylhydrazine. Parker [1046] uses hydroquinone instead of phenylhydrazine, as the former reagent is more convenient in use and nontoxic. In the first stage of the determination, the iron is reduced by metabisulfite, and is retained in the reduced state by hydroquinone. When the sample solutions are neutralized by ammonia, the pH may suddenly rise, and a bulky, gelatinous precipitate is formed, which is difficult to filter and wash. This can be avoided if the solution is neutralized by sodium carbonate. Nitrates interfere, since they oxidize the hydroquinone. Accordingly, if the sample to be solubilized is not readily soluble in hydrochloric acid alone, nitric acid should not be introduced; it is preferable to use bromine water or hydrogen peroxide, which are subsequently expelled by evaporation to dryness and sintering.

To ensure quantitative precipitation of aluminum, the solution is boiled for 45 min after the appearance of the precipitate, and the volume of the solution is maintained at about 500 ml. If Fe : Al = 10 : 1, satisfactory results are obtained by one precipitation (an analysis of a mixture of 1 g Fe and 100 mg Al gave 98−100 mg aluminum).

Palmer [1042] recommended the following procedure for the determination of aluminum in iron, steels and slags.

The acid sample solution is brought to the boil and 5 g $KHSO_3$ are introduced, then the solution is boiled for another 1–2 min. The solution is then diluted with water to 400 ml and heated to boiling. Five g of succinic acid, 4 g of urea, and 5 g of hydroquinone are introduced into the boiling solution. The mixture is boiled for another 1–2 min and is then removed from the hotplate. Sodium bicarbonate (solid salt or solution) is cautiously added to the appearance of a permanent precipitate, which is redissolved by the addition of 15% HCl. The solution is diluted to 600 ml, and brought to the boil. The boiling is continued until 45 min after the appearance of the precipitate, while the volume of the solution is maintained at not less than 500 ml. The precipitate is immediately filtered and washed with a 2% solution of ammonium nitrate. The precipitate is ignited at 1200°C and weighed as Al_2O_3.

Titanium is coprecipitated together with aluminum. In this case the ignited precipitate is then fused with $KHSO_4$, titanium is photometrically determined by hydrogen peroxide, and the result is suitably corrected. The determination takes about 2 hours. The relative error is about 2%.

If the sample contains zinc, ammonium chloride must be added during the precipitation. The solutions must then be neutralized very carefully to avoid a large increase in pH, which would result in the formation of gelatinous precipitates.

According to Willard and Tang [1273], precipitation by succinic acid with urea will separate 0.1 g of Al from 1 g Mn and Cu, and 0.1 g Li and Co; by reprecipitation 0.1 g Al can be separated from 1 g Ni, Co and Zn.

It is seen that aluminum can be separated from divalent metals much better by this method than by precipitation with ammonia.

Precipitation as diphenate. The ammonium salt of diphenic acid (diphenyl-2,2'-dicarboxylic acid or *o,o'*-dibenzoic acid) reacts with aluminum to form a sparingly soluble basic salt [3]. The solubility of the precipitate in water is $2.3 \cdot 10^{-5}$ mole per liter. The precipitate is practically insoluble in a 0.2% solution of the reagent, alcohol and ether; it is readily soluble in acids and alkalis. Ammonium diphenate precipitates other trivalent metals as well, but does not precipitate alkali and alkaline-earth metals, Fe(II), Mn, Co, Ni, Zn and Cd. Complete precipitation of aluminum diphenate is observed at pH 3.2–5.0. At pH > 5.0, $Al(OH)_3$ is partially precipitated together with the basic diphenate. At optimum pH values, aluminum is so completely precipitated by ammonium diphenate that the amount of aluminum remaining in solution, determined photometrically with aluminon, was less than 0.03 mg.

The presence of small amounts of ammonium salts favors the formation of a denser, more readily filtered precipitate. The optimum amounts of ammonium salts (chloride or nitrate) are 0.15 g per 0.03 g Al; larger amounts increase the solubility of the precipitate. The optimum precipitation temperature is 20–40°C.

The coprecipitation of Co, Fe(II), Ni and Mn with aluminum diphenate is insignificant; reprecipitation is required for separation from zinc [3].

In the presence of divalent metals the pH value should not be higher than 4.4; at higher pH values the coprecipitation increases. The following method is used to precipitate aluminum [3].

To a solution of not more than $0.1 g$ Al_2O_3 in nitric acid or hydrochloric acid are added 10–20 ml of a 1% solution of ammonium chloride or ammonium nitrate. The solution is diluted to 75 ml with water, 3–4 drops of methyl orange solution are introduced, and the pH is adjusted to 4.4 with acid or ammonia. The solution is heated until it just boils, and aluminum is precipitated from the gently boiling solution by a 3.4% solution of ammonium diphenate (4.5 ml per $0.01 g$ Al_2O_3). The precipitant is introduced from a burette drop by drop for 10 min, with vigorous stirring. The beaker is covered by a watch glass, and left for one hour on a boiling water bath. When the solution has cooled to 40–45°C, the precipitate is filtered through a white ribbon paper and washed, first by decantation, and then after it has been transferred onto the filter, 5–6 times more. The precipitate is ignited at 1100–1200°C and weighed as Al_2O_3.

For the preparation of ammonium diphenate, see [3].

Precipitation as cinnamate. Aluminum is precipitated from weakly acid boiling solutions by ammonium cinnamate as the sparingly soluble basic salt [321], with the following structure

$$(C_6H_5CH = CHCOO)_2 \diagdown \!\!\!\!\!\! {\diagup} Al,$$
$$HO \diagup$$

During precipitation of aluminum, the pH of the solution is adjusted to 5.0–5.1. The composition of the precipitate remains constant, irrespective of the amount of excess precipitant, and the precipitate can be used as the gravimetric form. The precipitate of basic aluminum cinnamate readily coagulates and can be rapidly filtered and readily washed. Up to 5–10g of NH_4Cl, NH_4NO_3 and $(NH_4)_2SO_4$ do not interfere. At pH 5.0–5.1, the following elements are also precipitated: Fe(III), Cr(III), Ti, In, Ga, Be, U, Sc [85,321, 322], Zr and Th [903]. Titanium and zirconium can be quantitatively precipitated by cinnamic acid from a more acid solution, at pH 1.9–2.1, and can thus be separated from aluminum [85, 85a]. Complete separation of titanium from aluminum can be attained at TiO_2 : Al_2O_3 ratios of 1 : 100 to 10 : 1.

In the presence of ammonium salts, divalent metals do not precipitate and can be separated from aluminum. However, if small amounts (5 g) of aluminum chloride are present, they may be partially coprecipitated. In the presence of 10g of ammonium chloride, aluminum is quantitatively separated from Mn, Ni and Zn; to effect separation from cobalt, the amount of ammonium chloride must be increased to 15g. As a rule, reprecipitation is unnecessary. According to Babachev [23], it is preferable to separate aluminum from divalent metals by cinnamic acid than by pyridine; this author suggests that the precipitation be carried out at pH 5.8–6.0.

Aluminum can be precipitated as the basic cinnamate by the following procedure [321].

To the acid sample solution, of volume 60 ml, are added 15 g NH_4Cl, and the solution is carefully neutralized with 1 : 1 ammonia to the appearance of a faint turbidity, which is redissolved by the addition of 2–3 drops of HCl (sp.gr. 1.12). The solution is diluted with water to 80–90 ml and heated to boiling. To the gently boiling solution are slowly added, drop by drop, 20 ml of ammonium cinnamate solution (5% calculated as cinnamic acid). When the precipitation is complete, the beaker is covered by a watch glass and the solution is gently boiled for 1–2 min. It is then left to stand on a water bath for 1– 1½ hr, with occasional stirring. The settled precipitate is then filtered by decantation through a white ribbon filter paper. The precipitate left in the beaker is washed twice by decantation with a hot washing solution (10g NH_4Cl, 20 ml of reagent dissolved in 100ml), using 20–25 ml of the solution each time. The precipitate is then transferred onto the filter and washed with hot wash liquor (20 ml of reagent solution and 280ml of hot water). The precipitate is dried, ashed, and ignited to Al_2O_3.

After the precipitate has been washed with the wash liquor and alcohol, it may be dried at 110°C in a Schott crucible.

The relative error of the method is 0.2–0.4% in the determination of 0.025–0.050 g Al_2O_3.

Precipitation in the form of other compounds. The acetate method, which is one of the earliest methods for the separation of aluminum, is now scarcely ever used, since the precipitation is not quantitative, the precipitate is gelatinous and difficult to filter, and tends to sorb foreign ions. Aluminum is usually precipitated as basic acetates at pH 5.2–5.6; however, according to Morachevskii and Bashun [270], the precipitation of aluminum is quantitative at pH 7.5–7.8 only. Under these conditions certain divalent metals such as zinc and cobalt are partially coprecipitated with aluminum. In the presence of iron as collector, aluminum can be precipitated at pH 5.8– 6.0 and can thus be separated from these metals. Other authors [921, 1024] also recommended the use of iron as collector to achieve quantitative precipitation of aluminum; the iron is then separated from aluminum by means of sodium hydroxide [1024].

The formate method is even more rarely used than the acetate method. Methods based on the separation of aluminum as basic salicylate and basic anthranilate are not used in analytical practice.

Precipitation as cryolite

The formation of the sparingly soluble aluminum complex, cryolite Na_3AlF_6, is the principle of very important methods for the determination of aluminum, including gravimetric, titrimetric, and potentiometric methods. The complex of composition Na_3AlF_6 is formed if NaF is present in an excess of more than 1.4%. If the excess is less, the composition of the complex corresponds to the formula $4AlF_3 \cdot 11NaF$ [312, 401]. Cryolite is markedly soluble in water; its solubility is 0.39 g per liter, or $1.36 \cdot 10^{-3}$ mole per liter [402]. Its solubility decreases rapidly in the presence of excess sodium fluoride. If this excess is 0.02 mole NaF per liter, the filtrate from cryolite is perfectly free of aluminum, even traces.

The most detailed study of the fluoride (cryolite) method for the determination of aluminum was carried out by Yakovlev [513]. The fluoride method is one of the most valuable methods for the determination of aluminum. It is highly selective, and aluminum can be simply and quite rapidly determined in multicomponent samples such as steels, alloys based on Cu, Ni and Co, containing various proportions of alloying elements. The fluorides of Mo, W, V, Nb, Zr, Ti and Fe are much more soluble than cryolite, and remain in solution; Co, Ni and Cr also do not interfere.

Some metals are kept in solution by adding masking agents. Examples of such agents are ammonium oxalate (for many metals) and ammonium citrate (for Ni, Cu and Ti). It is best to use a mixture of a saturated solution of ammonium oxalate and a 40% solution of ammonium citrate. In the presence of Complexone III, sodium fluoride quantitatively precipitates aluminum from a weakly acid solution, whereas Cu, Fe, Co, Ni, Zn, Sn, Bi, Sb and Mn remain in solution [41].

Cryolite must be precipitated from sulfuric acid solutions. If it is precipitated from hydrochloric acid solutions, V, Ti and Fe are coprecipitated. For this reason the solutions to be analyzed are usually first evaporated with sulfuric acid to the appearance of SO_3 fumes. In the presence of appreciable amounts of Fe(III), the sparingly soluble iron salt $2FeF_3 \cdot 5NaF$ (solubility 2.4 g/liter) is coprecipitated. Attempts have been made to reduce iron to ferrous iron by sodium sulfite [397, 402], KI, sodium thiosulfate, hydrogen sulfide, hydroxylamine, hydrazine and zinc amalgam. Zinc amalgam proved to be the best reducing agent for iron in the cryolite

method [513]. If the sample solution is shaken with zinc amalgam for
1–2 min, ferric iron is quantitatively reduced. The fluoride complex of
iron is much less soluble than cryolite; for this reason, the iron complex
can be completely removed from the filter by prolonged washing (10–
15 times) with a 0.5% solution of NaF, and preliminary reduction of iron
is unnecessary [332]. In practice, this experimental variant proved to be
more convenient than reduction by zinc amalgam, and is therefore widely
employed.

It is important that the solutions to be analyzed be correctly neutralized.
If the acidity of the solution is too low, Ti, Fe and Zr may coprecipitate;
if the acidity is too high, there is danger of contamination of the solutions
by the silica leached out of the glassware [513]. If the solution to be analyzed contains titanium, the solutions must be neutralized in the cold; if
hot solutions are neutralized, there is danger of separation of $Ti(OH)_4$,
which is difficult to dissolve in acids. In the presence of titanium and iron,
the precipitate should not be allowed to settle for more than 20 min, or the
results will be too high [513].

Tananaev and Talipov [402] determined aluminum by weighing it as
cryolite. The precipitate was washed with a 0.5% solution of NaF, then
with 50% ethanol, and dried at 120–130°C; the cryolite was weighed.
The filtration takes a long time; it can be accelerated by centrifugation.
A similar method was also proposed by Pen'kova [332] and Pender [1054].
However, centrifugation is not a suitable technique in routine determinations,
and another disadvantage of the method is the high consumption of ethanol.
Moreover, cryolite of constant composition is formed only if the experimental conditions are rigorously adhered to [401]. Accordingly, precipitation
as cryolite is suitable for preliminary separation of aluminum, but not its
determination.

The cryolite precipitate is dissolved and the aluminum in solution determined by the following method. Cryolite is readily dissolved in 40 ml of a
mixture of the following composition: 150 ml of a saturated solution of
H_3BO_3, 250 ml HCl (sp.gr.1.19) and 600 ml of water.

A number of authors precipitate aluminum by ammonia after the dissolution of cryolite. In this case, fluoride ions must be eliminated by evaporating the solution with sulfuric acid to the appearance of SO_3 fumes, and
the determination is thus more time-consuming. Yakovlev [513] recommended that aluminum be precipitated as hydroxyquinolate, without
evaporation of the solution. Fluorine does not interfere with the precipitation of hydroxyquinolate, provided that the solution is acidified before

precipitation to complex the fluorine with boron. Under these conditions aluminum hydroxyquinolate is quantitatively precipitated in the presence of 0.05 g F$^-$ with 8–10 times excess of boric acid. Vasil'ev [77] worked under somewhat different conditions, and determined aluminum by the hydroxyquinolate method only, in the presence of 0.003 g F$^-$ with a 30-fold excess of H_3BO_3.

Aluminum can be determined by the cryolite method using the following procedure [377a, 513].

To the solution are added 5–7 ml of sulfuric acid (sp.gr. 1.84), and the mixture is evaporated to the evolution of SO_3 fumes. The salts that separate out are redissolved by heating in 50–70 ml of water. The insoluble residue (SiO_2, WO_3 and Nb_2O_5) is filtered off. The solution is cooled, and neutralized with ammonia until the hydroxides begin to precipitate. The precipitate is then redissolved by adding sulfuric acid (sp.gr. 1.84) drop by drop, with 5 drops in excess. Twenty-five ml of masking solution (a 1 : 1 mixture of a saturated solution of ammonium oxalate and a 40% solution of ammonium citrate) and 60–80 ml of a 3.5% solution of NaF are added, and the solution is stirred. After 20 min the cryolite precipitate is filtered onto two fine-pored filter papers, and washed 10–15 times with a 0.5% solution of NaF. The precipitate on the filter is dissolved in 50–70 ml of a boiling H_3BO_3/HCl mixture (150 ml of a saturated solution of H_3BO_3, 250 ml of HCl (sp.gr. 1.19) and 600 ml of water). The filter is washed 5–6 times with hot water. The solution is neutralized by ammonia (sp.gr. 0.91) to methyl red, and then with HCl (sp.gr. 1.19) until the yellow color turns pink, and then 5 drops are added in excess. The solution is cooled to 30–40°C, and 20 ml of a 2.5% solution of hydroxy-quinoline in 4% acetic acid and 10 ml of a 20% solution of ammonium acetate are introduced. The mixture is stirred, and 25 ml more of ammonium acetate solution and 5 ml of ammonia (sp.gr. 0.91) are introduced. The solution with the precipitate is heated to 60–70°C, and is held at this temperature for 10–15 min to coagulation of the precipitate. The precipitate is left to settle for 30 min at room temperature, and is then filtered through two white ribbon papers and washed 8–10 times with a wash liquor (12.5 ml of a 20% solution of ammonium acetate and 2 drops of ammonia are added to 1 liter of water). The filter paper together with the precipitate is placed in a tared porcelain crucible and covered with 2–3 g of anhydrous oxalic acid. The precipitate is dried, ashed at 400°C, ignited at 1100°C, and weighed as Al_2O_3. One determination takes 4–5 hours.

If the aluminum content is 1%, the relative error is up to 5%; if 5–6% Al are present, the relative error is 2%.

The cryolite method is the standard one for the analysis of alloy steels, certain ferrous alloys, and metals.

The method can be used for the determination of aluminum in steels [403, 920], intermetallic phases containing Ni, Co, Cr, Fe, Ti, Zr, V, Mo, W and Nb [512], ferro alloys [29, 40, 332a], silico-zirconium [39], aluminum bronzes [41], brass, slags, and for the separation of aluminum from titanium [391] and beryllium [1104, 1126].

Aluminum can be gravimetrically determined as the double potassium salt [301], but this method is much inferior to the cryolite method owing to the higher solubility of the precipitate, and is not used in practice.

Determination as phosphate

The method is based on the formation of the sparingly soluble aluminum phosphate $AlPO_4$ in a weakly acid medium. The method is not as accurate as the other gravimetric methods: hydroxyquinoline, benzoate, succinate, etc. The composition of the precipitate is not constant but depends on the conditions of precipitation: the pH of the solution and the concentration ratio $Al:P_2O_5$ [971]. The precipitate always contains some P_2O_5, which is difficult to remove by washing. If the precipitate is washed with water, some of it may dissolve [613] or hydrolyze [89]. The best washing liquor is a hot 5% solution of ammonium nitrate. Lundell and Knowles [940] showed that accurate results can be obtained in the determination of milligram amounts of aluminum if the precipitation is conducted in the presence of paper pulp, and if the precipitate is washed until free of chlorides. If larger amounts (0.1–0.2 g) of Al_2O_3 are to be determined, the results are 2–3% too high owing to the strong sorption of P_2O_5. Aluminum phosphate begins to separate out at pH 3.5 [783]; it is usually precipitated from solutions at pH 5–5.4 [89, 940, 1281]. In the presence of calcium, the precipitation must be carried out in more acid solutions, pH 4–4.5, since calcium phosphate precipitates at pH 5.7–6.0 [783]. Luff [937] pointed out that lower pH values must also be maintained when precipitating from solutions containing zinc. Ray et al. [1102] obtained satisfactory results in the determination of aluminum phosphate at pH 3.7–3.9. Precipitation from alkaline solutions did not give satisfactory results, since $Al(OH)_3$ is partially precipitated under these conditions; moreover, at elevated pH values the phosphate is gradually converted to the hydroxide [971]. When precipitating aluminum as phosphate, it is very important to use a 5–10-fold excess of the precipitant; if the excess is too small, the results will be too low [940]. Aluminum phosphate is readily dissolved by mineral acids and by citric, tartaric, oxalic and other acids [1251], but not by acetic acid.

Ferric iron, Ti, Zr, and partially V, are also precipitated under the conditions employed for the precipitation of aluminum. Chromium is strongly sorbed on the aluminum precipitate and must be previously separated. The effect of chromate is not very strong [780], so that the effect of chromium may be reduced if Cr(III) is oxidized to Cr(IV). Depending on the experimental conditions and on the amounts present, Mn, Zn and Ca can also be coprecipitated with $AlPO_4$. The precipitation of Fe(III) is prevented by converting it to Fe(II). Thiosulfate may also be employed for this purpose [531, 652, 745]. According to Hillebrand et al. [89], thiosulfate is not very suitable for the reduction, since the sulfurous acid formed in the process dissolves $AlPO_4$ partially, and therefore the results are too low. These workers proposed that iron be reduced by hydrogen sulfide.

When aluminum is determined in steel, Ray et al. [1102] remove most of the iron by extracting with ether from a solution $6N$ in HCl. The residual iron and certain other elements are kept in solution by a mixture of thioglycolic acid and ammonium thiocyanate. Titanium and zirconium are preliminarily removed by sodium hypophosphite and bromine water; Cr, V, Mo, Sn, Mn, Zn, Ni and Co, if present in amounts in which they usually occur in steels, do not interfere. According to these authors, 500-fold amounts of Cr, V, Mn, Ni and Co do not interfere with the precipitation of $AlPO_4$ at pH 3.7—3.9. The separation from Fe, Cr, V, Sn, Mn, Zn, Mo, Ni and Co is so complete that in most cases reprecipitation need not be carried out. The pH must be rigidly controlled if accurate results are to be obtained. The best results are obtained at pH 3.7—3.9; at lower pH values the precipitation is incomplete, while at pH values above 3.9 the results are too high owing to contamination of the precipitate by other elements. Heating will accelerate the coagulation of the precipitate. Precipitates should not be left overnight.

The $AlPO_4$ precipitate should be ignited at 1200°C [968, 1124]; at higher temperatures the precipitate is partially decomposed. It is hygroscopic and should be rapidly weighed.

Precipitation as cupferronate

Cupferron, the ammonium salt of nitrosophenylhydroxylamine, is a very important reagent in the analytical chemistry of aluminum. It is used only rarely for the determination of aluminum, but is often employed for its preliminary separation and for removal of interfering elements. Even though a large number of metals are precipitated by cupferron, it can be used to separate aluminum from many metals if the medium is suitably adjusted. Cupferron quantitatively precipitates aluminum at pH 4.6. Beryllium, gallium, yttrium, cerium and erbium are also precipitated under these conditions; Cr(III), Tl(III), In, Th and U(VI) are partially precipitated. The metals Fe, Ti, V, Zr, Sn, Nb and Ta can be separated from aluminum by precipitation in acid medium ($1N$ acid). Aluminum in turn can be separated from Mn, Mg and alkaline-earth metals by precipitation at pH 4.5—4.6.

The precipitate of aluminum cupferronate has no constant composition, and for this reason it is converted to Al_2O_3 by ignition. When metals are separated from aluminum by precipitation in a more acid medium, the products of oxidation of cupferron, on which aluminum may be partly sorbed, are precipitated together with the cupferronates. A better separation is achieved if the cupferronates are extracted. For this reason, the separation of metals by precipitation as cupferronates, as well as precipitation of aluminum cupferronates, are limited in application. But methods involving extraction of cupferronates are widely employed (for details, see Chapter III).

N-Benzoylphenylhydroxylamine can be employed in place of cupferron [1152]. The conditions for the precipitation of aluminum and other metals by this reagent are the same as in precipitation by cupferron. The reagent has the advantage that it is more stable to heat, light, and atmospheric oxygen than cupferron. If the precipitation is carried out from a hot solution, crystalline precipitates are obtained, which the author claims are not contaminated by the reagent; the optimum precipitation temperature is about 65°C. Quantitative precipitation of the reagent takes place at pH between 3.6 and 6.4. According to Schome [1152], the aluminum precipitate can be dried at 110°C and weighed; the conversion factor to aluminum is 0.04064.

Triethanolamine salt of 4-stilbenylnitrosohydroxylamine (styrylcupferron) reacts with aluminum at a pH of about 7 to give a sparingly soluble compound of composition $[C_6H_5CH = CHC_6H_4N(O)NO]_3Al$, which can be used for the gravimetric determination of aluminum [688]. To determine aluminum, hydroxylamine hydrochloride is added to the hot solution, the pH is adjusted to about 7, 250 ml of a hot solution of styrylcupferron are added, and then a mixture of ethanol with pyridine (4 : 1) in an amount corresponding to 15% of the total volume of the solution. The precipitate is filtered, washed, and dried at 110°C. The conversion factor to aluminum is 0.03627. During the determination of 4–12 mg Al the error is less than 1.5%. The advantages of the method are that the precipitate is easily filtered and has a constant composition; besides, the conversion factor is small.

Precipitation in the form of other compounds

Embelin, $C_{17}H_{26}O_4$, reacts with aluminum at pH 4.0–4.5 to form the compound $Al_2(C_{17}H_{24}O_4)_3 \cdot C_{17}H_{26}O_4$, which is sparingly soluble in water and can be used for the gravimetric determination of Al [1101]. The precipitate can be ignited to Al_2O_3 or else dried at 105–110°C after it has been filtered through a No.4 Schott crucible. Embelin is a suitable reagent for the separation of aluminum from beryllium, since the latter precipitates at pH 6.7–7.0.

Salicylhydroxamic acid reacts with aluminum at pH 6.0–7.5 to give a white, sparingly soluble precipitate of composition $Al(C_7H_5O_3N)(OH)$, which is suitable for the gravimetric determination of Al [1065]. The metals Cu, Fe, Ti, V and Mo are determined by extracting their hydroxamates with chloroform from concentrated solutions of sulfuric acid. If the pH is adjusted to the correct value after the precipitant has been added, phosphate ions do not interfere.

bis-2-Pyridylglycol at pH 5.2 quantitatively precipitates aluminum [564]; the precipitate can be ignited to Al_2O_3 and aluminum quantitatively determined. The reagent also forms precipitates with Be and U(VI).

It has been suggested [1220] that aluminum be determined by hexamminecobaltichloride in the presence of sodium fluoride or ammonium fluoride. Beryllium does not interfere if the ratio Be : Al < 1 : 2.

Aluminum can be quantitatively precipitated by tannin from weakly acid, neutral, or ammoniacal solutions as a bulky sorption compound of tannin and $Al(OH)_3$ of variable composition. The precipitate can be filtered more readily than when ammonia is employed as precipitant. The precipitates are ignited to Al_2O_3. The disadvantage of working with tannin is that very bulky precipitates are obtained with small amounts of aluminum.

The methods used to separate aluminum from a number of other elements by means of tannin are inferior to many other methods, and tannin is no longer used in practice to separate or determine aluminum.

If aluminum and lithium salts are allowed to react in ammoniacal and alkaline solutions, the compound $Li_2O \cdot 2Al_2O_3$ is formed; this compound can be used as a gravimetric form of aluminum [1027, 1087], but the method is scarcely ever employed.

TITRIMETRIC METHODS

Complexometric, acidimetric, alkalimetric, and other titrimetric methods have been proposed for the determination of aluminum. Of these, only complexometric methods are widely employed.

Complexometric titration

Complexometric methods are highly accurate, almost as accurate as gravimetric methods, and much less time-consuming. They are therefore widely employed for the determinations of aluminum in various materials.

Aluminum reacts with Complexone III to form a complex of composition

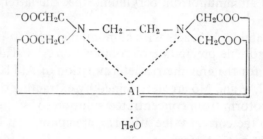

As a trivalent metal with a coordination number of 6, aluminum is bound by the main valence bonds to the hydroxyl groups, and forms coordination bonds with the amino groups and a water molecule. Such complexes are formed mainly in a weakly acid medium. At pH > 8, the main species are hydroxocomplexonates of type $[Al(OH)(H_2O)Y]^{2-}$ and $[Al(OH)_2Y]^{3-}$ [773]. The hydroxocomplexonates are formed in insignificant amounts even at pH 5−6 [1248].

Aluminum complexonate is a stable complex with an instability constant of $10^{-16.13}$ [490, 1154]. But, in spite of this, the process of formation of the complex is quite slow, owing to the tendency of aluminum to form a hydrated ion. Due to the high packing density of the water molecules in the inner sphere of the hexaquoaluminum complex, the exchange reaction between it and Complexone III is very slow at room temperature. At the boiling point, the exchange reaction becomes rapid. Moreover, aluminum forms polynuclear hydroxo complexes of type $Al_2(OH)_3^{3+}$, $Al_3(OH)_6^{3+}$, etc., up to $Al_n(OH)_{3(n-1)}^{3+}$ [551, 1028, 1118, 1248], which also react slowly with Complexone III. To ensure that the reaction of aluminum with Complexone III is quantitative, the reaction mixture must be heated to boiling. Direct titration by Complexone III is carried out in boiling solutions, and the back-titration in the cold, but the solution must be boiled after the introduction of Complexone III. If the back-titration variant is employed, Complexone III is introduced into acid (pH 1−2) or alkaline (pH 12−13) solutions, but not into weakly acid (pH about 4).

Direct titration by Complexone III

Titration in the presence of PAN and copper complexonate. Of the direct methods, the best is titration in the presence of 1-(2-pyridylazo)-2-naphthol (PAN), with small amounts of copper complexonate* [719].

* Copper-PAN complex has been proposed by a number of authors as indicator in this titration.

Aluminum is titrated in boiling solutions at pH 3. The aluminum displaces copper from the complexonate, and the copper liberated reacts with PAN to form a red complex. After all the aluminum has reacted, Complexone III at the end point of the titration decomposes the copper complex of PAN, and the red solution turns yellow (copper complexonate is somewhat more stable than the PAN-copper complex, corresponding to $pK_{inst} = 18.8$ [347] and 18.3 [634]).

The temperature of the solutions being titrated must be kept close to the boiling point. The best procedure is to titrate the boiling solutions, reheat them to boiling, and repeat the titration. According to Flaschka and Abdine [719], the way the sample solutions are neutralized is highly important. It is preferable to neutralize acid solutions to pH 4 with ammonium acetate, and then to add acetic acid to pH 3. If the acid solution is neutralized with ammonia to pH 3 and acetate buffer solution (pH 3) is then introduced, the results will be too low. The reason for this is the local increase in pH during neutralization with ammonia and the formation of polynuclear hydroxo complexes of aluminum, which react slowly with Complexone III.

Appreciable amounts of K, Na, Ca and Mg do not interfere with the determination of aluminum. Up to 30 mg of manganese in 100–150 ml of the solution being titrated are permissible. Iron is quantitatively titrated, so that it is possible to determine the sum Fe + Al [719]. Chromium should be present in small amounts only [430], since in boiling solutions chromium complexonate is rapidly formed. Titanium interferes [430] and must be eliminated, usually by extracting titanium cupferronate with chloroform. Zinc is also titrated; this forms the principle of the determination of the overall content of aluminum and zinc, followed by a correction for the zinc [165, 416]. Up to 1.8 g of sulfate ions do not interfere; in the presence of larger amounts the results are too low [430]. Therefore, the introduction of sulfates during the solubilization of samples should be avoided.

The end point of the titration is very sharp, and the method is one of the most highly accurate complexometric methods for the determination of aluminum. It has been used to determine aluminum in steels [1884, 1185], ferroalloys [146], titanium [59], magnesium [416] and zinc [165] alloys, silicates [625], carbonate rocks [74], and products of the titanium industry [430]. See also [73, 102, 133, 506].

The acid solutions are preliminarily neutralized with ammonia to pH 0–1 (but not higher). Bromophenol Blue indicator is added, and the neutralization is continued with a 10% solution of ammonium acetate, added drop by drop until the solution turns grayish blue. Five ml of acetic acid are then added, and the resulting pH of the solution

Table 8. Methods for direct complexometric determination of aluminum

Indicator	pH	Notes	References
Aluminon + methylene blue	4.4	Solutions are titrated at the boiling point	[871]
Alumocresone + methylene blue	2	The same; at the end point the violet solution turns blue	[268]
2-Hydroxy-3-naphthoic acid	3	Solutions at 50°C are titrated in UV light until the blue fluorescence turns green	[904]
Eriochrome Cyanine R	3.5—4.1	Solution is titrated at the boiling point to color change from red-violet to yellow	[800, 1215]
Pyrocatechol Violet	3.5—4.5 4.5—5	At the end point the dark blue solution turns yellow	[792] [480, 533]
Gallein	3.5	The solution is titrated at the boiling point in the presence of sulfosalicylic acid, which increases the stability of aluminum complexonate. The color change is from lilac to red	[868]
Naphthylazoxine S	6	Titration is continued until solution turns pink, 1—2 drops of 0.05 M Cu(NO₃)₂ are added, and the solution is further titrated until yellow color turns orange-pink	[738]
Xylenol Orange*	3—4	Titration in boiling solutions until red-violet coloration turns yellow	[752]
Methylthymol blue	3—4	The boiling solutions are titrated until the blue solution turns yellow	[752]

* To obtain a sharper color change at the end point, a 3 : 1 mixture of the indicator with methylene blue or methylthymol blue can be employed.

should be 3 (universal indicator paper); if necessary, more acetic acid or ammonium acetate are added. Five or six drops of a 0.1% ethanolic solution of PAN and 2ml of an approximately 0.005 M solution of copper complexonate are added. (To prepare the last solution, 20ml of 0.05 M $CuSO_4$ are titrated by 0.05 M Complexone III in the presence of 3—4 drops of PAN solution and 2—3ml of acetate buffer solution (pH 3) until the solution turns green; this solution is then diluted to 200ml with water.) Water is added to a final volume of about 100ml, and the solution is heated to boiling. The hot solution is titrated by 0.05 M Complexone III to the appearance of a pure yellow color. The solution is again brought to the boil and the titration is completed. The end point is reached when, after boiling for 30 sec, the yellow solution no longer turns orange [416, 719].

Titration in the presence of hematoxylin. If aluminum is preliminarily complexed with lactic or glycolic acid, it can be titrated directly with a hematoxylin indicator at pH 5.5—6 from hot solutions [1167]. The color change is very sharp. The method was tested on Ti—Al, Ni—Al, and other alloys. According to [533], the determination at pH 4.5—5 in solutions at 60°C is less accurate.

Titration with the indicator system vanadium complexonate − organic reagent. The principle of this method, proposed by Sajo [1135], is the same as that of Flaschka and Abdine [719]. Vanadium is displaced from its complexonate by a number of elements, including aluminum. Organic reagents that can be used to form a colored compound with the liberated vanadium include diphenylcarbazone, Pyrocatechol Violet, morin, carminic acid; the titration is continued until the violet coloration disappears (at pH < 5) or until it turns pale pink (at pH > 5).

Titration in the presence of Chrome Pure Blue [247, 495]. The solution is heated to boiling and titrated at pH 2 in the presence of the indicator and small amounts of copper until the violet coloration disappears.

Titration with Chrome Azurol S. Direct titration with Chrome Azurol S (at pH 4, at pH 5 near the end point), as recommended by Theis [1227], gives an indistinct color change at the equivalence point, and is of no interest [268, 625, 1066].

Other methods for the direct titration of aluminum will be found in Table 8.

Back-titration of Complexone III

Titration with Xylenol Orange. Titration with Xylenol Orange indicator is the most accurate method for the determination of aluminum by back-titration. The color change at the equivalence point is very sharp.

At pH > 3 aluminum reacts with Xylenol Orange to form a bright red complex, which is more stable than aluminum complexonate; for this reason direct titration with this indicator is impossible [822, 823, 894, 896]. Salts of Pb, Th and Zn can be employed in back-titration. Zinc salts are preferable, since the color change at the equivalence point is then very sharp (yellow through orange to red). The optimum pH value is 5.0—5.5; below pH 5 the color change is less sharp. A lead salt solution should not be employed as titrant if the titrated solution contains large amounts of sulfate ions, since the sparingly soluble $PbSO_4$ will then be formed; it is then preferable to use a zinc solution for the titration. When titrating with a solution of thorium, the pH should be kept at 2.3—3.5; at higher pH values the titration is not feasible, since thorium gives a more stable complex with the indicator than with Complexone III [822, 823]. The titration with a zinc solution should be carried out at pH 5.5; at lower pH values the color change is less sharp. All the three metals used as titrants, Th, Pb and Zn, form more stable complexonates than aluminum, their pK_{inst} values are 23.2, 18.04 and 16.50, respectively, whereas for aluminum complexonate $pK_{inst} = 16.13$ [347]. Nevertheless, aluminum is not displaced from its complexonate by these metals, owing to the high sensitivity of Xylenol Orange to Th, Pb and Zn, and the insignificant displacement rate compared to the rate of formation of the colored complexes between Th, Pb and Zn, and Xylenol Orange.

The following do not interfere with the titration in the presence of Xylenol Orange: appreciable amounts of alkali metals and ammonia (up to $10 g$ $NaNO_3$, KCl and NH_4NO_3; in the presence of $10 g$ $Na_2SO_4 \cdot 10H_2O$ the results are somewhat lower). Large amounts of Mg, Ca, Sr, and Ba do not interfere; neither does manganese if the titration is carried out by thorium.

The interference of many cations can be eliminated if Sajo's technique [1130, 1131] is employed: the excess Complexone III is titrated by a solution of zinc acetate, and then the aluminum complexonate is decomposed by heating with fluoride, and the mixture is again titrated by zinc acetate solution. The consumption of the titrant during the second titration is equivalent to the content of aluminum. Sajo used benzidine as indicator, and the ferricyanide-ferrocyanide redox system; if this indicator is replaced by Xylenol Orange, a highly satisfactory and quite specific method is obtained, which is extensively employed in laboratories.

The method is based on the reaction between aluminum complexonate and sodium fluoride:

$$AlY^- + 6NaF = Na_3AlF_6 + Na_3Y^-$$

occurring in a weakly acid medium in which complexonates of other metals do not react with NaF.

When sodium fluoride is employed, the amount of Fe^{3+} in the solution being titrated should not exceed 50 mg, or else the yellow color of iron complexonate will interfere with the determination of the equivalence point. The Fe : Al ratio must not be higher than 10 : 1. Ferrous iron interferes with the determination of aluminum and is therefore first oxidized by nitric acid or ammonium persulfate [598].

Titanium behaves similarly to aluminum, but its reaction with Complexone III is not quantitative. Therefore, the suggestions of certain authors [62, 166, 679, 680] that the content of titanium be determined on an aliquot by the photometric method, and a suitable correction introduced, are only practicable if the amount of titanium is small. The titration of the overall content of Al+Ti is only possible if the amount of TiO_2 is less than 4 mg [229]. Up to 5 mg Ti can be masked by the addition of H_2O_2 (1 ml of a 1% solution) before the addition of Complexone III [854]. In the presence of higher contents of titanium aluminum is determined after the introduction of phosphate buffer [166], when titanium is precipitated as phosphate and aluminum alone is titrated. However, the presence of phosphate ions impairs the sharpness of the color change at the equivalence point. Accordingly, if titanium is present in appreciable amounts, it is best eliminated before the determination, for example, by extraction as cupferronate. The reports of different authors [31, 934] that small amounts of titanium do not interfere are doubtful. In fact, if the contents of aluminum are high (30–50%), the effect of titanium is indeed insignificant, but this effect may become important if the amounts of aluminum present are small.

Manganese forms a stable complexonate, which does not react with fluoride. However, if large amounts of manganese are present, the color change becomes less sharp. Up to 5–10 mg Mn are practically without effect [23, 598], but larger amounts must be separated.

Silica gel interferes with the sharpness of the equivalence point and must be separated for this reason.

Chromium (III) forms a blue-violet complexonate, which makes the location of the equivalence point more difficult. According to [854], up to 30 mg Cr(III) are without effect on the results of the determination; if 40 mg are present, the equivalence point is difficult to identify, while in the presence of 60 mg the titration becomes impossible. Chromates and dichromates do not affect the results of the determination of aluminum. Therefore, the common practice is to oxidize Cr(III) to Cr(VI); up to 60 mg Cr(VI) do not interfere. Large amounts of chromates must be separated from aluminum, for example, by passing the sample solution in which chromium has been oxidized through KU-2 cation exchanger in the H^+-form. Then CrO_4^{2-} passes through the column, while the aluminum is sorbed on the resin, and is then eluted by acid and determined as above. Divalent nonferrous metals (Cu, Cd, Co, Ni, Pb) do not affect the determination of aluminum, except that some of them (Cu, Co, Ni) form colored complexonates which interfere with the location of the equivalence point. Calcium in amounts as large as about 60% does not interfere [71], but it must be remembered that some of the fluoride ions are bound by this element. If substantial amounts of calcium are present, the amount of the introduced sodium fluoride must therefore be increased. According to [854], 0.5—5.0% Al can be very accurately determined, at least in the presence of 3% Co, 3% Ni, 1% Cu, and 3% V; in the presence of molybdenum the titration is impossible as it is difficult to locate the equivalence point.

Aqueous solutions of Xylenol Orange are unstable, and it is preferable to use a 1 : 100 mixture with sodium chloride, which is stable for a long time. A mixture of Xylenol Orange with methylene blue is employed to improve the sharpness of the color change [1173]. In this case, the color change at the equivalence point is from green to red-violet.

Titration in the presence of Xylenol Orange can be used to determine aluminum in alloys [712], titanium alloys [1173], ferrotitanium [63], magnesium alloys [429], aluminum bronze [260], nickel-aluminum alloys [263], binary aluminum-copper alloys [345], aluminum-zirconium alloys [434], aluminum-iron alloys [345], aluminum-titanium alloys [665], ternary alloys of aluminum with zirconium and nickel [295], bauxite, nepheline ores and concentrates [16, 71, 558, 877], kaolin [147, 680], various minerals, ores and rocks [23, 71, 166, 229, 372, 386, 1083], refractories [71, 135, 669, 685], slags [259, 447], and welding materials [1077].

The following procedure [71] is recommended for the determination of aluminum.

The sample solution is diluted with water to 100—200 ml, 0.05 M Complexone III is introduced in excess (allowing for elements other than aluminum which react with Complexone III under these conditions). The solution is neutralized by 1 : 5 ammonia to pH 2—3 (thymol blue or universal indicator paper) and is boiled for 3—4 min. Twenty ml of acetate buffer solution (250 g of ammonium acetate and 20 ml of glacial acetic acid per liter) and 3—4 drops of a 0.1% aqueous solution of Xylenol Orange are added (or a small amount of a dry mixture with NaCl). The excess Complexone III is titrated by 0.05 M Zn(NO$_3$)$_2$ until the yellow solution turns red-violet. Twenty ml of a 4% solution of NaF are added and the mixture is boiled for 2–3 min. When cool, the liberated Complexone III, in an amount equivalent to the amount of Al present, is titrated by $Zn(NO_3)_2$ solution.

If the solutions being titrated do not contain appreciable amounts of sulfates, it is preferable to use $0.05\,M\,Pb(NO_3)_2$ as titrant during the back-titration

Titration with methylthymol blue. This indicator is an analog of Xylenol Orange and can be used under the same conditions. Methylthymol blue reacts with the usual titrants, e.g., solutions of thorium, lead, and zinc, to form colored complexes at somewhat higher pH values than Xylenol Orange. The optimum pH values are 3–6 for thorium [422], and 6.0–6.5 for lead and zinc [163, 423].

Medvedeva and Mazel' [253] titrate at pH 6.7–7.2.

Titration by zinc with dithizone indicator. The method was first used by Wänninen and Ringbom [1248]. The excess Complexone III is titrated by a solution of zinc at pH 4.5 with dithizone as indicator in a water-ethanol medium. The ethanol must be added to retain the dithizone in solution. The color of the titrated solution at the equivalence point changes sharply from greenish-violet to purple-red. This method and that with Xylenol Orange give the sharpest color changes.

According to the authors, the accuracy of the method is a function of the excess Complexone III present, but later Nydahl [1020] showed that the titration error is negligible in the presence of an excess Complexone III of between 1 and 160%. This is because the equilibrium reaction between zinc and aluminum complexonate, as postulated by Wänninen and Ringbom [1248], is, in fact, almost nonexistent. Nydahl [1020] showed that aluminum complexonate must be considered as an inert compound, unable to react with zinc. Titration by zinc in the presence of dithizone is highly accurate and reproducible. According to Nydahl [1020], during the determination of 0.02–2.0 millimoles of aluminum, the amounts found were 99.93–100.04% of the amounts taken, with a standard deviation of 0.063–0.072%.

The optimum pH value is 4.5, and the titration may be conducted at pH 4.4–5.0 [1020]. At higher pH values the color change is less sharp and the consumption of zinc solution is somewhat higher. This is explained by the formation of hydroxo complexes of aluminum, accompanied by the liberation of Complexone III. The low results obtained at lower pH values are explained by assuming that the reaction between aluminum and Complexone III does not proceed to completion [1248]. Wänninen and Ringbom [1248] showed that Complexone III should be introduced before the buffer solution. If the reagents are added in the reverse order, the results are too low, owing to the formation of hydroxo complexes. During the preparation of the solution for titration, the neutralization technique is also very important (see also the PAN method). After Complexone III has been introduced, the acid solutions must be neutralized by ammonia to pH ~ 2 (thymol blue), and boiled, and then the buffer solution added drop by drop. The solution must be held at the elevated temperature for 3 min before the titration [1020]. The pale yellow color of thymol blue does not interfere with the location of the equivalence point. Gottschalk [773] obtained satisfactory results at pH 4.1; there are reasons for believing [701] that heating is unnecessary. However, it is preferable to accept the recommendations of Nydahl, with minor modifications: Complexone III is added to the acid sample solution, which is then neutralized to pH 2−3 (with thymol blue indicator). The acetate buffer solution (pH 4.5−4.7) is added, the solution is boiled and kept boiling for 2−3 min, and then cooled under the water tap.

The ethanol concentration should be 40−50% [1248]; at lower concentrations the color change is indistinct. The optimum amount of dithizone is 2−2.5 ml of a 0.025% solution in ethanol. When such amounts of dithizone are titrated by 0.001 M zinc, the error due to the indicator is 0.12 ml of indicator; this volume must be deducted from the volume of the zinc solution consumed.

Cations which react with Complexone III under these experimental conditions interfere with the determination of aluminum: Cd, Co, Cu, Ga, In, Fe(II), Fe(III), Th, Pb, Ni, Ti, V(IV), Zn, REE, also cations which form stable complexes with dithizone (Hg, Ag, platinum metals), cations and anions which oxidize dithizone (Ce(IV), Cr(VI), Mn(VII), Mo, V(V), Zr, persulfates, periodates, nitrites). The effect of small amounts of oxidants can be counteracted by introducing 100 mg of ascorbic acid.

Fluorides, phosphates, organic hydroxy acids, sulfides and cyanides interfere because they react with aluminum or zinc. Alkali and alkaline-earth metals do not interfere in concentrations up to 0.1 M; the presence of chlorides, sulfates, perchlorates and acetates is permissible in concentrations up to 0.2 M. The concentration of nitrates must not exceed 0.1 M. It is more convenient to eliminate the effect of many cations by the addition of NaF, according to Sajo [325, 326, 1288].

Many workers determine the overall content of aluminum and iron and introduce a correction for iron after this element has been determined on an aliquot of the solution [369, 567, 623, 751]. The dithizone titrimetric method can be employed to determine aluminum in steels, in metallic uranium and uranium alloys [833, 1091], cement [623], silicates and rocks [223a, 557, 567, 707, 751, 1244, 1288], acid waters [639, 654] and other materials.

The determination of aluminum by titration with zinc in the presence of dithizone is one of the most accurate methods. Its disadvantage is that ethanol must be employed. The consumption of alcohol may be reduced to 10–15 ml per titration if the volume of the solutions to be titrated is reduced by evaporation.

Titration by iron solution in the presence of salicylic or sulfosalicylic acid. The method was proposed by Milner and Woodhead [976], and has now been widely accepted, largely owing to the work of Bashkirtseva and Yakimets [42–46, 351]. The titration is usually carried out at a pH of about 6. The sample solution is boiled with Complexone III and cooled before the titration. If the titrated solutions have not been previously cooled (pH 5), the color change is not sharp owing to the displacement of aluminum from aluminum complexonate by iron [976]. Milner and Wood-head [976] obtained very accurate results by titrating 2.5–60 mg of aluminum (error 0.2–0.8%) at pH 6.5. However, Bashkirtseva and Yakimets [45] found that if interfering ions are present, it is preferable to carry out the titration at pH 4.8, since the effect of extraneous ions is then much weaker than at pH 6. At pH 4.8 magnesium does not interfere, while at pH 6 only 80 mg of this element can be present. One hundred ml of the sample solution may contain 50 mg of Ca at pH 4.8, but only 5 mg Ca at pH 6. If 0.5–0.6 g of sodium sulfosalicylate are added per 100 ml of solution, up to 30 mg PO_4^{3-} do not interfere. Alkali metals, sulfates, nitrates, and acetates do not interfere. The solutions are boiled with Complexone III, and are then best neutralized while hot (70–80°C) to prevent hydrolysis [46].

Excess Complexone III is introduced into the sample solution, the solution is neutralized with ammonia to a pH of about 2 (universal indicator paper), heated to boiling, and boiled for 2–3 min. It is then neutralized with ammonia to phenolphthalein, the pink color is discharged by the addition of 1:1 HCl, and then 2–3 drops of the HCl solution are added in excess. Ten ml of acetate buffer solution (pH 6; 500 g of ammonium acetate and 20 ml of glacial acetic acid per liter) are added, the solution is boiled for 3 min and cooled, and then 2 ml of a 10% solution of sulfosalicylic acid are added and the solution is titrated by 0.05 M FeCl$_3$ until the lemon-yellow color turns golden-yellow [473].

The complexometric determination of aluminum by iron solution in the presence of sulfosalicylic acid is widely used in laboratory practice. Examples of such applications are the determination of aluminum in ferroalloys [160, 588, 589], bronzes [354, 976], zinc alloys [976], aluminum-thorium alloys [977], aluminum-silicon alloys [161], Al-Sb-Ga alloys [104], slags [182, 350], nepheline concentrates [183], alumina materials [108], rocks, silicates, refractories [267, 277, 373, 975], TiO$_2$ [288], soils [209], aluminate solutions [43, 44], vulcanizates [55], and rubber latex [82], etc.

Table 9. Complexometric determination of aluminum by back-titration methods

Indicator	Titrant	pH	Color change at equivalence point	Notes	References
Alizarin S	Th(NO$_3$)$_4$	3.5 4.3*	Orange to red (not sharp)	In calculations the empirical titer is used (about 1% higher than the theoretical)	[381, 479, 593, 721, 784]
PAN	CuCl$_2$ or CuSO$_4$	3	Yellow to red	Titration carried out in boiling solutions	[72, 1267]
		4–6	Yellow to red	Titration carried out in cooled solutions	[349, 635, 922, 1151]
4-(2-Pyridylazo)-resorcinol (PAR)	Pb(NO$_3$)$_2$	6.5–7.0	Yellow to yellow-orange	In presence of ammonium salts the solution is boiled before titration	[914]
Hematoxylin	AlCl$_3$	6.0	Yellow to pink-purple	Solutions titrated at 70°C	[1223]
Pyrocatechol Violet	CuSO$_4$ AlCl$_3$	5–6	Yellow to greenish blue	Required pH produced by acetate buffer solution, urotropin, or pyridine	[383, 905, 919, 1049, 1211]
		5–6	Light blue or pale green to dark blue	Titration carried out at 60–65°C	[464]
Chrome Azurol S	CuSO$_4$ or Cu(NO$_3$)$_2$	4–6	Yellow to blue	Organic solvent (CCl$_4$ or dichlorobenzene) added to improve sharpness of color change; the end point indicated by blue color in organic layer	[310, 315, 440, 484]

* Solutions heated to boiling

Table 9 (continued)

Indicator	Titrant	pH	Color change at equivalence point	Notes	References
Eriochrome Cyanine R	ZnSO$_4$	6–6.3	Yellow to lilac-red	Titrated solutions at 70–80°C	[788, 795, 1132]
	Fe(III)	4.1		Titration carried out at 40°C; end point located photometrically at 582mμ	[800]
Glycine Cresol Red	CuSO$_4$	6	Green to violet; if fluoride is introduced, from green to blue	Hot solutions titrated; fluoride introduced, solution boiled, and liberated Complexone III titrated in hot solutions	[1170, 1171]
Ammonium thiocyanate	Co(NO$_3$)$_2$	5–7	Pink to greenish-blue	Titration in 1 : 1 water-acetone medium	[932]
Zincon	ZnSO$_4$	8–9			[873]
Calcein W	CuSO$_4$	4.8		End point indicated by quenching fluorescence of Calcein W in UV	[807, 1240]
Omega-Chrome Green BLL	Th(NO$_3$)$_4$	3.5–3.6	Yellow to violet	Titration in solutions at 1–3°C in presence of ethanol	[516]
Arsenazo	ZnCl$_2$	5.5–6.2	Yellow to red		[183]
Acetylacetone	FeCl$_3$				[962]
Bindschedler's Green	FeCl$_3$	3.5	Red to green		[737, 1249]

Table 9 (continued)

Indicator	Titrant	pH	Color change at equivalence point	Notes	References
1-Hydroxy-2-naphthoic acid or 2-hydroxy-3-naphthoic acid	$FeCl_3$	3.0–4.0	Yellow to blue		[1043, 1158]
Mixture of diphenyl-carbazone and com-plexonate V(V)	$ZnSO_4$	5.0–5.5	Yellow to red		[745]
Same in presence of methylene blue	$ZnSO_4$	5–6	Yellow to red	Titration carried out in solutions at 40°C	[587]
Sodium pyrocatechol-3,5-disulfonate (Tiron)	$FeCl_3$	5		Titration carried out in solutions at 80–90°C; end point located photometrically at 620 mμ	[684, 687]
Naphthylazoxine S	$Cu(NO_3)_2$ or $Zn(NO_3)_2$	4.5–5	Pink to yellow		[664, 727]
Mixture of diphenyl-carbazone and o-phenanthroline	$Hg(NO_3)_2$	5–6		Titration in nitric or sulfuric acid solutions	[548]

Table 9 (continued)

Indicator	Titrant	pH	Color change at equivalence point	Notes	References
bis-(Glycine)methylene-dichlorofluorescein	CuSO$_4$	6.5		During analyses of mixture of Al with Cu, Ni, Co, Zn, Cd, V(IV) and Cr(III), all elements titrated together; mixture treated with fluoride and aluminum determined by back-titration	[562]
Pyrogallolcarboxylic acid	BaCl$_2$	12	Yellow to violet-red		[900]
2-Hydroxybenzene-1-azo-4,1-sulfophenyl-3-methyl-5-hydroxy-pyrazole	ZnSO$_4$	8–9			[118]

Titration in the presence of redox indicators. The indicator consists of ferricyanide-ferrocyanide redox system and some organic reagent such as benzidine [228, 244, 252, 302, 337, 853, 876a, 1130, 1131, 1133, 1134], dimethylnaphthidine [21, 718, 720, 742, 980, 1006, 1007], diphenylamine [242] or Variamine Blue [703, 1256]. The end point can also be found potentiometrically [1183]. The titrant is a solution of a zinc salt. The optimum pH value is 5.0–5.5. At the end point of the titration, the Zn^{2+} ions bind ferrocyanide to form a sparingly soluble compound, the redox potential of the ferrocyanide-ferricyanide system increases, and the colored oxidation product of the organic reagent is formed. Sajo claims that the selectivity of the method can be increased by adding sodium fluoride (cf. p. 66).

The determination by means of the Cu^{2+}/Cu^+ redox system, thiocyanate and Variamine Blue B is based on the same principle [848]. A $0.1 M$ solution of $ZnCl_2$ containing $0.005 M$ $CuCl_2$ is the titrant. The redox system contains Cu^{2+} ions and traces of Cu^+ ions. This system, in the presence of ammonium thiocyanate and Variamine Blue B, serves as indicator. At the equivalence point free Cu^{2+} ions appear, as well as traces of Cu^+ in $CuCl_2$ solution. Cuprous copper reacts with ammonium thiocyanate to form the sparingly soluble CuSCN. As a result, the redox potential of the system increases, and Variamine Blue B becomes oxidized with the formation of a violet color. The optimum pH value during the titration is 5–5.5.

Titration by zinc solution with Eriochrome Black T as indicator. Methods with titration in alkaline solution are less specific than titrations in weakly acid solution. The titration of aluminum by zinc solution in the presence of Eriochrome Black T as indicator has been most completely studied. The indicator reacts with aluminum to form a stable complex, which cannot be decomposed by Complexone III, so that back-titration only is feasible, and even then the formation of the Eriochrome Black T-aluminum complex must be prevented, by titrating below 10°C [633, 642] or even at 2–5°C [1150]. Solvents such as ethanol, methanol, isopropanol, acetone or formamide are added to improve the sharpness of the color transition at the equivalence point, since they reduce the stability of complexes between aluminum or iron and Eriochrome Black T [465, 614, 625, 876]. Most investigators conduct the titration at pH 10; this pH value is produced by means of ammonia buffer solution or pyridine [786], while solutions of zinc, and sometimes solutions of magnesium or manganese, are employed as titrants. The Eriochrome T titration method has a low selectivity. Owing to all the disadvantages just described, the method cannot be recommended.

Other back-titration methods. Table 9 lists other methods for the determination of aluminum by back-titration.

Of the many methods proposed for the determination of aluminum by back-titration, the most widely used are those with Xylenol Orange, dithizone, and sulfosalicylic acid as indicators. For comparative studies of complexometric methods, the reader is referred to [268, 625, 800, 951, 1066]. In a specially interesting study [1066], the content of aluminum in aluminum acetate was determined by 10 different complexometric methods, and the results were compared with those given by the hydroxyquinoline method. It was found that the accuracy of the complexometric method is close to that of the gravimetric hydroxyquinoline method, provided that PAN with copper complexonate, Eriochrome Black T, dithizone, Xylenol Orange, and methylthymol blue are used as indicators. Despite the fact that satisfactory results have been obtained with Eriochrome Black T, it cannot be considered a satisfactory indicator for the reasons just given.

Physicochemical methods for locating the equivalence point. The equivalence point can be located not only visually, but also by photometric [684, 800, 1240], potentiometric [757, 1080, 1081, 1183], amperometric [24, 83] and high-frequency [553] titrations by Complexone III.

Determination of aluminum in presence of other elements

The best method for determining aluminum in the presence of other elements is the fluoride method (see above).

Successive titrations at different pH values are widely employed, especially during the analysis of aluminum-iron mixtures. The iron is titrated first at pH 1–2, with sulfosalicylic acid as indicator. The pH is then adjusted to 5–6, and the excess Complexone III is titrated by a solution of an iron salt, in the presence of the same indicator. Many similar methods have been described; these involve the use of different indicators for iron, or aluminum is titrated by different methods. The overall content of aluminum + iron is sometimes determined, and then the iron is found in an aliquot and aluminum found by difference. However, methods with only a small difference in the pH values at which the two elements are determined should not be employed. For example, Endo and Takagi [509] titrate iron at pH 2 by salicylic acid, and then aluminum is titrated at pH 3 with copper + PAN as indicator. During the determination of aluminum and chromium present together in solution, advantage was taken of the difference in the stabilities of their complexonates at different pH values as a function of the duration of heating, since chromium complexonate is formed after fairly prolonged boiling only. Table 10 shows the different techniques for determining aluminum in the presence of other metals.

Table 10. Complexometric determination of aluminum in the presence of various cations

Metals	Titration method	References
Al, Fe	Direct titration of Fe in the presence of salicylic or sulfo-salicylic acid at pH 1–2. Complexone III is introduced, and the excess Complexone III is titrated at pH of about 6 by a solution of zinc, in the presence of Xylenol Orange. Aluminum complexonate is decomposed by fluoride, and the complexone liberated is titrated in the same manner (detn. of Al)	[23, 229, 667]
	The same, but without NaF	[23, 259]
	Back-titration of excess Complexone III by Pb solution in the presence of Xylenol Orange at pH 5–5.5 (detn. of Al+Fe). Aluminum complexonate is decomposed by fluoride and aluminum is titrated in the same manner (detn. of Al); Fe is found by difference	[1085]
	Back-titration of excess Complexone III by a solution of Th in the presence of Xylenol Orange at pH 2.7 (detn. of Al+Fe). Fe is colorimetrically determined by o-phenanthroline in an aliquot; Al is found by difference	[1077]
Al, Cr	The solution is heated with an excess of Complexone III for 5 min at 40–50°C; when cool, the solution is titrated by $FeCl_3$ solution in the presence of Tiron or sulfosalicylic acid (detn. of Al). Complexone III is again introduced, the solution is boiled for 5 min, cooled, and titrated in the same manner (detn. of Cr)	[931]
	The solution together with Complexone III in excess is boiled for 5 min, cooled to 40°C, and titrated by a solution of $FeCl_3$ in the presence of sulfosalicylic acid (detn. of Cr + Al). The solution is acidified to pH 1, the liberated Complexone III is titrated at 60°C in the same manner (detn. of Al); the Cr content is found by difference	[930]
	Back-titration by a solution of Zn at pH 5 in the presence of a ferricyanide-ferrocyanide redox system; the equivalence point is found by potentiometry (detn. of Al + Cr). After treatment with fluoride, the liberated Complexone III is titrated in the same manner (detn. of Al)	[1183]

Table 10 (continued)

Metals	Titration method	References
Al, Ti	Back-titration of excess Complexone III by a solution of Bi in the presence of Xylenol Orange at pH 1 (0.1 M HNO$_3$); in this way Ti is determined. The content of Al is determined in the same solution by back-titration by a solution of Zn in the presence of the same indicator at pH 5—6	[386]
	Complexone III is introduced to bind Al, Ti, and other metals. The excess Complexone III is titrated by a solution of Zn in the presence of Xylenol Orange, titanium complexonate is decomposed by tartaric acid, and the liberated Complexone III is titrated in the same way (detn. of Ti). KF is introduced into the solution from the titration, aluminum complexonate is decomposed, and the liberated Complexone III is titrated (detn. of Al)	[63]
Al, V	Back-titration of excess Complexone III by a solution of zinc in the presence of Xylenol Orange (detn. of Al + + V(IV)). Al is determined in another part of the solution by adding 30% H$_2$O$_2$ to decompose vanadium complexonate and titrating the excess Complexone III by a solution of zinc in the presence of Xylenol Orange; V is determined by difference	[712]
Al, Zn	Direct titration at pH 4.8—5.5, with Cu-PAN complex as indicator (detn. of Al + Zn). In another part of the solution Al is masked by NH$_4$F, and Zn is titrated in the same manner; Al is found by difference	[165]
	Back-titration of excess Complexone III by a solution of Zn in the presence of Eriochrome Black T (detn. of Al + + Zn). In another part of the solution Al is bound by a mixture of NH$_4$F with tartaric acid, and Zn alone is titrated. The content of Al is found by difference	[64]
Al, Zr	Direct titration in the presence of Xylenol Orange in $2N$ HCl (detn. of Zr). The pH is adjusted to 3.5; back-titration of excess Complexone III by a solution of Th in the presence of Xylenol Orange gives the content of Al	[434]
Al, Y	Direct titration in the presence of Xylenol Orange, after Al has been masked by sulfosalicylic acid (detn. of Y). The sum Al + Y is determined in another part of the solution by back-titration of excess Complexone III by a Zn solution with dithizone. The content of Al is found by difference	[780a]

Table 10 (continued)

Metals	Titration method	References
Al, Sc	Direct titration at pH 2 in the presence of Xylenol Orange (detn. of Sc). The content of Al is determined in the same solution by back-titration of excess Complexone III by a Zn solution at pH of about 6, with the same indicator	[112]
Al, Ga	Back-titration of excess Complexone III by a solution of $ZnCl_2$ at pH 6.5–7.5 with Eriochrome Black T as indicator (detn. of Al + Ga); addition of NaF followed by titration of the liberated Complexone III by $ZnCl_2$ solution (detn. of Al)	[71a]
Al, Fe, Cr	Direct titration in the presence of sulfosalicylic acid (detn. of Fe). Back-titration of excess Complexone III by a solution of Fe at pH 6 (detn. of Al). Boiling with excess of Complexone III for 5–8 min, cooling, followed by back-titration as before (detn. of Cr)	[929]
	Direct titration at pH 1–1.5 in the presence of sulfosalicylic acid (detn. of Fe). In the second part of the solution a complexometric determination at pH 5 of Al and Fe is carried out after the oxidation of Cr(III) to Cr(VI); Al is found by difference. In a third part of the solution Cr is determined by the volumetric method after oxidation to chromate	[23]
Al, Fe, Ti	Direct titration of Fe at pH 1–1.5 in the presence of sulfosalicylic acid. In another part of the solution Fe + Ti is titrated in the presence of H_2O_2. The content of Al is determined in the same solution by back-titration at pH 5 in the presence of Xylenol Orange. The content of Ti is found by difference	[23]
	Direct titration of Fe at pH 1 in the presence of sulfosalicylic acid. In the same solution Al + Ti is determined by back-titration of excess Complexone III by a solution of Fe in the presence of sulfosalicylic acid at pH 4–5. Ti is photometrically determined by H_2O_2 on a separate aliquot. The content of Al is found by difference.	[661]
Al, Fe, Mn	Direct titration of Fe at pH 1 in the presence of sulfosalicylic acid. Mn is determined in another solution by direct titration in the presence of Rochelle salt and small amounts of sodium diethyldithiocarbamate at pH 8.5–9 with Acid Chrome Dark Blue as indicator. The sum Fe + Al + Mn is	[350]

Table 10 (continued)

Metals	Titration method	References
Al, Fe, Mn (cont.)	determined in a third solution by back-titration of excess Complexone III against a solution of $CuSO_4$ + PAN at pH 6. The content of Al is found by difference	
	Direct titration of Fe in the presence of sulfosalicylic acid at pH 1−1.5. Al + Mn is found in the same solution by back-titration of excess Complexone III at pH 6 by a solution of Fe with sulfosalicylic acid. Mn is determined in a second solution by direct titration at pH 8.5−9 in the presence of Rochelle salt with Acid Chrome Dark Blue indicator. The Al content is found by difference	[351]
Al, Fe, Co	Direct titration of Fe at pH 1 in the presence of sulfosalicylic acid. The pH is adjusted to 6−7 and Co is titrated by murexide; the solution is acidified to pH 4−5 and the excess Complexone III is back-titrated by $FeCl_3$ solution to determine the aluminum content	[660]
Al, Fe, Mn, Mg, Ca	Direct titration of Fe at pH 2 in the presence of salicylic acid. Aluminum is determined by back-titration of excess Complexone III by $CuSO_4$-PAN solution at pH 3 in a boiling solution. The pH is adjusted to 4.5, the solution heated to 60°C, and Mn is determined by a similar titration. The pH is adjusted to 10, a similar titration is carried out, and the sum Ca + Mg is determined. In a separate solution Ca is precipitated as oxalate, Complexone III is introduced, and a similar titration is carried out to determine Mg	[72]

Determination of aluminum by other complexones

Complexones other than Complexone III have also been proposed for the determination of aluminum. The most important of these is 1,2-diaminocyclohexanetetraacetic acid (DCTA) [1084, 1169]. For this aluminum complexonate $pK_{inst} = 17.6$ [1154]. DCTA has certain advantages over Complexone III, and is a very promising reagent. Unlike Complexone III, DCTA instantaneously reacts with aluminum even at room temperature, while neutral salts (NaCl, KNO_3, etc.) present in large amounts do not interfere. This fact was utilized to simplify the analysis of various multi-cation systems. For example, it proved possible to determine aluminum in copper alloys, without any preliminary separation, by masking copper with thiourea [1082]. The overall Al + Fe content is determined on an aliquot by titrating the excess DCTA by a solution of lead nitrate in the presence of Xylenol Orange at pH 5–5.5. The iron is determined on another aliquot in the same way, after the aluminum is masked by fluoride. The aluminum content is found by difference. If other substances that react with DCTA (Pb, Zn, Ni, Mn, Sn) are present in the alloy, their sum is determined on a third aliquot, after copper has been masked by thiourea, and iron and aluminum by fluoride. The content of copper is determined on a fourth aliquot by titration in the absence of thiourea, while iron and aluminum are masked by fluoride. Fluoride masking is carried out in the cold, which is more convenient than boiling with fluoride when Complexone III is employed. Přibil and Vesely [1084] showed that more cryolite is dissolved by boiling.

The use of DCTA has simplified the analyses of systems containing Fe, Al and Cr [1085]; it is now possible to determine aluminum (and iron) in the presence of Cr(III), and also Al, Fe and Cr(III) in the presence of chromate. Aluminum can be titrated in the cold in the presence of Cr(III), which reacts with DCTA on boiling. Chromium (III) can be determined in the same solution after the solution has been boiled with excess DCTA. DCTA is not oxidized by chromate in a weakly acid medium, and for this reason Al and Cr(III) can be determined in the presence of large amounts of chromates. These determinations cannot be carried out by Complexone III, since this reagent is partially oxidized by chromate at the boiling point. Přibil and Vesely [1085] recommend the following method for the determination of aluminum and chromium.

Determination of Al and Cr(III). To an aliquot of the sample solution containing not more than 25 mg of chromium, DCTA is added in an amount sufficient to bind all the aluminum with some DCTA in excess. The pH is adjusted to 5–5.5 with urotropin (indicator paper), the solution is diluted to 250 ml, Xylenol Orange is added, and the solution is titrated by $0.05 M$ $Pb(NO_3)_2$ until the faint yellow-green color of the solution turns intense violet (determination of aluminum). More DCTA is added to bind all the chromium, with some in excess, and the solution is boiled for 10 min, cooled, and diluted to 500–600 ml with water. The pH of the solution is adjusted to 5–5.5 with urotropin, and the titration is carried out as in the determination of aluminum (determination of chromium). The color change in the second determination is as sharp as in the first.

Determination of Al and Cr(III) in the presence of chromate. The yellow color of the chromate has no effect on the sharpness of the color change. Zinc sulfate is the titrant (lead nitrate forms a $PbCrO_4$ precipitate, which slowly dissolves in DCTA). DCTA in excess is added to the solution containing Al, Cr(III) and Cr(VI), and aluminum is titrated by $0.05 M$ $ZnSO_4$ as above. An excess of DCTA is again added, the solution is boiled for 10 min, and Cr(III) is determined. In both these titrations the color change is very sharp.

DCTA has been used to determine aluminum in high-alloy heat-resistant alloys [600], copper alloys [1082], silicates [704, 1087], chromium ores and refractories [507], manganese ores [509], and basic slags [509].

Wilkins [1268] determined aluminum by hydroxyethylenediaminetetra-acetic acid in the presence of manganese (which forms a very weak complex with the reagent in an acid medium) by back-titration with a copper solution at pH 5 in the presence of a fluorescent indicator (methylcalcein or methylcalcein blue).

In practice, nitriloacetic acid has no advantages over Complexone III [1249].

Cheng [633] reported that the complexone Hel-242, with properties very similar to those of Complexone III, is also suitable for titration and masking purposes.

Determination by hydroxyquinoline

In these methods aluminum is precipitated as hydroxyquinolate, which is then dissolved in an acid, and the bound hydroxyquinoline determined by the bromatometric method.

For conditions of precipitation of aluminum hydroxyquinolate, see p. 27. The use of titration markedly reduces the duration of the determination. Moreover, the accuracy is quite high, since iodometry, a highly accurate volumetric method, is used in the last stage of the determination. These methods are therefore very valuable. In the absence of interfering elements, the usual determination procedure is as follows.

Aluminum hydroxyquinolate is precipitated, the precipitate is filtered through a medium-density filter paper and washed (p. 30). The precipitate is dissolved in a hot mixture of 30 ml HCl (sp.gr. 1.19) and 50 ml of water, added in small portions. The filter paper is washed with water, the solution and the washings are collected in a 300-ml conical flask with a ground glass stopper, with a mark at 150 ml. The solution is slowly titrated by $0.1 N$ or $0.2 N$ $KBrO_3$ containing about 20 g of KBr per liter. Just before the end point is reached, a few drops of a 0.1% solution of methyl red are added, and the titration is slowly continued until the solution turns yellow. Two more ml of bromate solution are then added. The flask is closed with the ground glass stopper, and the contents shaken, and after 2–3 min about 2 g KI are added and dissolved by swirling the solution. The liberated iodine is titrated by $0.1 N$ $Na_2S_2O_3$, with starch as indicator added at the end of the determination. One ml of $0.1 N$ $KBrO_3$ is equivalent to 0.2248 mg Al or 0.4248 mg Al_2O_3 [644].

Interfering substances must be removed before the precipitation of aluminum hydroxyquinolate or masked during the precipitation.

Classen et al. [645] improved the selectivity of the method by precipitating aluminum hydroxyquinolate from cyanide-containing and Complexone III-containing ammoniacal solution at pH 8.5–10.

Under these conditions the stability of aluminum hydroxyquinolate is much higher than that of aluminum complexonate, so that aluminum is quantitatively precipitated. With this method, aluminum can be separated from a large number of metals, due to the formation of cyanide complexes and complexonates. Iron should be reduced to Fe(II). Ferricyanides are reduced by sulfides [793], sulfites [645, 702, 916, 1109, 1250], or simply by boiling alkaline solutions [585, 1061, 1196]. Sulfites are the best reducing agents for this purpose. Between 2 and 20 mg Al can be deter-mined in the presence of 0.5–1 g of the following elements: Ag, As(III), As(V), Au, Cd, Ce(III), Ce(IV), Co, Cu, Fe(II), Fe(III), Ge, Hg(I), Hg(II), La and REE, Mg, Mn, Mo(VI), Ni, Pb, Pd, Pt, Sb(V), Se(IV), Se(VI), Sn(IV), Te(IV), Te(VI), Tl(I), Tl(III), W(VI), Zn and alkaline-earth metals. Up to 50 mg Y do not interfere; in the presence of larger amounts, about 0.5 mg Al is lost. Chlorides, sulfates, nitrates and perchlorates do not interfere.

Under these conditions Be, Bi, Ga, Hf, In, Nb, Sb(III), Ta, Th, U and Zr precipitate. Only 80–90% Al is determined in the presence of scandium. Chromium(VI) is incompletely precipitated as hydroxyquinolate; the interfering effect is insignificant in the presence of up to 20 mg Cr(VI). The effect of Cr(III) is much greater. If less than 20 mg Cr are present, Cr is oxidized to Cr(VI), and its interference thus eliminated. Moreover, chromium can be bound as complexonate by boiling for 5 min with Complexone III, and iron should be reduced by boiling with sulfurous acid. Fluorides present in amounts of up to 1 mg do not interfere; in the pres-ence of larger quantities the results are too low, even if a large excess of boric acid is present. Orthophosphates do not interfere, unless more than 100 mg Fe are also present (phosphates interfere with the complete reduc-tion of iron). Vanadium is incompletely precipitated; its effect is weaker at pH below 9. Titanium is quantitatively precipitated as hydroxyquinolate at pH $\leqslant$ 9; at pH $>$ 9 the precipitation is incomplete.

The solution containing 2–20 mg Al is diluted to about 100 ml, 5 ml of a 20% solu-tion of citric acid are added, followed by a small excess of 7 M NH$_4$OH. Three grams of KCN and 1 g of anhydrous sodium sulfite are added, the solution is stirred until the salts have dissolved and diluted to 150–250 ml. The solution is slowly heated to 80–90°C, and held at this temperature for 2 min. One gram of Complexone III is then added and the solution is held at 80–90°C for 2 min. It is then cooled to 70°C, 0.70 ml of a 2.5% solution of hydroxyquinoline (25 g of the reagent are dissolved in 29 ml of

6 M HCl, and the solution diluted to 1 liter with water) is added, with vigorous stirring, for each mg Al present, and then 20 ml in excess. The solution is heated to 80–90°C and held at this temperature for 30 min. It is then cooled to 50°C and filtered by decantation through a medium-density filter paper. The precipitate is washed with several small portions of warm (50–60°C) wash liquor (cf. p. 30) by decantation. The consumption of wash liquor should not exceed 100 ml. Towards the end, the precipitate is washed twice with 5–10 ml portions of cold water. The precipitate is then dissolved, and the determination completed as described above for the precipitation from weakly acid solutions [645].

The analysis of standard specimens of various materials (steels, magnetic alloys, bronzes, brasses, magnesium and zinc alloys, glasses) by this method gave very accurate results in the work of Classen et al. [645]. Determinations of specimens containing between 1 and 7% Al (approx.) yielded results which differed by only 0.01–0.03% from those obtained by gravimetry.

A similar method was used by Mohr [986] for the determination of aluminum in copper alloys. Reutel [1109] analyzed zinc and zinc alloys using cyanide in combination with tartaric and citric acids as masking agents. If magnesium is present in the alloy, it is coprecipitated with aluminum; it must accordingly be determined, and the results corrected (the presence of magnesium in trace amounts can be neglected). If the Zn : Al ratio does not exceed 100, the results are very accurate.

According to Gheocalescu [753], aluminum hydroxyquinolate is precipitated by 30–35 ml of 0.4 N hydroxyquinoline. The solution with the precipitate is diluted to 250 ml in a volumetric flask, and filtered through a dry filter paper into a dry flask. One hundred ml of the filtrate are acidified by 30 ml of 1 : 1 HCl, and the content of hydroxyquinoline is determined by bromatometry. Aluminum is determined by the difference between the volumes of the hydroxyquinoline solution and of the titrant.
The experimental error is not more than 0.33%.

The hydroxyquinoline bound as aluminum hydroxyquinolate may also be oxidimetrically titrated by a solution of V(V) [958] or Ce(IV) [963]. However, these methods are more time-consuming than the bromatometric determination of aluminum, and the accuracy is lower than that attained in iodometry. Aluminum hydroxyquinolate is also determined by potentiometric [69], amperometric [127], and high-frequency [98] titrations.

Determination by fluoride

These methods involve the titration of aluminum by fluoride ions in the presence of Al^{3+} ion as indicator. Alternatively, the excess fluoride can be back-titrated by a solution of an aluminum salt [153].

The following procedure may be used in direct titration.

To the solution to be analyzed are added 20 ml of acetate buffer solution at pH 1.7 (50 ml of 1 N sodium acetate and 55 ml of 1 N HCl in 250 ml), 1 ml of a 0.1% solution of ascorbic acid to reduce Fe(III), and 3 drops of a 0.1% solution of Xylenol Orange. The mixture is boiled and kept boiling for 1–2 min. The hot solution is titrated against 0.01 N NaF until the raspberry-red color turns pale yellow. During the titration solid NaCl must be introduced to saturation. If the sample to be analyzed is fused with sodium carbonate, the neutralization of the flux by hydrochloric acid produces a quantity of NaCl sufficient to reduce the solubility of cryolite.

Other indicators employed in titrations include hematoxylin [149, 368], salicylate, ferric thiocyanate [400], and the fluorescent indicator salicylal-o-aminophenol [817]. The excess fluoride added can also be back-titrated by a solution of potassium aluminum alum, with alizarin as indicator [33]. The relative error of these methods is about 1%.

Acidimetric methods

Aluminum is first converted to the hydroxide or a soluble alkali aluminate, and then treated with an excess of NaF or KF to bind it into a fluoride complex (cryolite). This reaction leads to the formation of hydroxyl ions, which are then titrated by acid:

$$AlCl_3 + 3NaOH = Al(OH)_3 + 3NaCl; \quad Al(OH)_3 + 6NaF = Na_3AlF_6 + 3NaOH.$$

In some methods tartrates or sodium gluconate are added to keep aluminum in solution. After the fluoride treatment a known volume of acid is added, and the excess acid titrated by a solution of sodium hydroxide. In another method [461] aluminum is precipitated as basic carbonate by passing carbon dioxide through the solution of alkali aluminate. The precipitate is dissolved in a standard solution of phosphoric acid taken in excess, and then the solution containing $Al(H_2PO_4)_3$ is titrated by alkali. Aluminum is sometimes precipitated as hydroxide, which is dissolved in a known volume of acid, and the excess acid is titrated by a solution of alkali. Numerous variants of the acidimetric method have been described.

Alkalimetric methods

These methods involve the complexing of aluminum by organic hydroxy acids, with liberation of hydrogen ions which are titrated by alkali. Both the sample solution and the solution of the hydroxy acid salt must be neutral; tartrates, citrates and oxalates are employed as hydroxy acid salts.

These methods, as well as acidimetric methods, are now scarcely ever used in laboratory practice.

Arsenate method

In these methods aluminum is precipitated as arsenate, which is dissolved in acid, and the bound arsenate is iodometrically determined [251, 436, 485–487]. Depending on the experimental conditions, the neutral arsenate $AlAsO_4$ or the basic arsenate $Al_5(OH)_3(AsO_4)_4$ is precipitated; precipitation of the basic arsenate is to be preferred, on account of the ease of filtration. Such methods seem to give accurate results, but the toxicity of arsenates presents a serious disadvantage.

Other methods

Bobtelsky et al. carried out a heterometric determination of aluminum by titration by solutions of hydroxyquinoline [576] and potassium phthalate [574]. The optical density of the solution is measured during the titration; owing to the formation of hydroxyquinolate or aluminum phthalate precipitates, the absorption by the solutions increases and attains a maximum at the equivalence point. The disadvantage of the method is that the titration is time-consuming (5–15 min with very vigorous stirring).

In another method [359], aluminum is hydrolytically separated with the aid of an iodide-iodate mixture, when iodine in an amount equivalent to that of the aluminum is liberated, and can be titrated by thiosulfate.

It has been suggested that aluminum be precipitated by sodium p-hydroxy-m-nitrophenylphosphinate, and the reagent contained in the precipitate bromatometrically titrated [258]. In another method aluminum is titrated by oxalate. The end point is determined graphically by the attenuated phosphorescence of the aluminum-morin complex [379]. A method [389] is described in which aluminum is precipitated as hydroxyquinolate and its nitrogen content determined by a nitrometric technique.

None of these methods is very promising.

Potentiometric titration

Determination by fluoride. The potentiometric determination of aluminum is based mainly on titration by fluoride. There are several variants of this method. In the original variant, potentiometric compensation titration, proposed by Treadwell and Bernasconi [1232], the sample solution is titrated by a solution of sodium fluoride in a CO_2 atmosphere in the presence of a few drops of an $Fe(III)–Fe(II)$ mixture, between a platinum and a calomel electrode. Ferric iron reacts with fluoride to form the complex Na_3FeF_6, which is similar to cryolite but less stable. During the titration Al reacts first and $Fe(III)$ next; as the $Fe(III)$ is bound into a complex, the value of the Fe^{3+}/Fe^{2+} redox potential suddenly changes, and this potential jump corresponds to the equivalence point. To ensure a better shift in equilibrium, the authors suggest that a 1 : 1 water : ethanol mixture saturated with NaCl be employed. The pH of the titrated solution must not be below 2.1, since then the fluoride complex is decomposed and the equivalence point can no longer be located. However, the Treadwell-Bernasconi technique is not very convenient in practice; the potential jump is not very sharp, and one titration takes 40–50 min. The method was later improved by other workers. It was shown that the potential jump becomes sharper if $Fe(III)$ is added in small amounts [407]. According to Talipov and Teodorovich [392], a large jump is noted if the ratio of ferric to ferrous ions is 3.5 : 1. According to Polyak [340] and other workers [441], the method can be improved and the duration of the titration shortened if more dilute acid is added at the moment of the incipient drop in the original potential. In this case, the potential of the system returns to its original value and remains constant up to the equivalence point.

Manchen [951a] and Tarayan and Ovsepyan [408] show that this method does not invariably yield correct results. It is preferable to adjust the pH to the required value at the beginning. According to Manchen [951a] the optimum pH value is 3.6–4.0 if the indicator solution employed is a mixture of 0.1 g $FeCl_3 \cdot 6H_2O$ and 20 g $FeCl_2 \cdot 4H_2O$ in 100 ml, acidified with a few drops of 0.1 N HCl, and pH 2.5–4.6 with an acidified mixture of 0.05 g $FeNH_4(SO_4)_2$ and 0.05 g $Fe(NH_4)_2(SO_4)_2$ in 100 ml. Tarayan and Ovsepyan [408] report lower pH values. According to these workers, the potential jump is observed at the exact equivalence point at pH 2.2–2.5; the initial pH of the solution being titrated should not be higher than 3.0–3.1. At pH values above 3, iron hydroxide may be formed, and the potential jump will occur too soon, before the aluminum has been fully bound as the fluoride complex. The authors suggest that the sample solutions (and also the standard solutions employed to determine the titer of the NaF solution) be neutralized to Tropaeolin 00 indicator (color change at pH 1.4–2.6).

According to Manchen [951a], correct results will only be obtained if, besides maintaining the pH at the required value, the sodium fluoride solution is added at a constant rate, such as 0.5 ml of NaF solution at one-minute intervals. The solution should be mixed by a magnetic stirrer.

The method of Treadwell and Bernasconi was used to determine aluminum in steels [136], in magnesium [340, 999] and copper [136, 1192] alloys, and in bauxites.

Chirkov [481] determines aluminum by a noncompensating potentiometric titration by fluoride, with an aluminum indicator electrode coupled with a nichrome electrode. The optimum pH value is 3–7, and the potential jump becomes sharper if the solution is saturated with sodium chloride [311, 412, 481]. The method of Chirkov has a number of advantages over that of Treadwell and Bernasconi: the titration takes less time and ethanol is unnecessary. Chirkov's method is widely employed in laboratory practice. It is used for the determination of aluminum in steel [248, 418], nickel [95], zinc [65] and magnesium [65, 66] alloys, slags [228], soils [8] and other substances. For studies carried out by this method see [151, 202, 311, 312].

Ferric iron, tetravalent titanium, and cupric copper interfere with the potentiometric determination of aluminum by fluoride. The following do not interfere: Mg, Ca, Mn, Zn, Ni, Cd, W, Mo(VI), Cr(III), Fe(II), Zr, Ti(III), V, Nb, Si. The effect of many elements can be eliminated by reduction with metallic zinc, when Cu, Sb, As and Sn precipitate, while Fe(III), Ti(IV) and Mo(VI) are reduced and do not interfere with the determination [95]. The effect of Ti(IV) can also be eliminated by reduction with amalgam [202]. Iron and certain other interfering elements can be separated by alkali [136, 412], while copper is separated by electrolysis [64].

Determination by other reagents. Aluminum can be potentiometrically titrated by a solution of sodium hydroxide. Polyak [340] determines aluminum in magnesium alloys by noncompensating potentiometric titration by sodium hydroxide, with an antimony indicator electrode and a saturated calomel reference electrode. The relative error is 3%, which is acceptable for rapid analyses. The procedure is technically simple; a complete analysis, including dissolution of the sample, takes 20 min. A similar method is described elsewhere [450a].

Alkali titration with an antimony electrode has been employed to determine aluminum in aluminate solutions [801, 986a].

Stammler and Pegnitz [1192] determined aluminum in copper alloys by the compensating potentiometric method by titration against sodium hydroxide between a quinhydrone indicator electrode and a saturated calomel reference electrode. Copper and lead are previously separated by electrolysis. Aluminum is precipitated by ammonia together with Fe and Mn, and

is separated from Zn and Ni in this way. The hydroxides are dissolved in
acid, and the aluminum is titrated by alkali, after Fe and Mn have been
masked with cyanide. The first potential jump, corresponding to the
neutralization of the free acid, occurs at pH 3.62; the second jump, corresponding to the reaction between aluminum and alkali, is observed at a
pH of about 6.7 (Figures 3 and 4). The difference in volume of the sodium
hydroxide solution at the two potential jumps corresponds to the aluminum
content.

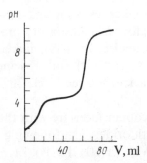

Figure 3. Titration curve
of an acid solution of
$AlCl_3$ (49.29 mg Al in
50 ml) by 0.1 N NaOH
[1192].

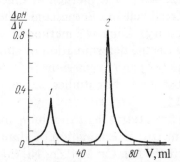

Figure 4. Differential titration
curve of the potentiometric titration of an acid solution of
$AlCl_3$ (49.29 mg Al in 50 ml)
by 0.1 N NaOH [1192]:

1) titration of free acid; 2) titration of $AlCl_3$.

A borax solution [504] can also be employed as titrant; Pt/Pt-black
serves as the indicator electrode. A potentiometric indication of the equivalence point can also be realized during titration by Complexone III.

Amperometric titration

In the amperometric titration of aluminum, the titrant is a solution of
a substance which either precipitates aluminum or forms stable complexes
with it. The titrant may be reduced on a mercury or a platinum electrode,
but titrants which are not reduced, such as NaF, can also be employed,
provided that a substance giving a diffusion current (in the present case
Fe^{3+} ions) is introduced as the indicator of the equivalence point. Aluminum can be amperometrically determined by the following titrants: sodium
or potassium fluoride [52,116,439,441–443,493,1239], hydroxyquinoline

[116, 286, 380], solutions of iron, calcium and vanadyl salts, with back-titration of excess Complexone III [24, 86, 764].

The disadvantage of titrations by hydroxyquinoline or urotropine [483] is that time is required for the formation of the aluminum precipitate.

Better results are obtained by using substances which rapidly react with aluminum, such as fluoride [1118a]. In this case neither aluminum nor fluoride participate in the electrode reaction, and small amounts of Fe^{3+} ions are introduced as indicator of the equivalence point (0.5 ml of 0.1 M $FeCl_3$). After all the aluminum has been converted to the sparingly soluble cryolite complex, the fluoride binds the Fe^{3+} ions into a complex, and the diffusion current due to Fe^{3+} disappears, indicating that the titration has terminated. The results must be corrected for the amount of fluoride that has reacted with Fe^{3+}. To reduce the solubility of cryolite, it has been suggested that this titration be carried out in a solution containing 50% ethanol. The method is used to determine aluminum in high aluminum alloys [493], copper alloys [439, 443], chromites [52], clays and chamottes [441, 442] with a rotating platinum electrode. It has been shown [52] that the end point is sharper if a mixture of Fe^{3+} and Cu^{2+} ions is used as indicator.

Babenyshev and Kuznetsova [24] determined aluminum by titrating excess Complexone III by a solution of $FeCl_3$, with amperometric indication of the end point. The sum of Al, Ca and Mg is determined by titrating the excess Complexone III by a solution of calcium nitrate at pH 8; in another part of the solution aluminum is bound by triethanolamine at pH 10 and the sum Mg+Ca is determined by titration against Complexone III. The content of aluminum is found by difference [86]. Aluminum can also be determined by titrating excess Complexone III by a solution of the vanadyl ion, VO^{2+} (pH 4 ± 0.5; acetate buffer solution); the indicator is the oxidation current of VO^{2+} on a platinum electrode at +0.6 V (with reference to a saturated calomel electrode) [764]. Fluorides and phosphates are added to improve the specificity of the method.

Conductimetric titration

The conductimetric determination of aluminum is based on the formation of stable complexes between aluminum and certain organic acids, or on the precipitation of aluminum as sparingly soluble compounds. Pasovskaya [328–331] proposed a number of methods based on the formation of acetate, tartrate, oxalate, and lactate complexes of aluminum. Magnesium and calcium do not interfere with such determinations, but ferric iron must be reduced to Fe(II) by ascorbic acid. The relative error of these methods is about 1%. Aluminum can be conductimetrically titrated [959] by a solution of the disodium salt of vanillinazine. Aluminum reacts with this salt in an almost neutral solution to form a 1:1 complex. Aluminum and zinc can be determined in the presence of one another, and then the second break corresponds to the terminal point of titration of aluminum. Khudyakova [464a] described a method for the automatic conductimetric titration in the determination of aluminum chloride and hydrochloric acid present together. Borax or sodium hydroxide solution can be used as titrants. The conductimetric curve will have two breaks, the first corresponding to the amount of hydrochloric acid, and the second to the amount of aluminum chloride.

Other authors carry out the conductimetric titration in a nonaqueous medium: aluminum is titrated by EDTA in dimethylformamide [148], or a solution of aluminum hydroxyquinolate is titrated by a solution of potassium methoxylate in a benzene:ethanol mixture [1236].

Coulometric titration

This method is described in only a few publications [200, 201, 838]. Kostromin et al. [200, 201] determine aluminum indirectly by titrating the bound hydroxyquinoline by electrically generated bromine. For the determination of aluminum in high-purity selenium [201], the selenium is distilled as SeO_2.

The dry residue is dissolved in dilute H_2SO_4; the pH is adjusted to about 9.6 by a borate buffer, the interfering elements are eliminated by extracting their diethydithio-carbamates with chloroform. Aluminum is then isolated by extracting its hydroxyquinolate by benzene. Benzene is evaporated, the hydroxyquinolate is dissolved in HCl (sp.gr. 1.19), and the bound hydroxyquinoline is determined by coulometric titration with $0.2 M$ KBr + $0.1 N$ H_2SO_4 as the supporting electrolyte. The end point is found by the amperometric method, with two platinum electrodes, to which a 200 mV voltage is applied. The background alone is first titrated and a correction introduced.

In the method proposed by Iwamoto [838] the H^+ ions formed during the formation of aluminum hydroxyquinolate are titrated against the electrically generated OH^- ions. The titration is carried out in a water-ethanol medium (1:1). The end point is found by determining the pH of the solution. Sodium sulfate, in a concentration of about $0.05 M$, is used as the supporting electrolyte. The relative error is up to 8%. Chloride ions interfere; magnesium, alkali metals, and alkaline-earth metals do not.

Other electrochemical methods

Cruse and Nettesheim [663] studied the determination of aluminum by high-frequency titration by sodium hydroxide, barium hydroxide, morin, and Eriochromecyanine R. The titration is time-consuming (30 min or more) and the accuracy is not very high; the method is therefore of no interest.

The determination of aluminum by high-frequency titration by NaOH, Ba(OH)$_2$, NH$_4$OH, triethanolamine, CH$_3$COOH, HF [1094, 1095, 1164] and hydroxyquinoline as titrants has no advantages over the more conventional methods.

A method for the separation and determination of aluminum and other components of a silicate solution, which is based on the different mobilities of the different ions, has been described [183]. The method is not very accurate: the relative error is 8—12% for aluminum contents between 8 and 12%.

PHOTOMETRIC METHODS

Photometric methods are widely employed in the analytical chemistry of aluminum. The Al^{3+} ion is not a chromophore, and for this reason colored reagents only can be used for the determination of aluminum. For reviews on photometric methods for the determination of aluminum, see [360, 421].

Determination by Aluminon

Absorption spectra and composition of complex. Aluminon* is most frequently employed for the colorimetric determinations of aluminum. The absorption maximum of the aluminum-Aluminon complex is at 528 mμ [758], 530 mμ [780], 535 mμ [776, 1287], and 540 mμ [545]. According to our own results, the absorption maximum is at 535 mμ (Figure 5). In the complex, the ratio of Al to Aluminon is 1 : 1. The salt-forming and the carbonyl groups of the quinonoid ring in the Aluminon molecule participate in the formation of the aluminum complex. The structure of the complex can be represented by the formula:

---

* Aluminon is the trisubstituted ammonium salt of aurintricarboxylic acid; it is a triphenylmethane dye. It was first suggested by Hammet and Sottery in 1925 [788a].

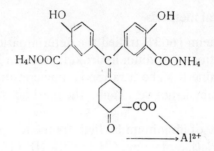

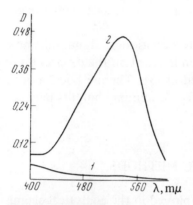

Figure 5. Absorption spectra of Aluminon and the aluminum-Aluminon complex:

1) $5 \cdot 10^{-5} M$ (pH 5.2) solution of Aluminon with reference to water; 2) aluminum-Aluminon complex with reference to a solution of the reagent (concentration of aluminum $5 \cdot 10^{-5} M$, pH 5.22).

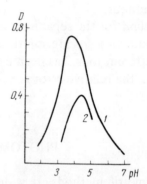

Figure 6. Effect of thioglycolic acid on the color of the aluminum complex at different pH values [780]:

1) solution containing 20 μg Al in 50 ml; 2) the same and 0.08% of thioglycolic acid.

Effect of pH. Different authors report that the maximum color intensity of the complex occurs at pH 3.2–3.8 [1287], 3.9–4.2 [1179, 1278], and at pH 3.8 [780, 1117]. In the presence of thioglycolic acid, added to eliminate the effect of iron and certain other metals, pH_{max} is shifted from 3.8 to 4.4 [780] (Figure 6). At pH_{max} the absorption by the reagent is weak, but as the pH decreases, the color of the reagent becomes much more intense. The pH corresponding to the maximum sensitivity of the method is at pH 3.8–4.0, but this range is unsuitable for analytical work owing to the superposition of the color of the reagent, the low stability of the aluminum complex and the danger of the liberation of free aurintricarboxylic acid. The very high pH values sometimes recommended in the literature, namely, pH 6.0 [1203] and pH 6.3 [1120], are also unsuitable, since at these pH values the sensitivity of the method is much lower.

The optimum pH range is 4.4–4.75. At pH 4.4 in the presence of glycolic acid the change in the color intensity with pH is negligible. But, if acetate buffer solution is employed to produce the desired pH of the medium, the buffer capacity is a maximum at pH 4.75 (pK of the acid 4.75), so that these pH values are more easily maintained constant (Figure 7). Aluminum is usually determined at pH 4.4 [19, 38, 1074, 1114, 1119], 4.75 [223, 249, 267, 449, 468, 780, 1086], or pH 4.4–4.75 [776, 1203].

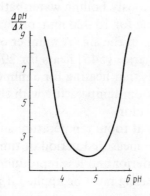

Figure 7. pH as a function of the acid content (Δx, g-equiv) in acetate buffer solution ($CH_3COOH : CH_3COONa = 1:1$).

Many workers determine aluminum at pH 5.0–5.3; the method then becomes less sensitive and the color varies more strongly with the pH than at pH 4.4–4.75, but if a high-capacity buffer solution is employed, the variations in the pH can be reduced to a minimum. At pH 5.0–5.3 the superposition of the color of the excess reagent is less and the aluminum complex is more stable, especially to heat; accordingly, this variant has been widely and successfully employed in the analyses of a great variety of different materials [105, 155, 179, 336, 382, 545, 555, 649, 659, 697, 829, 938, 939, 955, 1037, 1053, 1162, 1194].

Formerly, the colored aluminum compound was produced in weakly acid medium, and then the color intensity of the excess reagent was reduced by increasing the pH of the solution to 7.1—7.5 with ammonia-ammonium carbonate or ammonia-boric acid [1003, 1202, 1203, 1278, 1282]. However, these techniques are not advisable, since in weakly alkaline media the color intensity of the aluminum complex decreases greatly with decrease in the color of the excess reagent.

The pH of the medium is best adjusted by acetate solutions. Biphthalate buffer solutions [262, 264] can never have the same buffer capacity as acetate solutions, because of the lower solubility of phthalates.

Development of the color of the complex. At ordinary temperatures the color of the complex develops slowly, but is more rapid if the mixture is heated (the heating must be carried out on a boiling water bath). Various heating times have been recommended (between 4 and 20 min). According to Luke and Braun [939] and Yuan and Fiskell [1287], solutions must be held for five minutes on a vigorously boiling water bath. Craft and Makepeace [659] recommend heating for 15—20 min, or heating for 10 min followed by holding for 10 min in the air. A number of authors heat for 10 min [776, 1024], while Banarjee [545] heats for 20 min. Our own laboratory experience indicates that heating for 5 min on a boiling water bath, followed by cooling to room temperature with the bath, suffices for complete development of the color.

Others let the color develop at room temperature and determine the color intensity after an exactly measured period of time, say $30 \pm \frac{1}{2}$ min [780]. This technique gives inferior and less reproducible results than that with heating. Accordingly, heating must be applied if accurate results are to be obtained.

The temperature of the solution at which the photometric determination is carried out has a small effect on the optical density [555, 1074, 1258]; with increase in temperature, the optical density increases. Accordingly, the temperature of the samples should be kept as constant as possible. The difference between the temperature at which the calibration curve is determined must not differ by more than $5°C$ from that of the solution being analyzed [555].

Volume of solution. The optimum volume of the solution in which the color is allowed to develop is 20—30 ml [545, 659, 1282]. The optical density decreases with increase in volume.

Amount of Aluminon. According to the results of Craft and Makepeace [659], the optimum amount of Aluminon is 3.3—5 mg; this is achieved by the use of 10—15 ml of Aluminon buffer solution, containing 1 g of Aluminon in 3 liters. According to Grant [776], the maximum optical density corresponding to a content of $60 \mu g$ Al is obtained with 4 mg Aluminon in 50 ml of solution.

Employment of protective colloids. Various protective colloids have been proposed for the stabilization of the Aluminon complex (lake): starch, gelatin, gum Arabic, glycerol. All these increase the stability of the Aluminon lake and do not affect its color intensity, but only up to a certain limit, above which the color development is suppressed. The optimum quantities of the protective colloids (in 100 ml of solution) are: starch up to 5 ml of a 2% solution [1284]; gum Arabic, 1—4 ml of a 1% solution [1029, 1287]; gelatin, 5 ml of 1% solution [659, 939], glycerol 20 ml. It is convenient to first mix the protective colloid with the solution of Aluminon and the buffer solution, and thus the determination is simplified.

Grade of Aluminon and preparation of Aluminon solution. As a rule, different color intensities are obtained if the reagent has been prepared by different workers. The calibration curves prepared with reagents marketed by different manufacturers, or even with different batches of the reagent from the same manufacturer, do not coincide. Accordingly, a new calibration graph must be prepared with each new batch of Aluminon. There are many published data on the preparation and properties of Aluminon [813, 1146, 1179, 1202, 1292, etc.].

Many workers recommend 0.1% or 0.2% solutions of Aluminon. However, these solutions are not very stable, and the results are less reproducible than those given by the so-called composite Aluminon solution, which is stable for several months. This solution is prepared as follows [939].

One liter of water and 80 ml of glacial acetic acid are placed in a two-liter beaker and 500 g of ammonium acetate are dissolved in the mixture. One g of Aluminon is dissolved in 50 ml, and this solution is combined with the buffer solution. Then 2 g of benzoic acid are dissolved in 20 ml of methanol (or ethanol), and this solution is poured into the buffer solution. The mixture is diluted to two liters with water.

Ten grams of gelatin are dissolved in 250 ml of water by heating on a boiling water bath; the solution is occasionally stirred until the gelatin has completely dissolved. The warm gelatin solution is poured into 500 ml water, with stirring, cooled to room temperature, and diluted to one liter with water. The solutions of Aluminon and gelatin are transferred to a 4-liter bottle with a glass stopper, shaken well, and stored in a dark place.

The color of Aluminon solutions weakens on standing, especially during the first few days. For this reason the composite Aluminon solution should be allowed to stand for three days before use. The solution can be kept for one month.

The gelatin may be replaced by 800 ml of glycerin [1053], and the composite Aluminon solution thus obtained is more transparent. Owen and Price dissolve benzoic acid in isopropanol rather than in methanol [1037]. The presence of isopropanol in the solution of Aluminon ensures a better color homogeneity in the solution used for the photometric determination, and thus a higher reproducibility of the results. The solution prepared according to Owen and Price* [1037] keeps better and can be used immediately after it has been prepared.

Beer's law. According to the results of different workers, Beer's law holds up to 50 μg [170, 649], 60 μg [776] and 70 μg [1203] of aluminum in 50 ml. These discrepancies are explained by the different experimental conditions employed; in the two first studies the determination was carried out at pH 5.3 and 5.0—5.2, respectively, and in the last two studies at pH 4.5.

The *sensitivity* of the reaction between aluminum and Aluminon is 0.05 μg/ml [284].

Effect of other cations. Aluminon is not a specific reagent for aluminum. Many elements form colored compounds with Aluminon. The pH range of the reaction between Aluminon and the different metals is as follows:

Element	pH	Element	pH
Mo(VI)	1—4	Pb	6—9
Cr(III)	2—5.5	Co, Zn, Mn	7—9
Th, Sn(II), V(IV), U(VI)	2—7	Cd	7—11
Al, Fe(III), Ce(III)	2—9	Ni, Mg	9—11
Ti(IV)	2—11	Ca	⩾11
Cu(II), Be	4—9		

Ferric iron, V(IV), Be and Cu(II) interfere most.

In most cases iron interferes. At 525 mμ, the Fe(III)-Aluminon complex has a color intensity which is 2—2.5 times weaker than that of the aluminum compound. The effect of small amounts of Fe(III) is eliminated by reduction to Fe(II) by thioglycolic acid [586, 630, 632, 727, 938, 939], ascorbic acid or hydroxylamine [179, 406, 1052, 1203, 1287].

Thioglycolic acid or ammonium thioglycolate reduces Fe(III) to Fe(II), and binds the Fe(II) into a colorless complex. According to Luke and Braun [939], the effect of 50 mg of Fe(II) can be eliminated by thioglycolic acid, provided that the solution is allowed to stand for 15 min before the

---

* Unlike Luke and Braun [939], these workers dissolve the benzoic acid in 750 ml of isopropanol instead of 20 ml of methanol.

photometric determination. The solution in the photometric cell may darken owing to oxidation by atmospheric oxygen. Accordingly, if considerable quantities of iron are present, the solution in the cell must be allowed to stand for a few minutes before its optical density is determined (to reduce iron) [1053].

In the presence of thioglycolic acid, the color intensity of the aluminum complex changes, so that equal amounts of the acid must be introduced into the standard solutions as well.

Ascorbic acid is widely employed. To reduce 5 mg of iron the theoretical amount of ascorbic acid required is about 10 mg, but in practice not less than 75 mg of acid are needed [106]. After the addition of ascorbic acid, the solution should be held for 5 min at 70–80°C. It must be noted that the presence of ascorbic acid somewhat reduces the color intensity of the aluminum-Aluminon complex.

If iron is reduced by hydroxylamine, aluminum is determined at pH 5.0–5.2 [180] or 4.8 [1192a]. According to Yuan and Fiskell [1287], hydroxylamine reduces the color intensity of the complex at pH 4.8, but has no effect at pH 3.5. Therefore pH 3.5 is recommended by these workers. If the reduction is effected by hydroxylamine, the permissible amounts of iron are less than when thioglycolic acid or ascorbic acid are employed. At pH 3.5, if the solution is boiled for 5 min, up to $150\,\mu g$ Fe do not interfere; the results become 3%, 6%, and 35% too high in the presence of 200, 500, and $1000\mu g$ Fe, respectively [1287]. According to [1192a], threefold amounts of iron do not interfere at pH 4.8.

It has been recommended [866] that iron be masked by Complexone III and by equivalent amounts of zinc. The results are satisfactory even in the presence of a 20-fold excess of iron.

Nevertheless, neither this method, nor the reduction by hydroxylamine are of any special interest. Thioglycolic and ascorbic acids eliminate the influence of iron most effectively.

Aluminum and iron can be determined in the presence of one another. The total absorption of the two elements is measured at $530\,m\mu$, and then iron is determined on an aliquot by the thiocyanate method and aluminum is found by difference [265]. According to another author [307a], optical densities are determined at 530 and $570\,m\mu$, i.e., at the λ_{max} for aluminum and iron complexes, respectively. The content of each element is then found from previously prepared calibration curves.

Aluminum and iron can be determined after the iron has been reduced by hydroxylamine [299]; the optical densities are measured at 430 and $540\,m\mu$. The contents of aluminum and iron are found from a special nomogram. In another method, both elements are determined when present together in solution, by the photometric method for iron with ferrocyanide and the determination of aluminum by Aluminon; a correction is introduced for the color produced by iron [466]. However, it is only very rarely that the sum of the two elements has to be determined.

The chromium-Aluminon compound is formed slowly in the cold; thus, if the determination is carried out in the cold, chromium does not significantly interfere. According to Short [1162], up to 1 mg Cr does not interfere at pH 4.9−5.0; according to Konkin [180] this amount is up to 0.84 mg. Roller [1120] found that if the optical density is determined after 30 min, 0.1 mg Cr(III) is equivalent to 0.001 mg of aluminum at room temperature, and to 0.008 mg after 11 hours. If the determination is carried out in a hot solution, chromium interferes strongly: $200 \mu g$ Cr(III) produce a color equivalent to that of $40 \mu g$ Al [1162, 1287]. The interference of Cr(VI) is less serious than that of Cr(III). A number of workers accordingly recommend that chromium be oxidized to Cr(VI). Craft and Makepeace [659] found that up to 2% Cr(VI) does not interfere, while up to 1% Cr(VI) does not interfere according to Codell and Norwitz [649]. However, subsequent checks showed that even small amounts of Cr(VI) interfere with the determination of aluminum [545, 780]. Thus, the error introduced by $25 \mu g$ Cr(VI) is equivalent to the presence of $2 \mu g$ Al [545].

Heavy metals can be masked by thioglycolic acid. The complex formed by copper with thioglycolic acid is so stable that it is possible to determine aluminum in copper alloys. If large amounts of copper are present, a green coloration is formed in the presence of thioglycolic acid, but if a green filter is used for the determination of the optical density, this does not seriously interfere. Small interferences due to copper are compensated by introducing equivalent amounts of copper into the standard solutions and in the blank solution [939, 1053]. Copper can also be bound as a colorless complex with thiourea [450].

In the presence of thioglycolic acid, 10,000-fold amounts of zinc [1053] do not interfere. However, if the content of zinc exceeds 10 mg, the masking power of thioglycolic acid decreases, since zinc binds thioglycolic acid, and the amount of thioglycolic acid present may prove insufficient to eliminate the effect of iron and copper [939].

Large amounts of cadmium may be masked by thioglycolic acid [1053]. In the presence of more than 0.1 mg lead or bismuth, a precipitate may separate out; in this case, the solution should not be heated for longer than 3 min, since if it is heated for a longer period of time, it becomes turbid [1053]. In the absence of thioglycolic acid, even large amounts of lead do not interfere with the determination of aluminum [1053]. Up to 10% of tin can be masked by thioglycolic acid [1053]. It must be remembered that both tin and antimony tend to hydrolyze at the pH of the determination of aluminum.

According to Luke [938], up to 100 μg of the following elements: Ge, As(V), Sb(V), Mo(VI), Hg(II), Tl(III), Tl(I), Cd, Zn, Ni, Fe(III), Fe(II), Mn, Mg, Si, W, Ta, Nb, Y, In, U(VI), Ga, Ce(IV), Ce(III), La, Sm, Nd, Pr, B, P, Ba, Sr, Ca do not interfere if the determination is carried out by the thioglycolic acid method and the optical density is measured at 515 mμ. Heavy metals (Pb, Bi, Sn, and Ag) produce turbidity; cobalt reacts with thioglycolic acid to give a brown coloration, which absorbs slightly at 515 mμ. The elements Be, Sc, Cr(VI), Cr(III), V(V), V(IV), Cu(II), Cu(I), Zr, Hf, Ti and Th react with Aluminon to give red lakes; with reference to color intensity, 100 μg Be are equivalent to 75 μg Al; 100 μg Sc are equivalent to 10 μg Al; 100 μg of the other elements are equivalent to 5 μg Al. Titanium and zirconium do not interfere, in practice, if present in amounts not exceeding 20 μg.

According to Fedorov and Sokolova [450], the interfering elements can be more efficiently masked by a mixture of thioglycolic and ascorbic acids than by thioglycolic acid alone.

Large amounts of ammonium salts somewhat attenuate the color intensity of the aluminum complex [223, 1162]. Sodium chloride also interferes (1 g NaCl reduces the optical density by 4–5% in the presence of 2–20 μg Al, while higher amounts of NaCl produce a correspondingly higher interference).

Effect of anions. According to Craft and Makepeace [659], the following anions do not interfere if present in the amounts listed below:

Anion	Content, mg/ml	Anion	Content, mg/ml
Cl^-	12	ClO_4^-	45
NO_3^-	60	SO_4^{2-}	36

In the presence of up to 0.01 mg PO_4^{3-} per ml, the error does not exceed 5–6% [659]. According to other data, 50 ml of solution must not contain more than 3 mg PO_4^{3-}, the resultant error is not more than 3% [780]. Yuan and Fiskell [1287] found that 20-fold amounts of PO_4^{3-} do not interfere, and Yoe and Hill [1282] state that up to 2 mg H_3PO_4 do not interfere; in the presence of phosphates in amounts higher than the above, the color intensity of the complex is reduced. Complex-forming ions (tartrates, citrates, fluorides, oxalates, Complexone III) interfere with the color development [1287]. If the solution contains hydrogen peroxide, it must be decomposed [1287]. Free chlorine should be reduced by thiosulfate [1163].

To the sample solution are added 2 ml of a 4% solution of thioglycolic acid, then the solution is neutralized to pH 3 with ammonia (test with universal indicator paper). The solution is transferred to a 100-ml volumetric flask, 15 ml of the composite Aluminon solution are added (for the preparation of the solution, see p. 97), then the flask is shaken and placed in a boiling water bath for 5 min. The flask is then immersed in a cold water bath. After the solution has reached room temperature, it is made up to the mark with water and shaken. It is then poured into a photocolorimeter cell and left to stand for 1–2 min (if the sample contained much iron). The optical density is then measured at 525 mμ with reference to a blank solution. The aluminum content is read off a calibration curve, plotted from data obtained under the same conditions [938].

It was proposed by Green [777] that aluminum be determined by a mixture of Aluminon and Ferron. According to this author, with the mixed reagent it is possible to determine aluminum over a wider concentration range than when Aluminon alone is used, and no protective colloids are necessary.

Determination by Eriochrome Cyanine R

Eriochrome Cyanine R, which was first introduced into analytical practice by Eegrieve [699], is the most sensitive reagent for aluminum

The properties of the reagent and of its complex with aluminum. Eriochrome Cyanine R is an acid-base indicator which undergoes several color changes [796]: at pH 1.8–2.8 from orange-red to pink-red, at pH 4.5–6.2 from pink-red to yellow, and at pH 9.0–12.5 from yellow to blue-violet. The absorption maxima of the various forms are shown below.

Color	λ, mμ	pH
Orange-red	480	1
Pink-red	515	3
Yellow	430	8
Blue-violet	580	13

The absorption maximum of the aluminum complex is given by different workers (λ in mμ) as: 520–540 [250], 528 [1218], 530 [523, 759, 880, 881, 1247], 532 [656], 530–535 [796], 535 [568], 538 [808], 540 [285b], 550 [1195]. The most probable value is 530–535 mμ (Figure 8). The

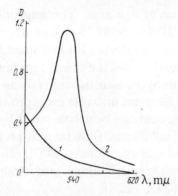

Figure 8. Absorption spectra of Eriochrome
Cyanine R (1) and its aluminum complex (2).

location of the absorption maximum is practically independent of the pH
and of the aluminum concentration. According to Hegedüs [796], the
shifts in the absorption maximum due to changes in the pH and in aluminum
concentration are small. Most workers [709, 808, 913, 1247, etc.] give the
formula of the complex as $Al(ER)_3$, but the aluminum : reagent ratio may
also be 1 : 2 [808, 974] or 1 : 1 [974]. Hegemann and Wilk [800] demon-
strated the formation of a 1 : 1 complex (pH 3.0–4.7) and a 1 : 2 complex
(pH 6.3) by the method of isomolar series. These authors found that for
the 1 : 1 complex, $\log K_{stab} = 7.62$ (pH 4.1). According to Hegedüs, the
product of the reaction between Eriochrome Cyanine R and aluminum has
a variable composition, and is an adsorptive compound. The molar extinction coefficient of the aluminum complex is $6.75 \cdot 10^4$ [656].

Many workers employed an aqueous solution of the reagent. However,
Eriochrome Cyanine R solutions are stable only when acid (pH up to 3.5)
[1212]. The reagent is particularly unstable in the pH range 4–7, which
corresponds to the color change from red to yellow. The reagent ages, but
its solution can be regenerated by acidification. It has been suggested that
the stability of the reagent can be improved by using its nickel salt [802].
When aluminum is determined in the presence of iron, better results are
obtained with the decolorized reagent [549].

The chromophore groups of Eriochrome Cyanine R are easily decomposed by a large number of reducing agents, such as sodium bisulfite. The
reduction of the chromophore groups is reversible. If the aluminum complex is decomposed by sodium bisulfite, it is reformed if one or two drops
of hydrogen peroxide are added. The chromophore groups of the reagent

are more stable in the presence of oxidants. The reagent is best stabilized by adding NaCl, NH_4NO_3 or HNO_3 [808]. Several authors [226, 568, 808, 809] have pointed out that such a nitrated Eriochrome Cyanine R has better complex-forming properties and the color is more stable. If nitrated Eriochrome Cyanine R is employed, nitrous acid is formed in the solution and the color of the reagent becomes unstable owing to the reduction of the chromophore group. The acid can be decomposed by the addition of urea and sulfanilic acid. The solutions of the reagent prepared in this manner are stable for 6 months or even longer [809]. The use of a composite reagent containing Eriochrome Cyanine, gum Arabic, and buffer solution has been proposed [243, 1228].

As in the case of Aluminon, the properties of Eriochrome Cyanine R vary with the method of preparation of the reagent [1247]. A new calibration curve must therefore be prepared for each new batch of the reagent.

Duration of color development and effect of pH. The conditions of formation of the aluminum complex of Eriochrome Cyanine R have been studied in greatest detail by Milner [972, 974] and Richter [1114].

The reaction between aluminum and Eriochrome Cyanine R takes time and the rate differs with the age of the solution. Richter showed that the reaction is quantitative only if the solution is heated and has an acid reaction. Thus, at pH 3.8, the color develops more rapidly than at pH 5.4; moreover, the results are more reproducible at the former pH value. For this reason Milner [972], Richter [1114, 1115], Walraf [1247], and others [597, 759, 1097] considered that aluminum is preferably determined at pH 3.8. However, most workers suggest that aluminum be determined at higher pH values (5.4–6.1), since at pH 3.8 the color of the reagent itself is much stronger than at a pH of about 6, while at the latter value the maximum color intensity of the complex is stronger than at pH 3.8 (Figure 9). Methods in which the advantages of both techniques are utilized are especially interesting. For example, colored compounds are formed in an acid medium at pH 3.8 [974] or even at pH 2.5 [754, 1218, 1257] within 20–30 min. The pH is reduced to 6 to reduce the color of the reagent present in excess, and the optical density is determined either immediately [974, 1257] or after 15 min [754]. According to [1217], the solutions must be held at pH 6 for 45 min, so that the color intensity becomes constant.

The development of the color can be accelerated if the solutions to be analyzed are boiled after the addition of all the reagents and are then rapidly cooled under running water [1114, 1247]. The temperature of the solution has practically no effect on its optical density. According to [549],

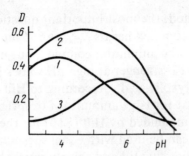

Figure 9. Color intensity of aluminum-Eriochrome Cyanine R complex as a function of the pH [831]:

1) solution of reagent measured with reference to water; 2 and 3) solution of aluminum complex measured with reference to water and solution of reagent, respectively.

the temperature coefficient at 15–30°C is 0.0035 per degree centigrade for optical densities of between 0.30 and 1.08. If the solution is boiled and then rapidly cooled, its temperature has practically no effect on its optical density [1247].

Acetate buffer solutions are used to produce the optimum pH-value. Buffer solutions containing malonic, tartaric or citric acids must not be used, since they decompose the dye. A mixture of aniline with aniline hydrochloride can be used as a buffer solution (pH 4.8–5.2), since absorption by the aluminum complex is unaffected by the presence of this buffer.

Application of Beer's law. According to Hill [808], Beer's law holds up to a concentration of 80 µg Al/50 ml; at high concentrations, deviations from Beer's law can be minimized by adding large amounts of Eriochrome Cyanine R.

Effect of cations. Eriochrome Cyanine R reacts with a large number of metals to form colored complexes. The pH values of the incipient formation of the colored complexes are given below [1212]:

Element	pH	Element	pH	Element	pH
U(IV), Mo(VI)	0	Al, Be, In, U(VI)	2.5	Co	7.0
Fe(III), Zr	0.5	Ce(III)	3.5	Mg	9.5
Ti	1.0	Cu(II)	5.5	Mn	10.0
Ga, Th	1.5	Ni, La	6.5	Ca, Ba	11.0
V(V)	2.0				

As in any other method, the most important question is the effect of iron. The iron-reagent complex has its λ_{max} at 556 mμ; its color is much less intense than that of the aluminum complex, but nevertheless iron interferes strongly. Its effect is stronger at high pH values [568], and is eliminated by ascorbic or thioglycolic acid. According to Hill [809], ascorbic acid is a better reducing agent for large amounts of iron than mercaptoacetate. This acid has been recommended by Hill [810] for the analysis of steels, while mercaptoacetate can be used with equal success for the analysis of iron ores [809]. Thioglycolic acid and thioglycolates somewhat reduce the color intensity of the aluminum complex, and this must be taken into account when constructing a calibration curve [809, 925].

Chromium reacts with Eriochrome Cyanine R to form a complex, but its color intensity is much weaker than that of the aluminum complex [1114]; the development of the color of the aluminum complex is thereby suppressed, since part of the reagent is bound by chromium. Chromium can be eliminated as chloride, but this lengthens the analytical procedure, and other methods for the elimination of chromium have been proposed. According to Hill [808], the addition of $FeSO_4$ and 8-hydroxyquinoline assists the formation of the aluminum complex in the presence of large quantities of chromium. A number of workers [926, 1247] compensate for the effect of chromium by introducing the same amounts of this element into the standard solutions. Lilie and Rosin [926] recommend that in determinations of aluminum in steels, several calibration curves be plotted for several concentrations of chromium. If the chromium content is 0–2%, satisfactory results will be obtained from calibration curve constructed in the absence of chromium.

For other cations, 100 μg Al can be determined in the presence of 0.5 mg Ni, Co, Mo, Mn and W [926]. According to other data [831], 4–150 μg Al can be determined in the presence of 3 mg Mn, 0.5 mg Cr and Sn, 0.3 mg Cu, and 0.15 mg P. Five μg Al can be determined in the presence of 0.25 g Zn without preliminary separation [831]. Large amounts of Cd, Pb and Sn do not interfere. The effect of copper can be eliminated by introducing sodium thiosulfate [250], and in this way it is possible to determine aluminum in copper-zinc alloys without any preliminary separation [250]. Arsenic interferes to a small extent [1195].

Separations must be carried out when analyzing multicomponent materials. Hill [808] separates a large number of elements which interfere with the analysis of iron ores from aluminum by fusing the sample with sodium carbonate. When the melt is leached, vanadium passes into the extract together with aluminum. The colored complex of V(IV) with

Eriochrome Cyanine R can be decomposed by adding a solution of hydroxy-quinoline in chloroform. If the content of vanadium is known, a correction can be made: 1% V is equivalent to 0.12% Al. Beryllium and zirconium give colored complexes; $2\,\mu g$ Be is equivalent to $1\,\mu g$ Al [656]. The beryllium and zirconium complexes can be decomposed by fluorides. Complex-one III decomposes all the complexes except those of these elements; thorium forms a complex at low pH values and high concentrations of Eriochrome Cyanine R only. Aluminum can be separated from many elements by means of NaOH [818].

Effect of anions. Large amounts of chlorides weaken the color intensity of the aluminum complex; sulfates have an even stronger effect [1114]. Fluorides and phosphates interfere strongly even in small amounts [522, 808, 1114]. Organic hydroxy acids also interfere with the formation of the aluminum complex.

Eriochrome Cyanine R reacts with aluminum in the presence of cetyltri-methylammonium chloride to form a blue ternary complex with $\lambda_{max} = 587\,m\mu$ and a molar extinction coefficient of $1.17 \cdot 10^5$ [361b]. The optimum pH of the medium is 5.3–6.3. Beer's law holds at $0.01–0.08\,\mu g$ Al/ml. With this method, P, Zr, Cu, Cr, Ga, Bi, Ta, Ti, Pt, Ge, Be, Th, and U interfere.

The determination of aluminum by Solochrome Cyanine RS, which is a reagent resembling Eriochrome Cyanine R, has been described [540, 541, 1071]. The maximum difference between the absorptions of the reagent and the complex is observed at $546\,m\mu$ and pH 6.1, but the optimum pH value is 5.8. Under these conditions the colors are very stable on standing, and the reproducibility is satisfactory. The above authors found that at pH 5, and also at pH 6.0–6.1, the stability and the reproducibility of the color are lower. All the metals that react with Eriochrome Cyanine R also react with Solochrome Cyanine RS to form colored complexes. Solochrome Cyanine R, sometimes found in the literature, is merely a synonym for Eriochrome Cyanine R.

Determination by Chrome Azurol S

Chrome Azurol S is the sodium salt of 2,6-dichlorodimethylsulfoxy-fuchsonedicarboxylic acid. The free acid under the name of "Alberon" was proposed by Mustafin et al. [164, 285] for the photometric determination of aluminum and beryllium. Chrome Azurol S was later widely used [18, 417, 592, 596, 772, 820, 1004, 1041, 1189] for the analysis of aluminum.

Properties of the reagent and its aluminum complex. Chrome Azurol S has the following absorption maxima [7, 592, 950, 1189]:

pH	$\lambda, m\mu$
6–7	430
~1	465
~3	495
12–13	590–600

Studies at pH 5.6–5.8 by the method of molar ratios and isomolar series showed that the aluminum : Chrome Azurol ratio in the complex is 1 : 2 [265a, 417, 427, 592, 820]. It has been pointed out [636] that this complex is formed at pH $\geqslant 3.5$, while the complex formed at pH $\leqslant 3.5$ has the composition 1 : 1. Srivastava et al. [1189] also mention the formation of a 1 : 1 complex (at pH 4). This complex has an absorption maximum at 570 mμ [636], and the logarithm of the formation constant of the complex is 5.20 [636]; according to other data [1189], log $K_{stab} = 4.3$ at pH 4. The 1 : 2 complex has a maximum absorption at 545 mμ [417, 592, 596, 636, 1041, 1189] (Figure 10); the logarithm of its formation constant is 9.64 [636], while the molar extinction coefficient of the complex is 59,300 [820]. Chrome Azurol S and Eriochrome Cyanine R are the two most sensitive reagents for aluminum.

The color of the aluminum complex with Chrome Azurol S develops rapidly; the maximum color intensity is attained within 10 min, and it then remains constant for a long time. It is independent of the temperature over a wide temperature range [1189]. The maximum color is observed at pH 5.8; as the pH decreases, the color intensity decreases slowly, while above pH 6.1 the drop in intensity is very rapid (Figure 11). Between pH 5.7 and 6.1 the optical density does not significantly vary with the pH. Moreover, the absorption of the reagent is weak at these pH values, while at pH 3–5.5 the reagent is strongly colored. Therefore, the optimum pH of the medium is 5.7–5.8.

At pH 4, the sequence of addition of the reagents does not affect the results of the determination [1189], while at pH 5.6–5.8 it is important [417, 592]. If the sample solution is weakly acid (pH 4–5), the color intensity is a function of the volume of the solution before the addition of Chrome Azurol S (the smaller the volume, the stronger is the color intensity). This effect is not noted if the reagent is introduced into an acid solution. Therefore, the optimum pH of the medium is not produced by using buffer solutions, but by adding 5 ml of 0.1 N HCl, then Chrome Azurol S, and finally 5 ml of 2 N sodium acetate solution. The pH of such solutions is 5.7–5.8.

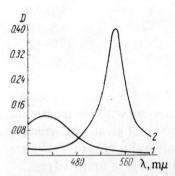

Figure 10. Absorption spectra of Chrome Azurol S and of the aluminum complex of Chrome Azurol S:

1) $1.6 \cdot 10^{-5} M$ solution of Chrome Azurol S with reference to water; 2) $0.8 \cdot 10^{-5} M$ solution of the aluminum complex with reference to water.

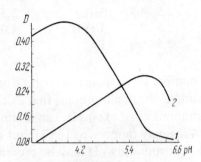

Figure 11. Dependence of the color of Chrome Azurol S and of the complex of aluminum with Chrome Azurol S on the pH:

1) 2 ml of a 0.1% solution of Chrome Azurol S with reference to water; 2) 0.01 mg of aluminum and 2 ml of a 0.01% solution of Chrome Azurol S with reference to a solution of the reagent.

The amount of the Chrome Azurol S present affects the color intensity of the aluminum complex; the optical density increases with the amount of the reagent. But, large quantities of the reagent must not be used, since then the blank solution is more intensely colored. In the presence of 2 ml of a 0.1% solution of Chrome Azurol S, up to 0.030 mg of aluminum can be determined; if more aluminum is present, correspondingly larger amounts of Chrome Azurol S must be added.

The sensitivity of the method is 0.006 μg Al per ml. If a photocolorimeter is employed, 0.5 μg Al in 100 ml of solution can still be determined.

If the optical density is plotted against the amount of aluminum, the graph is rectilinear up to 100 μg Al/100 ml [1189].

Effect of cations and anions. The pH values at which colored complexes of certain metals are formed with Chrome Azurol S are given below [683].

Metal	pH of stabil-ity of complex	Metal	pH of stabil-ity of complex	Metal	pH of stabil-ity of complex
Zr(IV)	1.5–5.5	Sc	3.0–6.5	Cu(II)	4.5–7.0
Hf(IV)	1.5–7.0	In	3.5–5.5	Y	5.5–7.5
Fe(III)	2.5–4.5	Pd(II)	3.5–8.0	Cd	9.5–11.5
Th	2.5–7.2	U(VI)	4.0–6.0	Mg	10–12
Ti	2.5–6.0	Be	4.5–7.5		

A large number of these metals form Chrome Azurol S complexes under the same conditions as aluminum, that is, they interfere with the determination of aluminum. For example, vanadium (V) does not greatly interfere, and up to 4 mg may be present [592], but only equal amounts of V(IV) are permissible. Chromium (VI) may be present in amounts of up to 10 mg [820], or only up to 1 mg [592]; this discrepancy is not very relevant, since if ascorbic acid is employed, Cr(VI) can be reduced to Cr(III). At pH 5, Cr(III) and Mo(VI) do not interfere if present in up to 20-fold amounts, but larger amounts attenuate the color intensity of the aluminum complex [164]. The effect of Cr(III) is weaker at lower pH values; thus, at pH 5.8, only equal amounts of Cr(III) may be present. At pH 5, 100-fold amounts of Zn, Mn, Co, Ni, As(V), V(V), Cd, Pb, Sb(III) do not interfere [164]. Calcium in 10,000-fold amounts and magnesium in 2,500-fold amounts do not interfere; alkali metals may be present in considerable amounts [417].

Stannic tin and W(VI) interfere by attenuating the color intensity of the aluminum complex; 1 mg W(VI) may cause a relative error of 2%.

The effect of iron in amounts of up to 3 mg/50 ml can be eliminated by reduction with ascorbic acid. Large amounts of iron which have been reduced by ascorbic acid reduce the optical density of the aluminum complex [18, 1041]; it is recommended, accordingly, that the same amounts of iron are introduced into standard solutions. Iron can be also be masked by thioglycolic acid [820]. At pH 6 the thioglycolic-iron complex is rather absorbant, and therefore the bulk of the iron is removed by extraction with methyl isobutyl ketone. Thioglycolic acid will mask up to 0.5 mg Fe in 50 ml. Cupric copper may be masked by thioglycolic acid and thiosulfate. The effect of molybdate ions can be eliminated by binding with phosphate ions [1041]. The effect of microgram amounts of titanium is eliminated in the same way. The effect of tungstate ions can be reduced by adding Ca^{2+} ions [1041]. Four mg of Ca^{2+} eliminate the effect of 2 mg W(VI) during the determination of 20 μg Al.

The following anions do not interfere: up to 3 g of Cl⁻ ions, nitrates, perchlorates, up to 0.3 g of sulfate ions, 50-fold amounts of phosphates, carbonates, sulfites, thiosulfates, and acetates. The following anions interfere: fluorides, tartrates, citrates, hydrogen peroxide.

Aluminum has been determined by Chrome Azurol S in steels [18, 78, 164, 378, 592, 596, 772, 820], copper alloys [164, 997], tin [371, 488], zinc alloys [265a], magnesium [1004], magnesium alloys [417], uranium alloys [709a], ferroboron [378], products of the titanium industry [417], iron ores, agglomerates, martensite slags [563], minerals and rocks [1113], and soils and natural waters [415a]. Chrome Azurol S is employed in the differential spectrophotometric determination of aluminum [265a, 417].

A procedure for the determination of aluminum by Chrome Azurol S is given below [417].

The sample solution, which should not contain more than 30 μg Al, is placed in a 100-ml volumetric flask. Then 2 ml of a 1% solution of ascorbic acid are added and the solution is neutralized to pH 5–6 with a 10% solution of NaOH (check with universal indicator paper). Five ml of 0.1 N HCl are added, then water to a final volume of about 50 ml, then 2 ml of a 0.10% solution of Chrome Azurol S, and 5 ml of 2 N sodium acetate are introduced, and the solution is made up to the mark with water. The optical density of the solution is determined after 10 min at 545 mμ, or in a photocolorimeter with a green filter, against a blank solution. The aluminum content is found from a calibration graph.

In the presence of cetyltrimethylammonium chloride Chrome Azurol S reacts with aluminum to form a blue ternary complex, which can be employed for the photometric determination of aluminum [361a]. The optimum pH for the formation of this complex is 5.8–6.0; the absorption maximum is at 620 mμ, and the molar extinction coefficient of the complex is $1.08 \cdot 10^5$. Beer's law is obeyed at 0.01–9.06 μg Al in 1 ml. The interfering ions are Be, Ga, Ti, U, V, Zr, Cu, Fe, Cr, Sn, Ta; the effect of Cu and Fe(II) is eliminated by o-phenanthroline.

Determination by Xylenol Orange

Xylenol Orange, 3,3-*bis*-(N,N-dicarboxymethyl)-(aminomethyl)-o-cresol-sulfonephthalein, reacts with aluminum to form an intensely red-colored complex, which can be utilized for the photometric determination of aluminum.

The maximum absorption by the complex occurs at pH 3.4±0.1 [420, 696, 1036] (Figure 12). At pH 3 the absorption maximum of the complex is at 555 mμ, while that of the reagent is at 435 mμ (Figure 13). At ordinary

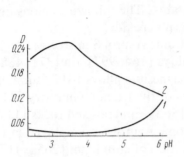

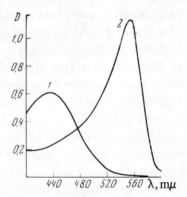

Figure 12. Optical densities of Xylenol Orange and its aluminum complex as a function of the pH:

1) 1 ml of a 0.1% solution of Xylenol Orange with reference to water; 2) 0.02 mg Al and 2 ml of a 0.1% solution of Xylenol Orange with reference to a solution of the reagent.

Figure 13. Absorption spectra of Xylenol Orange and of the complex of Xylenol Orange with aluminum:

1) $5 \cdot 10^{-5} M$ solution of Xylenol Orange with reference to water; 2) $5 \cdot 10^{-5} M$ solution of the complex with reference to water.

temperatures the color development of the complex is very slow, and about 4 hours are needed for full development. The complex formation is accelerated on heating, and the maximum color intensity is obtained by heating the solutions on a boiling water bath for 3 min. Otomo [1036] studied the absorption spectra at different pH values and different reagent : aluminum ratios, and found that a 1 : 1 complex with $\lambda_{max} = 555$ mμ is formed at pH < 3. The complex formed at pH > 4.5 has a reagent : aluminum ratio of 2 : 1 and $\lambda_{max} = 505$ mμ. At the latter pH small amounts of the 1 : 1 complex are also formed. The absorption by solutions of the complex with $\lambda_{max} = 555$ mμ gradually decreases on standing; during the first two hours the color intensity decreases by about 10%. At the isosbestic point (at 536 mμ) the absorption remains constant for several hours.

At pH 3.4 ± 0.1 Beer's law holds at concentrations of 5–25 μg Al in 25 ml [1036]. The sensitivity of the method is 0.001 μg Al per cm^3 (according to Sandell) for $D = 0.001$. The molar extinction coefficient of the complex at pH 3.4 ± 0.1 is 21,100 [1036].

Otomo [1036] found that the following elements do not interfere with the determination of 0.5 mmole (about 13 μg) of Al (if the permissible error is 5%): Ba, Ca, Cd, Cu(II), Hg(II), Mn(II), Pb, Tl(III), U(VI), Co(II), Mg. Equimolar amounts of Bi, Ce(III), Cr(III), Fe(III), La, Nd, Ni, Pd, Sn(II), Th(IV), V(IV), Y, Zn, Zr(IV), Ti(IV) interfere. Small amounts of Fe(III), Cu(II), Bi, Ni, Pb, Cr(III) are masked by thioglycolic acid [696]. Chlorides and nitrates do not interfere. Large amounts of sulfates, phosphates, tartrates and citrates produce a considerable decrease in the absorption. Fluorides, nitrilotriacetic acid and Complexone III interfere strongly.

Xylenol Orange has been employed for the determination of aluminum in uranium [67], copper alloys [261], nepheline concentrates and nepheline-apatite ores [17] and naturally occurring pigments [246]. Kazakov and Pushinov [154] determined aluminum by Xylenol Orange in the presence of beryllium which had been masked by fluoride. The presence of fluoride ions affects the optical density of the aluminum complex to a certain extent, and for this reason the same amounts of fluorides should be introduced into the standard solutions and into the blank solution. Molot et al. [266] used Xylenol Orange to determine aluminum and iron in the presence of one another. He determined the iron at pH 2.6, when the rate of formation of the aluminum complex is insignificant. The colored aluminum compound was obtained by heating at 100°C for 15 min.

A modification of the Xylenol Orange method involving the use of Complexone III. To increase the selectivity of the determination of aluminum by Xylenol Orange, Complexone III is sometimes added to mask several elements [420]. Complexes of aluminum with Complexone III and Xylenol Orange have about the same stabilities. In solutions which contain Complexone III, the complex of aluminum with Xylenol Orange is not formed, but if Complexone III is added to a solution which contains an already formed aluminum-Xylenol Orange complex, the color intensity of this complex decreases by about 10% only. This effect can be compensated for by adding the same amounts of Complexone III to the standard solutions. As in the variant without Complexone III, the most intense color develops at a pH of about 3.5. At pH 4.5−5.5 the color intensity does not, in practice, vary with time, while at pH 3−4 the intensity decreases on standing. At all pH values the color intensity does not significantly change when the solution is left to stand for 30 min. It is, therefore, advisable to measure the optical density within 30 min after the addition of all the reagents.

During the determination of 20 μg Al at pH 3, the effect of the following metals is eliminated by Complexone III:

Metal	Excess of metal	Metal	Excess of metal	Metal	Excess of metal
Fe(III)	1.5	In	6	Pb, Ce(III)	200
Cu(II), Th	2	Bi	12.5	Co, Mo(VI)	250
Be	2.5	Gd	35	La	400
Zn, Ni	5	Pr(III), Nd	100		

At pH 3–3.5 the following metals do not form colored compounds with Xylenol Orange and do not interfere with the determination of aluminum: alkali metals in large amounts, Sr (10,000 : 1), Ca or Mg (2,000 : 1), Cd (500 : 1), Mn(II) (250 : 1), W(VI) (100 : 1), Fe(II) (5 : 1). Cations which do not react with Xylenol Orange at pH 3 present in amounts higher than those just indicated weaken the color intensity of the aluminum complex. Some elements react with Xylenol Orange to form complexes with a stability comparable to that of the aluminum complex of Xylenol Orange. Their complexes are formed simultaneously with the aluminum complex, and these elements therefore weaken the color intensity of the aluminum complex. Such an effect at pH 3 is shown by Fe(III), Zn, Y, Th, Cu, In, Bi and Ni at ratios which exceed those shown above. A number of elements react with Xylenol Orange to form very stable complexes, which are not decomposed or are decomposed only partially by Complexone III, so that the results are too low. The following elements interfere if present in the same amounts as aluminum, or less: V(V), V(IV), Cr(III), Zr, Ti(IV) and Ga.

The variant involving the use of Complexone III has been proposed [420] for the determination of aluminum in materials used in the titanium industry.

Pritchard [1088] determined aluminum in soil extracts by Xylenol Orange after Fe(III) and Fe(II) had been masked by Complexone III.

The following procedure [420] can be employed to determine aluminum by Xylenol Orange.

The sample solution in a 100-ml volumetric flask is neutralized by a 10% solution of NaOH to about pH 3 (universal indicator). One ml of a 1% solution of ascorbic acid is added, then 2 ml of a 0.1% solution of Xylenol Orange, and 5 ml of buffer solution at pH 3 (mixture of 179 ml 1 N HCl and 821 ml of a 1 N solution of glycine). The solution is made up to 40 ml with water, heated for 3 min on a boiling water bath, and then cooled under the tap. Three ml of 0.025 M Complexone III solution are added, the solution is made up to the mark with water and is then shaken. The optical density is determined after 30 min at 555 mμ in a spectrophotometer, or in a colorimeter with a green filter, in a 3- or 5-cm cell, against a blank solution.

The content of aluminum is read off a calibration curve, constructed under similar conditions.

Determination by Methylthymol Blue

Methylthymol Blue, 3,3-*bis*-(N,N-dicarboxymethyl)-(aminomethyl)-thymolsulfophthalein, is an analog of Xylenol Orange. The two reagents are similar in many respects, and Methylthymol Blue reacts with aluminum under the same conditions. Methylthymol Blue reacts with many substances to form colored complexes [422, 423, 426]. Many of these are decomposed by Complexone III, and this reagent has practically no effect on the color of the complex of aluminum with Methylthymol Blue. Therefore, as in the case of Xylenol Orange, Complexone III is employed to enhance the selectivity of the method [420, 431].

The maximum color intensity is at pH 3.5; in the presence of large amounts of Methylthymol Blue it is at pH 3 (Figure 14). The color of the complex develops slowly in the cold and instantaneously if the solution is heated to boiling. The color intensity then decreases somewhat in the course of time, but this decrease is insignificant, and reproducible results can be obtained if the optical density is determined a definite period of time after the addition of the reagents. Methylthymol Blue has an absorption maximum at 435 mμ, while that of its aluminum complex is at 585–590 mμ (Figure 15). The absorption by the reagent is negligible at the absorption maximum of the complex; at equal molar concentrations, the color intensity of the complex is more than 100 times stronger than that of the reagent. This means that the reagent can be used in the determination of small amounts of aluminum. The molar extinction coefficient of the complex at pH 3 and 590 mμ is $1.9 \cdot 10^4$; the composition of the complex is 1 : 1 [420]. Beer's law is obeyed at medium concentrations of aluminum.

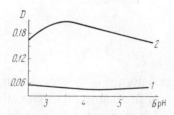

Figure 14. Dependence of the optical density of solutions of Methylthymol Blue and its aluminum complex on the pH:

1) 1 ml of a 0.1% solution of Methylthymol Blue (with reference to water); 2) 0.02 mg Al and 1 ml of a 0.1% solution of Methylthymol Blue with reference to solution of the reagent.

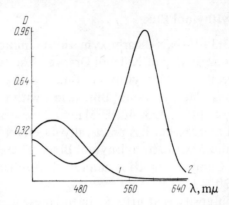

Figure 15. Absorption spectra of Methylthymol Blue and its aluminum complex:

1) $5 \cdot 10^{-5} M$ solution of Methylthymol Blue with reference to water; 2) $5 \cdot 10^{-5} M$ solution of its aluminum complex with reference to water.

In the presence of 3 ml of $0.025 M$ Complexone III at pH 3, the following amounts of foreign metals may be present:

Metal	Excess of metal	Metal	Excess of metal
Be, Ni, Cr(III), Th	2.5	La	250
Fe(III)	3	Y	300
Cu(II), Mo(VI)	3.5	Cd	400
Bi	25	Gd, W(VI)	500
Fe(II)	45	Tl(I)	1,000
In	80	Mg	2,000
Zn, Co	100	Ca	2,500
Mn, Ce(III), Nd,		Sr	10,000
Pr(III)	200		

Large amounts of alkali metals do not interfere. The metals Ga, V, Zr and Ti interfere even if present in small amounts. The effect of Fe(III) in amounts of up to 0.9 mg at pH 3 is eliminated by ascorbic acid, but at higher pH values, for example, pH 5, this is no longer possible. Up to 3 ml of a 1% solution of ascorbic acid is without effect on the color intensity of the aluminum complex, but larger amounts weaken it somewhat. Methylthymol Blue is used for the determination of aluminum in metallic titanium [431] and in various materials of the titanium industry [420].

Aluminum can be determined by the following procedure [420].

The solution in a 100-ml volumetric flask is neutralized by a 10% solution of NaOH to pH 2–3, water is added to a volume of approximately 30 ml, then 1 ml of a 1% solution of ascorbic acid, 2 ml of a 0.1% solution of Methylthymol Blue, and 5 ml of buffer solution at pH 3 (mixture of 179 ml of 1 N HCl and 821 ml of a 1 N solution of glycine). The mixture is heated for 3 min on a boiling water bath and cooled, then 3 ml of 0.025 M Complexone III are added, and the solution is made up to mark with water.

After 30 min the optical density is determined at 590 mμ in a spectrophotometer or in a photocolorimeter with a green filter against a blank solution. The content of aluminum is read off a calibration curve.

Determination by Chromoxane Violet P

Chromoxane Violet P was suggested as a reagent for the photometric determination of aluminum by Mustafin et al. [230, 231, 232, 284a]. Both the reagent and the complex absorb at 500 mμ, but this does not interfere, since the color of the reagent is very weak. The molar extinction coefficient of the complex is 52,500 [231], its composition is 1 : 1, and its apparent $K_{inst} = 4.4 \cdot 10^{-7}$ [231]; the maximum color intensity is observed at pH 5. In the cold the color takes three hours to develop, but the development at the boiling point is instantaneous. It is preferable to work with a freshly prepared reagent solution, since if aged solutions are employed, the results obtained are rather too low. The sensitivity of the method is 0.01 μg Al per ml. Beer's law is obeyed over a wide range of aluminum concentrations.

The metals Fe(III), Ti(IV), Be, Th, Cu(II), V(V) and Mo(VI) react with Chromoxane Violet P under approximately the same conditions as aluminum. Alkali and alkaline-earth metals, Ni, Co, Zn, Cd, Pb, W(VI), Cr(III), Cr(VI) and Bi do not react. Zinc and cadmium do not interfere even when present in a 100,000-fold excess. Nickel and cobalt interfere, but this interference is caused by the color of their ions only, and can be eliminated if they are introduced into the standard solutions as well. Molybdenum(VI) and tungsten(VI) do not interfere when present in 8,000-fold and 1,000-fold amounts, respectively, but larger amounts attenuate the color of the aluminum complex. The effect of Fe(III) is eliminated by ascorbic acid, and that of Cu(II) by thiosulfate.

Chromoxane Violet P is employed to determine aluminum in metallic Cd, Zn, Ni and Cu and their alloys [284a] and in natural waters [232]. Aluminum and iron can be determined in the presence of one another [232].

The reagent has not yet been used in laboratories to any extent, mainly because it is not readily available. It would appear to be suitable for the colorimetric determination of aluminum.

Another reagent which has been proposed for the determination of aluminum is Alumocresone [218] (ammonium salt of trimethyldihydroxy-fuchsonetricarboxylic acid; Chromoxane Violet P is the sodium salt of this acid).

Determination by Pyrocatechol Violet

Pyrocatechol Violet reacts with aluminum in an acid medium to form a blue complex; the reagent alone is yellow under these conditions. The absorption maximum of the reagent is at 445–450 mμ [529, 1128], while that of the complex is at 580 mμ [285a, 396, 1276] (Figure 16), at 615 mμ [529], or at 620 mμ [1128]. These discrepancies can be explained by the shift in the position of the maximum with increase in the concentration of aluminum; the maximum is first produced at shorter wavelengths (570–600 mμ), and shifts toward longer wavelengths as the Pyrocatechol Violet : : aluminum ratio increases to 1 : 1. The absorption maximum of the complex corresponds to a weak absorption by the reagent itself.

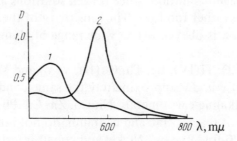

Figure 16. Absorption spectrum of Pyrocatechol Violet and its aluminum complex [1276]:

1) a $2 \cdot 10^{-3}$% solution of the reagent with reference to water; 2) solution containing 80 μg Al$_2$O$_3$ in 1 ml of a 0.2% solution of the reagent in 100 ml, with reference to water.

Different data have also been reported for the composition of the complex. According to Anton [529], aluminum and Pyrocatechol Violet are present in the complex in the ratio of 1 : 1.5. The authors of [396] report that the composition of the complex is 1 : 2. In fact, several complexes are formed. Ryba et al. [1128] showed that a monometallic and a bimetallic complex exist. The constant of complex-formation of the former is $10^{19.13}$, while the overall constant of complex formation of the bimetallic species is $10^{24.08}$. According to the data of different workers, the molar extinction coefficient of the complex is $6.8 \cdot 10^4$ at 580 mμ [396], $2.5 \cdot 10^4 - 2.9 \cdot 10^4$ [1128], or $4.31 \cdot 10^4$ [285a].

The optical density of the complex reaches a maximum at pH 5.9 [1276] and remains constant as the pH is further increased (Figure 17). As the pH increases, the color intensity of the solution of the reagent at first increases slowly, and from pH 6.4 increases rather rapidly. Aluminum is best determined at pH 5.9–6.1. This pH can be produced by using an acetate buffer solution; since the capacity of this buffer is low over this pH range, the solutions must be previously neutralized to this pH value. Under optimum conditions the maximum color intensity of the complex is attained within one hour and remains unchanged for several hours thereafter; if the solution is heated, the reaction rate is accelerated. Beer's law is obeyed up to 80 μg Al in 100 ml in the presence of a sufficient excess of the reagent (2 ml of a 0.15% solution) [1276], while according to other data [396] the law holds up to a content of 50 μg Al per 25 ml of solution.

A large number of divalent, trivalent and tetravalent metals react with Pyrocatechol Violet. The divalent metals react at higher pH values than aluminum, and their effect is accordingly negligible. The maximum color intensity of trivalent and tetravalent metals is observed at the following pH values [396, 1128]:

Metal	pH	Metal	pH
Zr	2.20	Th	3.70
Bi	2.4	Ga	4.05
Sn(IV)	3.5	In	5.50

The absorption maxima of these metals are close to the maximum for aluminum; the molar extinction coefficients of their complexes are also close to that of aluminum: from $2.5 \cdot 10^4$ to $2.9 \cdot 10^4$ [1128]. Thus, all these metals interfere with the determination of aluminum; Ti, Cr, V and borates also interfere. Ferric iron reacts with Pyrocatechol Violet to form an unstable complex, a product of oxidation of the reagent, i.e., it interferes as well.

Of the divalent metals, copper interferes if present in concentrations of more than 2 ppm. Up to 5 μg of Zn, Mg, Mn, CaO and CeO_2 in 100 ml, and

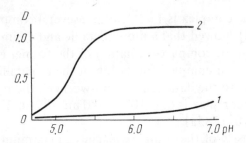

Figure 17. Dependence of the absorption of the complex of aluminum with Pyrocatechol Violet on the pH [1276]:

1) reagent; 2) solution containing 80 μg Al_2O_3.

up to 1 mg UO_3 do not interfere; 1 mg Be and 5 mg of lanthanum oxide lead to an error of about 75%. Of the anions, phosphates and fluorides present in amounts of more than 50 μg interfere with the determination of 80 μg Al.

The relative error of the method is 0.7–6% for Al contents of 12–24%.

Pyrocatechol Violet as a reagent for the photometric determination of aluminum has no advantage over the conventional reagents for aluminum. It is not now used in analytical laboratories.

Determination of aluminum by other reagents of the triphenylmethane series

Acid Chrome Pure Blue B (Chromoxane Pure Blue) reacts with aluminum in weakly acid and neutral media to form a violet compound. The reagent itself is golden-yellow at pH below 8, and violet colored at pH above 8. The maximum absorption of the complex is at 540–550 mμ and pH 6* [494]. The aluminum : reagent ratio in the complex is 1 : 3 [494], and the molar extinction coefficient is 20,000 [497]. The color of the complex is constant for 8 hours. Beer's law is obeyed with 1–12 μg Al in 25 ml. The metals Zn, Mg, Ni, Cd, Pb, Mn, Co and 5-fold amounts of In and Ga do not interfere. Cupric copper does not significantly interfere, and its effect may be eliminated by the addition of thiosulfate. Ferric iron interferes strongly, but its effect can be eliminated by the addition of ascorbic acid [196, 494, 496]. Sulfates suppress the development of the color of the complex [196]. The method has been employed for the determination of aluminum in magnesium and zinc alloys with a relative error of 6% for aluminum contents between 2% and 9%; it has been suggested for the determination of aluminum in sulfur, for aluminum contents higher than $3 \cdot 10^{-4}$%.

* Other publications of these authors [196, 497] report λ_{max} = 570–580 mμ and pH_{max} = 5.4.

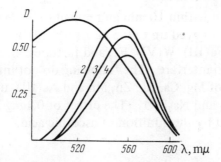

Figure 18. Absorption spectra of sulfochrome and its aluminum complex [332a]:

1) $5 \cdot 10^{-5} M$ solution of reagent at pH 4.0; 2–4) $5 \cdot$
$\cdot 10^{-5} M$ solution of its aluminum complex at pH 4.1,
3.9 and 3.4, respectively.

Sulfochrome has very recently been proposed as a reagent for the photometric determination of aluminum [237, 332c]. Sulfochrome is the diammonium salt of 3,3'-dimethyl-4-hydroxyfuchsone(4')-5,5'-dicarboxy-2",4"-
disulfonic acid. The aluminum : sulfochrome ratio in the complex is 1 : 1.
The maximum absorption of the reagent is at 520 mμ and that of the complex at 560 mμ (Figure 18). The complex is formed at pH 3–6 and its
optimum pH of formation is 3.8–4.1. The color development takes 2 hr in
the cold, but only a few minutes at 50–80°C. The molar extinction coeffi-
cient of the complex is $3.8 \cdot 10^4$, and the apparent $K_{diss} = (3.10 \pm 0.93) \cdot 10^{-6}$.
Beer's law holds for 1–45 μg Al in 50 ml. Sulfochrome reacts with Fe, Cr,
Be, Zr, V, Ga, Sc; very faint colorations are given by Ti, W, and Mo.

Sulfochrome is a sensitive reagent for aluminum. Its disadvantage is that
the absorption maxima of the reagent and the complex are close to one
another, and the reagent itself is strongly colored under the experimental
conditions of the determination of aluminum.

Cherkesov et al. [476] determined aluminum in cadmium salts by *gallein*
at pH 4. The aluminum complex formed under these conditions can be
extracted by isobutanol. The distribution coefficient of the complex between the organic solvent and the aqueous solution is three. The complex
is almost quantitatively recovered by three extractions.

Cadmium in 100,000-fold amounts does not interfere with the determination of aluminum.

Pyrogallol Red reacts with aluminum to form a colored complex with an
absorption maximum at 525 mμ and pH 4.8–5.2; this has been utilized for
the spectrophotometric determination of aluminum [395]. The maximum

color intensity is attained within 10 min and remains constant for about two hours. Beer's law is obeyed up to 0.5 μg Al per ml. Even small amounts of Zr, Ga, In, Fe(III), W(VI), F⁻, and tartrates, and considerable amounts of V(V) and Co interfere. The following do not interfere: 0.1 mg Cu(II), Ti and Be, 1 mg of Mg, Ca, Ba, Zn, Mn and As(III), up to 50 mg KNO_3, NaCl, $(NH_4)_2SO_4$ and $Na_2S_2O_3$. The effect of 0.1 mg Fe(III) and 1 mg Cu(II) is eliminated by the addition of ascorbic acid.

Determination by hydroxyquinoline

The chelate compound of aluminum with 8-hydroxyquinoline is soluble in organic solvents, giving an intense yellow coloration. The determination of aluminum is based on the photometric determination of the extract. Hydroxyquinoline is one of the most widely used and one of the most important reagents for the determination of aluminum, and is not inferior to Aluminon.

Properties of the complex and experimental conditions of its extraction. According to the data of different workers, solutions of aluminum hydroxy-quinolate in organic solvents have absorption maxima at 380 [1289], 385 [670, 957], 388 [646], 389 [811, 864], 390 [994, 1264], and 395 mμ [750, 983, 1191]. Our own data show that the absorption maximum is at 395 mμ. The absorption maximum of the reagent is in the UV, at 372 mμ [1201]. Below 370 mμ the reagent absorbs very strongly, between 395 and 410 mμ it absorbs very weakly and, if the extraction conditions are kept constant, the error produced by the color of the reagent can be eliminated. The aluminum: hydroxyquinoline ratio in the complex is 1 : 3. The pH values for the quantitative extraction of aluminum hydroxyquinolate have been given as 4.5–11.5 [750, 910], 4.9–9.4 [1264], 4.2–9.8 [771], and 5–10 [994]. The very narrow pH ranges reported by Moeller [983] are erroneous. The optimum conditions for the extraction are pH 5 in acid media and pH 9 in alkaline media (Figure 20).

According to Kambara and Hashitani [858], the parameter "pH of half-extraction" is important in describing the extraction process. In the case of aluminum hydroxyquinolate these values are 4.05 and 11.27. The pH values of a practically complete extraction (99.9%) are one pH unit higher for an acid solution (i.e., pH 5.05) and 0.75 lower for an alkaline solution (i.e., pH 10.5). Wiberley and Basset [1264] state that the amount of the extracted reagent increases more rapidly at high pH values, and therefore the result of the blank determination becomes higher. According to these workers, this value is constant at pH 5.0–7.0, which is therefore considered

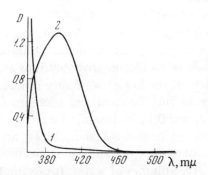

Figure 19. Absorption spectra of 8-hydroxyquinoline (1) and its aluminum complex (2).

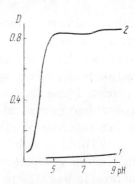

Figure 20. Effect of pH on the extraction of aluminum hydroxyquinolate [1264]:

1) blank solution; 2) solution containing 160 μg Al.

to be the optimum range. Gentry and Sherrington [750] found that the complete extraction of 0.05 mg of aluminum at pH 8.9 takes place at a concentration of hydroxyquinoline of 1%. The data of Wiberley and Basset [1264] show that the complete extraction of 0.16 mg of Al at pH 6.3 requires 2 ml of a 2% solution of hydroxyquinoline. There is evidence [869] that better results are obtained if the extraction is carried out from solutions preheated to 60–70°C. However, this merely complicates the course of the determination. Numerous studies indicate that satisfactory results can be obtained at room temperature as well. A number of workers reported the values of optimum volumes of sample solutions as a function of the amounts of aluminum they contain [705]; however, the order of magnitude of the concentration of aluminum is usually not known in advance, so that these data cannot be utilized. It is important, however, to have equal volumes of sample and standard solutions (10 to 30 ml).

The organic solvent employed in the extraction is usually chloroform. Other solvents which have been proposed include benzene [316, 1209], carbon tetrachloride [352, 534], trichloroethylene [581, 933], and isoamyl alcohol. According to [856], benzene extraction is more complete than chloroform extraction and the separation takes place more rapidly. In addition, the benzene extract has a higher optical density than the chloroform extract, for equal amounts of aluminum hydroxyquinolate. Raines and Larionov [352] conducted a comparative study of various solvents. They found that the sensitivity of the method is 0.002 mg for CCl_4, $CHCl_3$, and C_6H_6, 0.005 mg for butyl acetate, and 0.02 mg for isoamyl alcohol; the last named solvent, which is sometimes recommended, is thus not very suitable. If the extraction is carried out with carbon tetrachloride, the solutions are more transparent. From the point of view of the lowest solubility in water, it is more convenient to use carbon tetrachloride, benzene and trichloro-ethylene, with solubilities of 0.08, 0.082, and 0.1 g in 100 ml of water, respectively [375]. The solubility of isoamyl alcohol is higher, 2.67 g, and the solubilities of butyl acetate and chloroform are 0.5 and 1 g, respectively.

If aluminum is to be extracted from large volumes of water, for example, during the analysis of water, the solubility of chloroform in water must be allowed for; for this reason it is recommended that individual calibration curves be constructed for each volume of the aqueous solution. The extracts often contain water and are turbid. Traces of water can be eliminated by shaking the extract with anhydrous sodium sulfate or filtering through filter paper. The latter technique is simpler, but some of the solvent is lost through evaporation; treatment with anhydrous sodium sulfate takes longer, but may prove more expedient, for example, when a large number of determinations are carried out simultaneously.

The color of aluminum hydroxyquinolate solutions is quite stable. Chloroform solutions of the complex darken in sunlight, with a consequent increase in the optical density and erroneous results [957, 984]. Linnell and Raab [927] studied the stability of chloroform solutions of aluminum hydroxyquinolate, and found that partial decomposition takes place even in the dark, and that it is accelerated by oxygen and sunlight.

According to Gentry and Sherrington [750], Beer's law is obeyed up to $30 \mu g$ Al/10 ml; similar results were reported by Wiberley and Basset [1264] (up to $160 \mu g$ in 50 ml). Other workers report somewhat wider ranges: $50–250 \mu g$ Al in 50 ml [811] and up to $120 \mu g$ Al in 20 ml [957].

Effect of foreign cations. More than 40 elements form colored complexes with hydroxyquinoline. For the pH of their precipitation (conditions of formation) see p. 27. Since the specificity of the method is low, the elimination of the interfering elements is highly important. They can be masked by cyanides. Iron is previously reduced by sulfide or sulfite. The use of cyanides is so effective that as little as 0.0001% Al (for example, in nickel) can be determined [137]. Hydrogen peroxide is employed to mask elements which form peroxide complexes [144, 646, 657, 867].

The use of a mixture of cyanide with hydrogen peroxide is even more effective [856, 864, 938]. Luke [938] studied the effect of 50 elements on the determination of aluminum by hydroxyquinoline in the presence of this mixture. Luke extracted $100 \mu g$ each of different metals and found that at 400 mμ the following elements do not absorb, or else absorb very little: Ge, As(V), Sb (V), Sn, Ag, Hg(II), Cu(II), Cu(I), Cd, Mo(VI), Tl(III), Tl(I), Zn, Ni, Co, Fe(III), Fe(II), Cr(VI), Cr(III), Mn, Mg, Si, W, Ta, Nb, Hf, Th, Zr, Ce(IV), Ce(III), V(V), V(IV), La, Sm, Nd, Pr, B, P, Ba, Sr, and Ca. Up to 10 mg of Fe(III), Fe(II), Ni, Zn, Cu(II), and Cu(I) absorb very little or not at all. The elements Be, Ti, Ga, Y, In, Bi, Pb, U(VI), and Sc interfere. The color produced by Be, Sc, Y, and Pb disappears if the chloroform solutions of these elements are passed through glass wool into another separatory funnel with 15 ml of ammonia buffer solution and shaken for 3 min. Classen et al. [646] mask the interfering elements by a mixture of Complexone III and cyanide. Aluminum is not extracted from alkaline solutions (pH 8–9) containing Complexone III and cyanide by solutions of hydroxyquinoline in chloroform. Aluminum can be extracted by an alcoholic solution of hydroxyquinoline after it has been left to stand at room temperature for 1 hr. Attempts to shorten the duration of standing by heating proved unsuccessful. After the solutions had been allowed to stand for 15, 30, 45, 60, and 90 min, the amounts of aluminum found were 50, 60–70, 85–90, 100, and 100%, respectively, of the aluminum introduced. Under these conditions, the complex of Fe(III) is not sufficiently stable, and the element interferes with the determination. Therefore, iron is first reduced by sulfite and is then converted to ferrocyanide by adding cyanide. The extraction of aluminum from alkaline solutions in the presence of Complexone III and cyanide will separate it from the following elements: As(III and V), Ba, Cd, Ca, Ce(III), Cs, Cr(III), Co, Cu, Ge, Au(III), La, Pb, Li, Mg, Mn, Hg(I and II), Mo, Ni, Pt(IV), Se(IV and VI), Si, Ag, Sc, Te(IV and VI), Th, Sr, Sn(IV), W, V(V), and Zn. The following anions are separated at the same time: borates, phosphates, iodides, bromides, fluorides. Under these conditions Bi, Ga, In, Nb, Sb(III and V), Ta, Ti, U(VI), and Zr are coextracted with aluminum [646].

If all these elements, together with aluminum, are reextracted from the extract by $2N$ HCl, hydrogen peroxide is added in excess, and the solution is again extracted with chloroform at pH 7.5–8.5. Nb, Ta, Ti, V and U(VI) remain behind as peroxide complexes in the aqueous layer and can be separated from aluminum. The metals Bi, Ga, Sb, and Zr can be extracted with chloroform as cupferronates from $2N$ HCl, while indium can be separated by extracting the diethyldithiocarbamate with chloroform at pH 3–5. Beryllium hydroxide does not interfere with the extraction of aluminum if 1–2 drops of a wetting agent are added (Tergitol, etc.). The small amount of beryllium which passes into the extract can be separated from aluminum by a complementary extraction at pH 5. If V(IV) is present in the sample solution, it begins to be extracted before aluminum, and its extraction rate is very slow. If 0.5–1 mg Ti is introduced, the extraction of V(IV) and Al is accelerated. Pentavalent vanadium is not extracted, but is treated with bisulfite and thus reduced to V(IV).

The optimum quantities of Complexone III and cyanide employed in the method are: 0.1 g of iron or steel requires 1 g of Complexone III and 3 g KCN; 0.5 g requires 3 and 5 g, while 1 g requires 8 g Complexone III and 8 g KCN, respectively.

Manganese must be bound by a 20-fold excess of Complexone III. In the method of Classen et al. [646], up to 25 μg Th do not interfere, but larger quantities interfere with the extraction of aluminum. In the presence of 100 μg Th aluminum is not extracted. In such cases large amounts of hydroxyquinoline must be added to precipitate thorium hydroxyquinolate. In the presence of scandium the amount of hydroxyquinoline must be correspondingly increased (1 ml of a 5% solution of hydroxyquinoline will extract 1 mg Sc), or aluminum will not be extracted. Classen et al. found that the relative error involved in the determination of 0.04–7% of aluminum in steels is 1–10%. For details, the reader is referred to the textbook by Sandell [360].

Margerum et al. [957] determined aluminum in thorium by masking the thorium by the sodium salt of 4-sulfobenzoarsonic acid in the presence of acetate buffer. The aluminum is extracted at pH 4.7–5.0, since under these conditions the extraction of thorium is negligible. A precipitate may separate out in acid solutions containing sulfobenzoarsonic acid; in the presence of the buffer mixture the solubility of this precipitate increases. Up to 0.25 g Th can be retained in 100 ml of solution. To prevent the thorium from being extracted by chloroform, the molar ratio of sulfobenzoarsonic acid to thorium must be at least 2:1.

If the masking is not effective, the interfering elements should be previously separated. Table 11 shows the techniques used for the masking and elimination of the elements interfering with the determination of aluminum by hydroxyquinoline.

Effect of anions. Large amounts of chlorides, nitrates and sulfates do not interfere with the determination of aluminum [750]. Bromides and iodides do not interfere [646]. Perchlorates do not interfere up to a concentration of $1M$. If SiO_2 is present as a true molecular solution, it does not interfere at $Al_2O_3 : SiO_2 = 1 : 4$. In the presence of polymerized SiO_2, when the above ratio exceeds $1 : 4$, the results may be 10% or even more too high. Before the determination of aluminum it is advisable to treat the solution to be

Table 11. Elimination of the influence of other elements during the determination of aluminum by hydroxyquinoline

Method	Elements	Notes	References
Adjustment of pH of solution	Fe	Extraction of iron hydroxyquinolate at pH 2, 2.5 or 2.8	[276, 697, 750, 869, 983, 1098]
The same	Cu	Extraction of copper hydroxyquinolate at pH 2.8	[869]
Masking with cyanide in alkaline medium	0.1 g Cu, Ni, Co, Zn, Cd, Cu, Ni	Extraction of aluminum from a solution of 2 g of ammonium nitrate, 1 g KCN, 1 ml of 1:1 ammonia in 45 ml of water	[750]
The same	Co, Zn	Washing chloroform extract with KCN solution	[282, 957]
Masking with mixture of KCN and Na_2S	Fe	Extraction of aluminum from solution containing 2 g KCN, 1 g Na_2S and 2 g ammonium nitrate. After the addition of KCN the solution is heated for 3 min at $50°C$, cooled, Na_2S is added and aluminum hydroxyquinolate extracted	[750]
Masking with mixture of KCN and ammonium carbonate	U	Correction for iron is introduced	[457]
Masking with mixture of ammonium carbonate and ammonia (pH 9.5–10)	U	To the solution, $2N$ in HCl, are added 20 ml of a 30% solution of H_2O_2; after 1 min, the solution is neutralized to phenolphthalein with ammonia, and the aluminum extracted	[534]
H_2O_2 (pH 7.5–8.5)	Ti, V, Nb, Ta, U(VI)		[646]
H_2O_2	Nb, Th, U, Ce	Extraction of aluminum from solution at pH 9.2 containing 5 ml of 30% H_2O_2	[811]
Masking with a mixture of KCN, H_2O_2, and sodium sulfite	Fe, Ti, U, V, Sn, Co, Ni, Mn, Cu, Cr, Zn, Cd, Mo, Zr	To the solution are added 2 g of tartaric acid and 1 ml of 3% H_2O_2. After 5 min 5 ml of saturated sodium sulfite solution are added, after another 3 min 1.3 g KCN. The solution is heated to $70–80°C$, cooled, 2 g ammonium nitrate introduced, the pH adjusted to about 8.9, and aluminum is extracted	[864]

Table 11 (continued)

Method	Elements	Notes	References
Masking with DCTA	Fe, Pb, Cu, Bi, Mg	To the solution are added tartrate, 5 ml of 0.1% DCTA solution, the solution is made alkaline, cyanide and hydroxyquinoline are introduced, and the aluminum extracted	[510]
Formation of complex with o-phenanthroline	Fe	Extraction of Al from solution at pH 5, containing 0.1 g $NH_2OH \cdot HCl$ and 2 ml of a 0.1% solution of o-phenanthroline	[670]
Complexing Fe by 2,2'-dipyridyl and complexing F^- by beryllium	Fe, F	Extraction of Al from solution to which 5 ml of masking solution (5 g NH_4Cl, 17 g CH_3COONa, 0.2 g of 2,2'-dipyridyl, 2 ml HCl, and 2 g $BeSO_4$ in 500 ml water) have been added	[1265]
Extraction of hydroxyquinaldates of interfering elements at pH 9.2	Cd, Co, Cu, Fe, Mn, Ce, Ni, Pb, Sb, Sn, Zn, Ti	Extraction of hydroxyquinaldates with 5 ml $CHCl_3$ and 5 ml of a 2.5% solution of 8-hydroxyquinaldine in 5% CH_3COOH; the aqueous layer is shaken twice with 5 ml of chloroform each time, and aluminum hydroxyquinolate is extracted	[811]
Extraction of diethyldithiocarbamates	Mn Fe, Cu, Ni In	At pH 5.2–5.4 From solutions at pH 4.5 At pH 4	[960] [478] [646]
Extraction of chloride complexes	Fe, Ga	Extraction of chlorides with methyl isobutyl ketone; pH is adjusted to 5.6–5.9, sodium thiosulfate added and aluminum hydroxyquinolate extracted; the extract is washed with 10 ml of the mixture (2 g KCN + 2 g Complexone III in 100 ml) at pH 10	[863]
Extraction of thiocyanates	Fe	Extraction of Fe thiocyanate with ether-tetrahydrofuran mixture from 4 N HCl	[1187]
	Ti	Extraction of titanium thiocyanate with ether	[99]

Table 11 (continued)

Method	Elements	Notes	References
Extraction of cupferronates	Zr, Ga, Bi, Sb	Extraction of cupferronates with chloroform from $2N$ HCl or H_2SO_4; Cupferron is decomposed and aluminum hydroxyquinolate extracted	[457a, 657]
Extraction of complexes with N-benzoyl-N-phenylhydroxylamine	Ti, V, Zr	Extraction from a solution which is $1M$ in $HClO_4$; the aqueous layer is shaken with chloroform and aluminum hydroxyquinolate extracted	[775]
Electrolysis on mercury cathode	Many metals	Acidity $1N$ in sulfuric acid, current strength 5 A	[1264]
Fusion with sodium carbonate	The same	Fusion with 2 g Na_2CO_3 at 1000°C for 10 min; leaching with water and extraction of aluminum hydroxyquinolate	[1047]

analyzed by sodium hydroxide to convert silica to the molecular form [109]. Fluorides, even when present in concentrations as low as 10 μg, interfere with the extraction of aluminum hydroxyquinolate; this interference cannot be eliminated by the introduction of boric acid [646]. Small amounts of fluorides (up to 500 μg) do not interfere with the determination of aluminum in thorium, since thorium binds fluorides into a stable complex [957]. According to Gentry and Sherrington [750], up to 0.15 g of phosphates does not appreciably affect the results of the determination of aluminum, but more than 200 μg of phosphoric acid interfere with the reduction of iron [646]. Up to 0.2 g of tartrate in 50 ml of the solution do not significantly interfere [750]; according to other data, 0.3 g of tartaric acid in 80 ml solution are permissible [869]. For this reason tartaric acid is used to mask small amounts of iron [869]; 0.3 g of tartaric acid will mask 5.6 μg of iron. Some workers introduce tartaric acid to retain aluminum in solution in an alkaline medium. In such cases, the same amounts of tartaric acid are also introduced into the standard solutions.

Methods for the simultaneous determination of aluminum and iron in the same solution have also been proposed [2, 281, 282, 458, 459, 949, 991, 992]. The hydroxyquinolates are extracted at pH 5.2–5.5; the optical densities are measured at 390 and 470 mμ, respectively. At 470 mμ aluminum hydroxyquinolate does not absorb, while iron hydroxyquinolate absorbs at both wavelengths.

Variants of methods based on the determination of aluminum by hydroxyquinoline. Parks and Lykken [1047] suggested that aluminum hydroxyquinolate be dissolved in chloroform or in 0.1 N HCl, and that the optical density of the yellow solution thus obtained be determined in the UV. If chloroform is used as solvent, the precipitate should be dried at 140°C; if 0.1 N HCl is employed, drying is not necessary. The optical density of the HCl solution is measured at 251.5 mμ; satisfactory results are obtained for aluminum contents exceeding 0.050 mg Al. Phillips and Merrit [1060] proposed that the optical density of the hydrochloric acid solution of aluminum hydroxyquinolate be measured at 360 mμ, since at this wavelength the changes in the spectrum due to variations in acidity between 0.006 M and 10.0 M are insignificant. At other wavelengths the concentration of the acid must be controlled.

Indirect methods for the determination of aluminum, based on the conversion of the hydroxyquinoline complex of aluminum to an azo dye by diazotized sulfanilic acid in an alkaline solution [524, 673, 681], and on the reduction by phosphotungstic, molybdotungstic, phosphotungstomolybdic silicomolybdovanadic, and silicomolybdotungstic acids by hydroxyquinoline [174, 728, 1226] are of little interest.

Determination by azo compounds

Determination by Stilbazo. Stilbazo, stilbene-2,2'-disulfonic acid-4,4'-bis-(azo-1'')-3'',4''-dihydroxybenzene, diammonium salt, was first proposed by Kuznetsov et al. [216, 217] as a reagent for the photometric determination of aluminum.

The aluminum complex of Stilbazo has an absorption maximum at 496 mμ [5, 251b], 500 mμ [1260], or 510 mμ [137]. The absorption maximum of the reagent is located in the UV. The reagent absorbs strongly at the absorption maximum of the complex (Figure 21), which is a serious disadvantage of Stilbazo as a reagent. Owing to the superposition of the strong color of the reagent, it is very difficult to determine small amounts of aluminum by Stilbazo. The composition of its aluminum complex is 1 : 1 [5, 1260]; its instability constant, found by optical measurements, is 5.96 · 10^{-6} [1260], while its molar extinction coefficient, according to the data of different workers, is 34,600 [5] or 38,000 [656]. Other workers quote lower values, for example, $\epsilon = 19,500$ [1260].

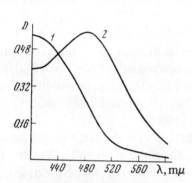

Figure 21. Absorption spectra of Stilbazo and its aluminum complex:

1) $2 \cdot 10^{-5} M$ solution of Stilbazo with reference to water;   2) $2 \cdot 10^{-5} M$ solution of the Stilbazo-aluminum complex with reference to the reagent.

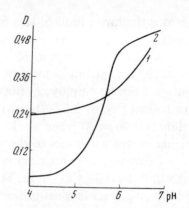

Figure 22. Color intensity of the Stilbazo-aluminum complex as a function of the pH:

1) solution of reagent with reference to water; 2) complex of aluminum with Stilbazo with reference to solution of reagent.

The absorption by the reagent is approximately constant between pH 4 and pH 5.8; it increases above pH 5.8. The absorption by the complex increases almost linearly with increase in pH between pH 4 and 6. The effect of the pH is less marked between pH 6 and 7 only (Figure 22), but over this pH range there is strong interference by foreign ions. Most workers determine aluminum at pH 5.4–5.6. Wetlesen [1259] proposes pH 4.8 for experiments. He states that interference by foreign ions is less marked at this pH. Since the pH strongly affects the color intensity of the complex, the pH of the medium must be rigidly controlled by a pH-meter.

The color of the aluminum complex develops at once, and is stable for a long time; the color intensity is independent of the temperature between 20 and 70°C [5]. The optical density of the solution increases with increase in the concentration of Stilbazo, and attains a maximum at the molar ratio reagent: aluminum = 3:1 [446]; according to other workers [204], this ratio is 4:1 [204]. It is recommended [204] that 5 ml of a 0.06% solution of Stilbazo be employed for aluminum contents of the order of 30 μg Al in 50 ml of solution. If the concentration of Stilbazo is increased, suspensions may be formed [912]. According to Iwasaka and Omori [137], the optical density is maximum at Stilbazo concentrations of 0.006–0.1%; 5 ml of the reagent is the lower limit of this range. Other authors work with very high reagent concentrations in solutions: $2.8 \cdot 10^{-4}M$ [150] and $3.1 \cdot 10^{-4}M$ [912], which corresponds to 18–20 ml of a 0.05% solution in 50 ml. In view of the very strong coloration of the reagent itself, such high concentrations must not be employed. Beer's law is obeyed at contents of 10–100 μg Al in 100 ml [5] (or 2.5–20 μg Al in 25 ml [1260]). The sensitivity of the method is 0.06 μg Al when working with 1 ml of solution; the sensitivity is higher in larger volumes of the solution [193].

The following elements react with Stilbazo under the same conditions as aluminum: Fe(III), Cu(II), In, Ti, Ga, Bi, Sb, Sn, Th, Mo(VI), W(VI); and the following react in a neutral or alkaline medium: Pb, Co, Ni, Zn, Cd, Be, Hg, Ca, Mg, Mn.

The following elements do not interfere with the determination of 30 μg Al at pH 5.65 [1260]:

Element	Excess	Element	Excess
Nb(V)	10	Ni, Zn, Cd, Pb	100
Cr(VI)	13	As(III)	300
As(V)	33	Ca, Mg, Mn, Co	1000

The results of other workers [203, 204] show somewhat larger amounts of lead (up to 500:1) and zinc (up to 1000:1) are permissible; antimony in ratios of up to 120:1 does not interfere. Up to 10 mg Ag in 100 ml solution do not interfere [150]. According to Arginskaya and Petrashen' [6], contents of up to 0.25% Mo and 0.05% W are permissible. These workers determine aluminum in the presence of 12.6% V(IV). Wetlesen and Omang [1260] state that both V(IV) and V(V) intensify the coloration if present in larger amounts than Al. Fifty μg of aluminum can be determined with satisfactory accuracy in the presence of 500 μg Cr(III) [1259]; SiO_2 does not interfere up to a ratio of 30:1 [203]. Up to 6.5 mg Cu may be masked by thiosulfate [1259]; the effect of small amounts of Cu (0.5 mg) can be eliminated by ascorbic acid [203].

The effect of Fe(III) can be eliminated by preliminary reduction. Hydroxylamine [6, 845] is suitable for the reduction of iron below pH 4.6 only, that is, in the pH range in which the use of Stilbazo is inexpedient. Thioglycolic acid, Complexone III, and maleic acid [845] eliminate the effect of iron and affect the aluminum complex. Ascorbic acid is considered the most suitable reagent for the reduction of ferric iron. The acid will eliminate the effect of 5 mg Fe(III) in 50 ml of the solution; larger amounts of iron reduce the color of the aluminum complex [5]. When aluminum is determined in steel, the iron must be previously eliminated, or else the calibration curve must be constructed on the background of aluminum-free steel.

Anions which do not interfere with the determination of aluminum include 1 g of sulfate ions, up to 0.1 g of chloride and nitrate ions; considerable amounts of neutral salts interfere [84]. The following interfere: oxalates, tartrates, citrates, nitrilotriacetic acid, Complexone III, fluorides, and phosphates. Phosphate ions in amounts of 0.5 mg do not interfere with the determination of 0.03 mg Al. Aluminum can be determined in the presence of fluorides if these are bound into a complex by introducing beryllium ions [289].

Stilbazo can be used to determine aluminum in steels [4, 5, 287, 444, 845], ferroalloys [251b], various materials of the lead and copper industries [203, 204, 446], metallic silver [150], glass [245], phosphoric furnace slags [111], and other materials.

Determination by Arsenazo. Arsenazo, the sodium salt of benzene-2-arsonic acid-(1-azo-7)-1,8-dihydroxynaphthalene-3,6-disulfonic acid, reacts with aluminum to form a violet complex, which can be used for the photometric determination of aluminum [198, 214, 215, 503]. The maximum color intensity of the complex is observed at $575-580$ mμ [214, 656] at pH 5.8 [198]. Under these conditions the absorption by the reagent is quite strong. The optimum pH value has been reported as $5.1-5.8$ [198, 214, 503] and $6.1-6.8$ [656]. The color develops within 15 min, and then remains constant for several hours [214, 215, 656]. If the amount of reagent is increased, the optical density increases, and the maximum is attained when 10 ml of the solution being determined contain 1.0 ml of a 0.05% solution. As the amount of the reagent is further increased, the optical density decreases [215]. The sensitivity of the photocolorimetric determination is 0.5 μg Al in 10 ml [215]. Beer's law is obeyed at Al contents of $1-8$ μg Al in 10 ml [215], or $0-100$ μg Al in 100 ml [656].

Under the experimental conditions employed in the determination of aluminum, Fe(III), Zr, Hf, Ga, In, Pd, Th, and Ti also react with Arsenazo to form colored compounds, and thus interfere with the determination. The effect of iron is eliminated by ascorbic acid. Copper (in amounts up to 10-fold excess) can be bound by urea into a colorless complex [214]. Zinc does not interfere if present in 25-fold excess [214]. Beryllium interferes strongly (0.7 μg Be is equivalent to 1 μg Al) [656]. Up to 10 μg Cr [656] and 40 μg W [503] do not interfere. Large amounts of alkali and alkaline-earth metals, magnesium and manganese do not interfere. Fluorides, phosphates, hydroxy acids, and other substances which bind aluminum into a complex, interfere. The effect of sulfates is not as strong.

Determination by 5-sulfo-4'-diethylamine-2',2-dihydroxyazobenzene. This reagent was proposed by Floerence [724] for the spectrophotometric determination of aluminum in the presence of beryllium. It is inferior to Aluminon, Chrome Azurol S, etc., in some respects, for example, the color intensity of the reagent at the absorption maximum of the complex. Its advantages are that it does not react with beryllium and that the complex of the reagent with aluminum is chemically inert, so that the interference of a large number of elements may be eliminated by Complexone III.

The absorption maximum of the aluminum complex is at 535 mμ (Figure 23), but at this wavelength the reagent itself absorbs strongly, and it is preferable to work at 540 mμ. At pH between 4 and 5 the optical density varies little, and beyond pH 5.5 the absorption decreases sharply (Figure 24). It is best to work at pH 4.7, which corresponds to the maximum capacity of the buffer solution.

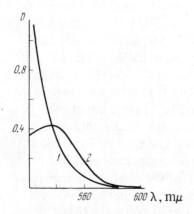

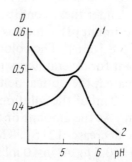

Figure 23. Absorption spectra of 5-sulfo-4'-diethylamine-2',2-dihydroxyazobenzene and its aluminum complex [724]:

1) 3.6·10^{-3}% solution of reagent with reference to water; 2) aluminum complex (6.7 μg) with reference to solution of reagent.

Figure 24. Effect of the pH on the color of 5-sulfo-4'-diethylamine-2',2-dihydroxyazobenzene and its aluminum complex [724]:

1) solution of reagent with reference to water; 2) complex of aluminum (6.7 μg) with reference to solution of reagent.

At pH 4.7 the composition of the complex is 1 : 1. At pH 4.7 and 540 mμ the molar extinction coefficient of aluminum is 41,000, and the sensitivity of the method (according to Sandell) is $0.00066\,\mu g/cm^2$. Beer's law is obeyed up to $0.6\,\mu g$ Al per ml. The relative standard deviation of the method is 0.5% [724]. At room temperature the rate of formation of the aluminum complex is slow, and it must be accelerated by heating at $40°C$ for 15 min. The complex is stable for at least 19 hours; its absorption decreases by only 2% in 120 hours.

Beryllium interferes if present in quantities exceeding $200\,\mu g$, since it forms a weakly colored complex with the reagent. The color intensity of the aluminum complex is also reduced. This reduction of color intensity is independent of the concentration of aluminum, and Beer's law holds for aluminum even at large concentrations of beryllium (15 mg in 25 ml). Aluminum can be determined in the presence of fairly large amounts of beryllium, if a correction for beryllium is introduced with the aid of preliminarily found curves. The elements Fe, Cr, Co, Cu, Ti, W, V, U, and Th react with the reagent to form colored complexes at pH 4.7. Except for Co and V, the color intensities of their complexes are much weaker than that of the aluminum complex. The effect of these elements, except Fe and Co, is eliminated by Complexone III, which decomposes their complexes. The decrease in the absorption of the complex of aluminum under the influence of Complexone III is independent of the concentration of aluminum, and no significant error is introduced if the absorption in the standard and sample solutions is determined within not more than 30 min. The iron complex is decomposed by Complexone III, but the results for aluminum are too low even in the presence of small amounts of iron. Complexone III has no effect on the strongly colored complex of cobalt with the reagent ($\epsilon = 31,000$; 540 mμ). If ferrocyanide is introduced, the effect of iron and cobalt can be reduced to a minimum. Small amounts of vanadium are bound into a complex by hydrogen peroxide.

If $5\,\mu g$ Al are determined by 2 ml of a 2% solution of Complexone III and 2 ml of a 0.1% potassium ferrocyanide solution, the effect of the following elements can be eliminated: $100\,\mu g$ per 25 ml of Ag, As(III), Be, Bi, Ca, Ce(III), Cu(II), Hg(II), Li, Mg, Mn(II), Mo(VI), NH_3, Nd, Ni, Pb, Pt, Sb(III), Sn(IV), Th, U(VI), W(VI), Zn, PO_4^{3-}, SiO_3^{2-}; $50\,\mu g$ per 25 ml of Ti(IV), $20\,\mu g$ per 25 ml of Fe(III), $10\,\mu g$ per 25 ml of Co, Cr(III) and Zr, and $500\,\mu g$ per 25 ml of chlorides, perchlorates, nitrates and sulfates. The effect of $20\,\mu g$ of V(V) in 25 ml can be eliminated by 0.1 ml of 3% H_2O_2; 1 mg Be eliminates the effect of $100\,\mu g$ of fluoride in 25 ml; citrates interfere strongly, even if present in the amount of only $10\,\mu g$ in 25 ml.

Determination by hydroxyanthraquinones

Determination by alizarin. Alizarin was the first photometric reagent for aluminum; it was proposed in 1915 by Atack [535]. However, owing to its low solubility in water, the reagent is not used to any extent, and Alizarin S (sodium alizarinsulfonate) is employed instead.

Alizarin S is an acid-base indicator with two color changes: at pH 3.7—5.2 (yellow to brown-red) and at pH 10—12 (violet to pale yellow) [657]. The dissociation constants of Alizarin S, as a dibasic acid, are $3 \cdot 10^{-6}$ and $3 \cdot 10^{-10}$, respectively [27]. Alizarin S and aluminum form a 1 : 3 complex [26, 535, 791, 1001]. Other data are also available on the composition of this complex. At pH 3.6—3.9 [1045] and at pH 4.6 [269] a 1 : 1 complex has been reported. If the complex is formed in an alkaline medium and is then acidified, it consists of one aluminum atom and two molecules of Alizarin S [269]; this complex is also formed in the presence of calcium [269, 1045].

The absorption maximum of the reagent is at 423 mμ [1045]. The absorption maximum of the complex is at 500 mμ [34, 1024] and 475—480 mμ [552, 1045]. The molar absorption coefficient of the complex is 3,700 [25] or 4,100 [269]; the coefficient in the presence of calcium is 10,100 [269] and 18,000 [657].

Various optimum pH values have been suggested: 3.6 [1001], 3.6—3.8 [1235], 3.92 [1024], 4.0 [791], 4.5 [1045], 4.4—4.65 [657], 4.6 [754]. It is undesirable to work at very low pH values (pH 3.6), because the lake rapidly precipitates and the color is slow to develop [791]. Oelschläger [1024] gives the optimum pH as 3.92; if a buffer solution with sufficiently high capacity is used and small amounts of phosphates introduced, a very stable color is obtained. According to Parker and Goddard [1045], the maximum color intensity is observed at pH 4.55; at pH 3—7 the color takes 3 hr to develop, but if the solution is heated the color development becomes more rapid. At pH 4.55 and 60°C the color is completely developed within 15 min [1045]. In the presence of calcium ions the maximum color intensity is also noted at pH 4.3—4.7; at pH 3.5, 3.9, and 4.5 at 60°C the color develops fully within 30 min.

The following conditions can be considered as optimum: pH 4.3—4.7 and heating at 60°C for 30 min, both in the presence and absence of calcium [1045].

Variations in the temperature of the sample solution between 20 and 30°C have no significant effect on its optical density [550]; volume changes of 25—45 ml during color development have no effect [550]. A number of authors recommended the use of protective colloids to improve the stability of the colored compound: glycerol [535, 1235], gum Arabic [1229], a mixture of starch and glycerol [1229]. The most recent studies indicate, however, that no protective colloids are necessary.

Beer's law is obeyed for contents of up to $80 \mu g$ Al in 100 ml, when 5 ml of a 0.01% solution of Alizarin S is employed [550, 656]. If the amount of Alizarin S is increased, this limiting value can be raised to $200 \mu g$ Al in 50 ml [530]. In the presence of calcium, the color of the aluminum complex is intensified and the absorption peak shifts toward the red. In the absence of aluminum the effect of calcium on the absorption spectrum of the reagent is insignificant. If the amount of calcium is increased to $500 \mu g$ in 10 ml, the color of the solution becomes more intense; if still more calcium is added, the color intensity remains constant. If more than $1,500 \mu g$ Ca are present, the solution becomes turbid. The optimum amount of calcium is $1,000 \mu g$ in 10 ml of the sample solution. The Li^+ ion also intensifies the color of the aluminum complex, but to a smaller extent. If the content of lithium is greater than 10 mg/ml, the color of the aluminum complex becomes weaker [584].

The following do not interfere with the determination of aluminum if present in concentrations of $25 \mu g/ml$: Na, K, Mg, Zn, Cd, Ni, As(III), Co. The following interfere, either by forming colored complexes with Alizarin S or by changing the absorption intensity of the aluminum complex: Fe(II), Fe(III), Cr(III), Sb(III), Bi, W(VI), Mo(VI), V(V), Cu(II), BO_3^{3-}, Ca, Li, Sn(IV), Ti(IV), Pb, Mn(II), PO_4^{3-}, SiO_3^{2-}. Sn(IV), Ti(IV), Pb, and Mn(II) give precipitates or turbidity in the final solution; B, Pb, and Si interfere to a very small extent only. Phosphates reduce the optical density of solutions. Beryllium somewhat intensifies the color intensity, $40 \mu g$ Be are equivalent to $1 \mu g$ Al [656]. Attempts to eliminate the effect of iron by citric [1001, 1283], tartaric, oxalic, and phosphoric [1001] acids proved unsuccessful. Iron can be masked by thioglycolic acid [754], cyanide and thiosulfate [743]. The effect of 5 mg Fe_2O_3/liter can be eliminated by introducing 0.1–0.2 g of solid sodium thiosulfate.

The interfering anions include fluorides, phosphates, citrates, and tartrates. Sodium sulfate does not interfere up to a concentration of $0.045 M$ [1045], but ammonium sulfate interferes more strongly. Sodium and ammonium chlorides rather intensify the color intensity of the aluminum compound (up to a concentration of $0.02 N$ their effect is negligible). Aluminum is best determined by Alizarin S in perchloric acid medium.

Determination by quinalizarin. Quinalizarin (1,2,5,8-tetrahydroxyanthraquinone) was proposed by Kolthoff [888] as a reagent for the photometric determination of aluminum. However, if Kolthoff's procedure is employed, the determination can be qualitative only, since the colored complex tends to precipitate. Quinalizarin was found suitable for the quantitative determination of aluminum only after a suitable solvent had been found to dissolve the colored lake. A mixture of butyl carbitol with water was suggested for this purpose [532]. The photometric determination of the colored solution is carried out at 570 mμ. Aluminum and quinalizarin form a 1 : 3 [532] or a 2 : 3 [601] complex. The optimum pH of the medium is about 5. Beer's law is obeyed between 0.05 and 2.0 μg Al/ml. Twenty-fold amounts of Mn, Cd, Pd, and Sn, 40-fold amounts of Zn, and 1,000-fold amounts of Mg do not interfere. The effect of Cu and Fe is eliminated by introducing 1 ml of a 1% solution of diethyldithiocarbamate within 20—25 min after the addition of quinalizarin.

To 1–2 ml of the sample solution containing 0.5–20 μg Al, a 40–50-fold amount of quinalizarin (with reference to aluminum) in the form of a 0.1% solution of ethanol (1.4 M in ammonia), and 1 ml of acetate buffer solution at pH about 5 (a mixture of 6 pts of 5 N acetic acid and 4 pts of 5 N ammonia) are added. Butyl carbitol is then introduced to a content of 40–50% in the final solution.

The optical density is determined at 570 mμ [532].

Quinalizarin sulfonate can be employed for the determination of aluminum. The reagent reacts with aluminum in an absolute methanol medium to form a violet complex, while a solution of the reagent itself is light yellow. The absorption maximum of the complex is observed at 560 mμ; the composition of the complex is 1 : 1; the optimum pH is 0.3–0.5 [1038]. The maximum color intensity is obtained after 40 min, and then the color intensity remains stable for a long time. The effects of light, sequence of the addition of the reagents, and variations in temperature between 15 and 35°C, are negligible. Beer's law holds up to 1.7 ppm. The Sandell sensitivity is 0.02 μg/cm^2. Foreign cations and anions do not interfere in the following ratios:

Ion	Excess
Be, Sc, Ti(IV), Y, Zr, P$_4$O$_7^{2-}$	0.2
Th	0.5
Cu(II), SO$_4^{2-}$	60

Hundredfold amounts of Au(III), Ba, Bi, Ca, Cd, Co, Cr(III), Fe(II), Ga, Hg(II), In, K, Li, Mg, Mn(II), Ni, Pb, Pd(II), Sb(III), Sn(II), Sn(IV), Sr, Zn, U(VI), BO$_3^{3-}$, Br$^-$, I$^-$, S^{2-}, NO$_3^-$, and citrates do not interfere. Three ml of water may be present in 50 ml of solution, but larger amounts of water interfere with the development of the color. The effect of Fe(III) can be reduced by a mixed solvent (10 ml ether + 40 ml absolute methanol), when 60-fold amounts of Fe(III) produce a +2% error.

Determination by other reagents

Many other organic reagents have also been proposed for the photometric determination of aluminum. Table 12 lists a number of reagents which are only rarely employed.

Methods based on the nephelometric determination of aluminum cupferronate [1034, 1122], photometric determination of the phosphorus contained in aluminum phosphate in the form of the reduced phosphomolybdic hetero acid [110, 811], and methods based on the reaction of the fluoride complex of aluminum and titanium in the presence of hydrogen peroxide [455] are of no interest.

Nazarenko et al. [292] studied the reaction between aluminum and polyhydroxyflavones, namely, quercetin and rutin, but did not recommend these compounds for the photometric determination of aluminum.

FLUORIMETRIC METHODS

Determination by salicylal-*o*-aminophenol

The fluorimetric determination of aluminum by salicylal-*o*-aminophenol (Manganon) is the most frequently used fluorimetric method [37, 56–58, 168, 224, 227, 435, 671, 1134]. The maximum fluorescence intensity of the aluminum complex is noted at pH 5.8 and 530 mμ [56]. Efimychev and Tumanov [435], and also Saylor and Ledbetter [1144] give the composition of the fluorescent complex of aluminum with the reagent as 1 : 2, its instability constant as $2.02 \cdot 10^{-7}$, and its molar extinction coefficient as 9,200 [435]. It was pointed out by Babko et al. [35] that a 1 : 1 complex is formed at pH below 6.5, and a 1 : 2 complex at pH above 6.5. The formation of a 1 : 1 complex was also reported by Dagnall et al. [671].

The sensitivity of the method is 0.0025 μg Al per 1 ml [58]. The fluorescence intensity is proportional to the concentration of aluminum up to 0.5 μg in 5 ml [58]. The fluorescence attains a maximum within 15 min, and then remains constant [56]; it is usually measured after 40–50 min. The content of aluminum is found from a calibration curve, or by the method of additions. The optimum amount of the reagent is 0.3 ml of a 0.1% solution in 10 ml [56]. The fluorescence intensity decreases if the

Table 12. Photometric methods for the determination of aluminum not in common use

Reagent	λ_{max}, mμ	pH$_{max}$	ϵ of complex	Effect of other elements or ions	Notes	References
Apigenin (4',5,7-trihydroxyflavone)	380 390*		23,800 16,500	Sc interferes; REE do not	Is used to determine Al in metallic Ce, in salts, and REE oxides	[1018a]
Ferron (7-iodo-8-hydroxy-5-quinolinesulfonic acid)	370	5.0		The following do not interfere: the same amounts of Mg, Mn; up to 1 mg Ca; Cu up to the ratio Cu:Al=1:4; Zn up to the ratio Zn:Al=1:7. The following interfere: Fe, U(VI), Zr, Th, Ni, Cr(VI), Mo(VI), Ti, Ga, Be	Beer's law holds up to 40 μg Al/25 ml	[50, 590, 675, 849, 915]
Hematoxylin	580	8.2–8.65		Many elements interfere	After lake has formed, the pH is reduced to 6.8–7.1 to attenuate the color of excess reagent	[781, 790, 824, 1202]
Sendachrome (3-carboxyquinone-3-carboxy-4-hydroxy-phenylmethide)	510	3.6–3.9		Zr, Th, Be, Ti, Cr(VI) interfere	The color develops within 5 min at 50°C and is stable for 6 hr; Beer's law holds up to 40 μg Al in 25 ml	[140, 390]
Calmagite**	570		42,000	Insignificant interference by 80-fold amounts of F^-, Cl^-, Br^-, I^-, ClO_4^-, BO_3^{3-}, NO_3^-, NO_2^-, SO_4^{2-}, SO_3^{2-}, $S_2O_3^{2-}$, CO_3^{2-}, CN^-, CNS^-, PO_4^{3-}, CH_3COO^-, formates and citrates	Complex has composition 1:3. It can be extracted by chloroform from solution at pH 6.8 (borate buffer solution) in the presence of methylcaprylammonium chloride	[1280]

Table 12 (continued)

Reagent	λ_{max}, mμ	pH$_{max}$	ϵ of complex	Effect of other elements or ions	Notes	References
Calcichrome	620 (reagent at 317) 317*	~6	7,200	Cu(II), Fe(III), Ti, V(V), Zr interfere	The color of the complex develops within 20 min and is stable for 24 hr. The optimum amount of the reagent is a two-fold excess. Sensitivity $2.3 \cdot 10^{-3}$ and $3.7 \cdot 10^{-3}$ μg Al/ml at 317 and 620 mμ, respectively	[836]
α-Methylformaurine-3,3'-dicarboxylic acid	510 (reagent at 360)	3.3	11,500	Effect of Fe(III) is suppressed by hydroxylamine		[15]
Sodium salt of o-carboxyphenylazochromotropic acid	590	~5	21,530	Most ions, including Mg, Ca, Sr, Ba, Cd, nitrates and sulfates interfere even in small amounts (of the order of 10^{-2} mg/ml)	The composition of the complex is 1:1. The color of the complex develops within 5 min at 50°C	[948]
Anthrapurpurin	500 (reagent at 334 and 420) 380* 390*		23,800 16,500			[616]
Diamine Pure Blue FFG						[492]

* Other values of λ.   ** 1-(1-hydroxy-4-methyl-2-phenylazo)-2-naphthol-4-sulfonic acid.

concentration of the reagent is above or below this value. The reagent is used as a solution in acetone (the presence of up to 10% acetone does not interfere; larger amounts greatly reduce the fluorescence intensity).

The following do not interfere: up to $100 \mu g$ Ag, Tl, Y in 10 ml; up to $10 \mu g$ Bi, Cd, V, Mo, Gd, $C_2O_4^{2-}$ in 10 ml; up to $7 \mu g$ Fe(III) and Cu(II) in 10 ml [168]. Larger amounts of Fe(III) and Cu(II) quench the fluorescence, and in this case aluminum is determined by the method of additions. Tin interferes and must be completely separated [228].

Babko et al. [37] reported that the sensitivity and selectivity of analyses of NaCl and $NaNO_3$ can be increased if the aluminum complex is extracted from solution by isoamyl alcohol at pH about 6.4. The fluorimetric method with salicylal-o-aminophenol was employed to determine aluminum in HCl, HF, H_2SO_4, HNO_3, H_2O_2, CH_3COOH [58], and $GeCl_4$ [57], and in lithium, rubidium, cesium [57] and lead [168] salts, in high-purity salts [224], and in high-purity tin [228]. The sensitivity of the method is $10^{-5}-10^{-8}\%$, and the relative error is about 20%.

Determination by 8-hydroxyquinoline

The intense green-yellow fluorescence of chloroform solutions of aluminum hydroxyquinolate in the UV forms the principle of one of the most important fluorimetric methods for the determination of aluminum [451, 651, 767, 779, 785, 827a, 1016, 1082, 1106, 1125, 1233, 1272]. The method is very sensitive: up to $10^{-4}\%$ Al can be determined, and the effect of other metals is less than in photometric determinations. Aluminum hydroxyquinolate is usually extracted with chloroform from solutions at pH 6–9. The fluorescence intensity is the same whether it is measured immediately or only after 24 hr [767]. The method is most sensitive if 2 ml of a 0.2% solution of hydroxyquinoline in 50 ml are used; if 2 ml of a 2% solution of the reagent are employed, the sensitivity of the method deteriorates somewhat, but its reproducibility improves [767]. The decrease in the fluorescence intensity which results from an increase in temperature from 10 to 40°C is insignificant. A chloroform solution of hydroxyquinoline also absorbs some UV light at concentrations above 0.05%, but the content of hydroxyquinoline in solutions for the fluorimetric determinations is usually lower. The content of aluminum is found from a calibration curve, plotted for the range between 0 and $5 \mu g$ Al/ml.

The following do not interfere [1016]:

Element	Content, $\mu g/ml$	Element	Content, $\mu g/ml$
Mn	0.3	Zn	5
Cu	1	Ca, Cr	10

According to [1082], up to 20-fold amounts of Co, Ni, Cd, Sn, Bi, Tl, Mo(VI), and U(VI) also do not interfere; Ce, Pb, Mg, Hg, Ag, Th, W [1233] do not fluoresce. The fluorescence given by aluminum hydroxyquinolate is quenched by Fe, Ti and V. The same amounts of titanium do not interfere; iron also has practically no effect up to 5 $\mu g/ml$. Vanadium interferes more strongly than titanium or iron. Gallium and indium do not interfere in amounts up to 15% and 3%, respectively, of the content of aluminum. Aluminum and gallium can be determined in the presence of one another [651] by using solutions of gallium and aluminum hydroxyquinolates of different sensitivities and different parts of the exciting radiation.

The following anions do not interfere: chlorides, sulfates, nitrates, and perchlorates; up to 100 μg per ml of SiO_2, up to 60 μg per ml of chromate, up to 30 μg per ml of phosphate, and up to 20 μg per ml of fluoride. Citrates and tartrates interfere. The relative error of the method varies between 5 and 15%, depending on the content of aluminum. The fluorimetric hydroxyquinoline method has been utilized in determinations of aluminum in steels and bronzes [767], magnesium [451], uranium [451], bismuth salts [451], phosphate rocks [779], vegetable materials [1125], water [1016], tungsten and tungsten oxide [672].

Determination by Pontachrome Blue-Black

Pontachrome Blue-Black R (Eriochrome Blue-Black B), the zinc or sodium salt of 4-sulfo-2-hydroxy-α-naphthaleneazo-β-naphthol, reacts with aluminum to form a complex of composition $AlOHO_2 \cdot (C_{20}H_{11}N_2SO_3Na)_2$ [1252]. Solutions of the complex in organic solvents (ethanol, amyl alcohol) give a red fluorescence in the UV with a maximum at 539 mμ [142]; pH_{max} 4.8−4.9 [141, 1252] (Figure 25). For an aluminum content of 0.05 mg in 50 ml of solution the maximum fluorescence intensity is noted when 1.5 ml of a 0.1% solution of the reagent in ethanol are added. If larger or smaller amounts of the reagent are added, the fluorescence intensity becomes constant within one hour, and then remains constant (at 24°C). Between 10 and 25°C the fluorescence intensity increases somewhat with decrease in temperature (at 10°C it is only 2% stronger than at

25°). The dependence of the fluorescence intensity on the aluminum concentration is almost linear up to 0.05 mg Al (with 1.5 ml of a 0.1% solution of the reagent).

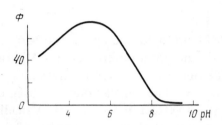

Figure 25. Dependence of the fluorescence intensity ϕ of the aluminum complex of Pontachrome Blue-Black R on the pH [1252].

The sensitivity of the method is 0.0001 μg/ml [1076]. The aluminum complex can be extracted by amyl alcohol from solutions at pH4.8—5.4 [142].

Iron, copper, chromium and cobalt interfere, and these elements are removed by electrolysis on a mercury cathode. Of the elements which are not thus separated, zirconium does not interfere, titanium does not significantly interfere, while V(V) strongly quenches the fluorescence [1252]. Gallium and large amounts of nickel also interfere. The metals Li, Na, K, Rb, Cs, Be, Mg, Ca, Sr, Ba, Cd, Pb and Tl(I) do not interfere; Se, La, Mn, Ni, Zn, Hg, In and Sb(III) do not interfere in concentrations up to 4μg/ml [1076]. Of the anions, up to 40μg/ml of BO_3^{3-}, SiO_3^{2-}, NO_3^-, PO_4^{3-}, AsO_4^{3-}, CO_3^{2-}, SeO_4^{2-}, SeO_3^{2-}, Cl^-, Br^-, I^- [1076] do not interfere. The reagent is suitable for the determination of as little as $5 \cdot 10^{-6}$% of aluminum in a one-gram sample.

Pontachrome Blue-Black is one of the most common and important reagents for the fluorimetric determination of aluminum. It has been used to determine aluminum in steels [1252], bronzes [1252], magnesium [142], antimony [294], minerals [1252], tungsten and tungstic oxide [672]. See also [141, 290, 672].

Determination by morin

Aluminum reacts with morin to form a complex of composition 1:1 [1214] with an intense green fluorescence. The maximum fluorescence is noted in the presence of 40 ml of ethanol and 12 ml of a 0.15% solution of morin in ethanol (final volume of solution 100 ml). The fluorescence intensity decreases if smaller or larger quantities of ethanol and reagent are employed [1261]. The fluorescence intensity increases with decrease in temperature, and becomes constant below 15°C [1214]. The measurement can be taken 15–20 min after the addition of morin. The optimum pH value is 3–3.3 [1261, 1269]. The stability constant of the complex is $2.96 \cdot 10^5$ [1214]. A fluorescence similar to that of aluminum is also given by Be, Ga, In and REE; Pb, Zn and Mo give a fluorescence if the solution has not been sufficiently acidified with acetic acid. Silver quenches the fluorescence. Iron and chromium give black precipitates with morin. According to Will [1269], aluminum present in 2 parts per billion can be determined in the presence of Fe(III) $\leqslant 10$, Cr(III) $\leqslant 30$, Cu(II) $\leqslant 20$, Mg $\leqslant 200$, $NH_4^+ \leqslant 500$, Ca, Co, Ni, Zn, Pb, $SiO_3^{2-} \geqslant 1,000$, $PO_4^{3-} \leqslant 3$, $F^- < 5$ parts per billion. Chlorides and nitrates, and up to 9 mg of sulfate in 100 ml do not interfere [1261]. The concentration of aluminum in the sample solution should be between 0.0005 and 0.005 mg in 100 ml [1269]; the relative error of the method is up to 5% [1214].

Morin is a highly sensitive reagent for the fluorimetric determination of aluminum. Its disadvantages are that it is not very specific and the blank values are too high. Also, its aluminum complex is sensitive to temperature changes. Therefore, morin is now rarely used to determine aluminum by the fluorimetric method.

Determination by 2-hydroxy-3-naphthoic acid

2-Hydroxy-3-naphthoic acid reacts with aluminum at pH 3 or above to form a complex with a bright blue fluorescence. At pH 5.8 the composition of the complex is 1 : 1; $K_{stab} = 2.37 \cdot 10^4$ [875]. At pH above 2 the reagent gives a green fluorescence.

To determine aluminum, 5 ml of acetate buffer solution (pH 5.8) and 10 ml of a $10^{-4} M$ solution of the sodium salt of 2-hydroxy-3-naphthoic acid are added, the solution is diluted to 100 ml with water, and the fluorescence intensity is measured after one hour; the wavelength of the exciting light should be 370 mμ.

Between 0.2 and 12.5 μg Al can be determined in this way. The elements Be, Sc, Zr, Ga, B, Tl, Cu, U, Th, Sb, Bi, In, Ti, Sn, As, Cr(VI), NO_2^-, SO_3^{2-} interfere.

This reagent has been suggested only recently and no laboratory experience is as yet available; it has no special advantages over the other reagents used for the fluorimetric determinations of aluminum.

Determination of aluminum by quercetin

Quercetin, 3,5,5,2',4'-pentahydroxyflavone, can be used for the fluorimetric determination of aluminum; the complex of aluminum with the
reagent gives an intense green fluorescence [101]. The optimum conditions
for the determination are: 3–5 ml of ethanol, 1–2.3 ml of a 0.1% solution
of the reagent in a 14 ml volume; pH 4.4–5.0; temperature 20–30°C. The
fluorescence intensity rises to a maximum immediately after the addition of
quercetin, and remains constant for several hours. A linear dependence of
the fluorescence intensity on the amount of aluminum is noted for aluminum contents between 0.0015 and 0.0040 mg. Aluminum in an amount of
0.003 mg can be determined in the presence of up to 400 mg KCl, NaCl and
KNO_3, up to 100 mg K_2SO_4, KBr, KI, Ba, 40 mg Ca, 45 mg Hg(II), 4.8 mg
Mg, 40 mg Mn, 6 mg Co, 2 mg Ni, 10 mg AsO_4^{3-}, 0.3 mg Sn(II), 0.03 mg PO_4^{3-},
0.01 mg V(V), 0.4 mg Cu(II), 0.08 mg Mo(VI), 0.003 mg Fe(III), and
0.03 mg Cr(III).

Quercetin is a nonspecific reagent for the fluorimetric determination of
aluminum.

Determination by N-salicylidene-2-amino-
3-hydroxyfluorone

The method has been proposed by White et al. [1262]. The optimum
conditions are: ethanol content 10%, a fourfold (molar) excess of reagent
with reference to Al, pH 5.1–5.4 (Figure 26). Between 0 and 30°C the
temperature has no significant effect on the fluorescence intensity. This
intensity varies linearly with the amount of aluminum between 0.001 and
0.04 mmole of Al. The visual sensitivity is 2 pts per billion.

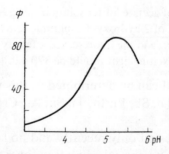

Figure 26. Dependence of the fluorescence intensity ϕ of the aluminum complex
of N-salicylidene-2-amino-3-hydroxyfluorone in 10% ethanol on the pH [1262].

The maximum excitation of the reagent takes place at 280 and 320 mμ, and the fluorescence maximum is at 385 mμ; and correspondingly for the complex at 445 and 525–530 mμ, respectively. The aluminum and the reagent in the complex are in the ratio of 1:2. The complex can be extracted by benzene. The effect of foreign ions is the same as that noted for other reagents. Iron, present in amounts equal to or greater than that of aluminum, reduces the fluorescence intensity and must be removed. Iron does not interfere at a molar ratio Fe:Al = 0.7:1.

The specificity of this reagent resembles that of other reagents commonly employed for the fluorimetric determinations of aluminum, but the reagent is more sensitive. It is more sensitive than salicylal-o-aminophenol, 2-hydroxy-3-naphthoic acid, morin, or Pontachrome Blue-Black R. Therefore, the reagent can be considered a promising one for the fluorimetric determination of aluminum.

Determination by salicylaldehyde formylhydrazone

Salicylaldehyde formylhydrazone forms a fluorescing complex with aluminum at pH 4.7–4.9; the aluminum : reagent ratio in the complex varies between 1:1 and 1:2 [817a]. The sensitivity of the reaction is 0.08 μg Al in 25 ml. The fluorescence intensity varies linearly with the amount of aluminum between 0.7 and 22 μg Al/25 ml. The following do not interfere:

Element or compound	Excess	Element or compound	Excess
Na, K, NH$_4^+$	10,000	Ag, Hg, Bi, U, Co	10
Hg, Ca, Cd, Mn	1,000	NaCl	200,000
Sr, Ba, Pb, Tl	100	Na$_2$SO$_4$	100

Ferric iron, Cr(III), Ni, Zn, and Cu interfere. The effect of 25 μg Cu in 25 ml can be eliminated by adding 0.3 ml of a 5% solution of thiosulfate. One ml of a 1% solution of thioglycolic acid will mask 2.5 μg Fe(III), 1 μg Cr, 5 μg Ni, and 20–100 μg Zn. The relative error is 1%.

Salicylaldehyde formylhydrazone is not a very sensitive reagent; many metals interfere with the fluorimetric determination of aluminum by this reagent. Alkali and alkaline-earth metals alone interfere to a small extent only. Therefore, the reagent is of no interest for the fluorimetric determination of aluminum.

Determination by lumogallion

Lumogallion, 2,2″,4′-trihydroxy-5-chloro-(1-azo-1′)-benzene-3-sulfonic acid, reacts with aluminum to form a fluorescent complex with a 1 : 1 composition [307b, 307c].

The maximum fluorescence intensity is noted at pH 5, after the solution has been heated for 20 min at 80°C. The fluorescence maximum of the complex is found at 576 mμ. The aluminum complex can be extracted by n-butyl and n-amyl alcohols or isoamyl alcohol. The fluorescence is stable for 2 hr. Beer's law is obeyed at 0.1−2 μg Al in 25 ml. The following elements interfere: Fe(III), Ni, Co, Cu(II), Sn(IV), Sc, Ti(IV), V(V), Cr(VI). Lumogallion has been employed for the fluorimetric determination of aluminum in seawater [307b].

Determination by other reagents

Pontachrome Violet SW, which is structurally similar to Pontachrome Blue-Black R, has been proposed as a reagent for the fluorimetric determination of aluminum. The method is highly sensitive, and the relationship between the fluorescence intensity and the concentration of aluminum is practically linear up to 20 μg Al in 50 ml. The disadvantages of the reagent are that the pH must be strictly controlled, and the fluorescence is quenched in the presence of even traces of iron [1252].

The zinc salt of 4-sulfo-2,2′-dihydroxyazonaphthalene need not be considered, owing to the high influence of the pH and of foreign ions [101]. Other fluorimetric reagents which have also been proposed include o,o′-dihydroxyazobenzene [1145], 8-hydroxyquinaldine [763], salicylalsemicarbazide [188]. Korenman and Grishin [187] studied hydroxyanthraquinone dyes and Golovina et al. [92] certain flavonoid dyes. A number of reagents with o,o′-dihydroxyazo groups have been prepared for this purpose [518]. Babko et al. [28] carried out a comparative study of certain reagents for the fluorimetric determination of aluminum.

The reagents and the water used in fluorimetric work must be of a very high degree of purity. The distilled water should be deionized. The reagents employed must be of the highest available degree of purity, or else common reagents must be preliminarily purified. It is recommended that buffer solutions be purified from aluminum impurities by passing them through a column filled with EDE-10P anion exchanger, or activated charcoal, with lumogallion to bind the aluminum. If the hydroxyquinoline method is employed, the buffer solution is shaken with a chloroform solution of hydroxyquinoline. All solutions should be kept in silica or polyethylene vessels.

POLAROGRAPHIC METHODS

Prajzler [1078] was the first to study the possibility of determining aluminum polarographically. Since then, many publications on direct, and also indirect, determinations of aluminum have appeared.

Direct methods

Prajzler [1078] noted a distinct aluminum wave on the background of $0.05 N$ $BaCl_2$, with a half-wave potential of -1.75 V, with reference to a saturated calomel electrode (s.c.e.). This potential is very close to the discharge potential of Na, K and Ba, and the wave given by aluminum is not well marked in the presence of large amounts of these ions. The hydrogen wave with $E_{1/2} = -1.58$ V at pH 3, with $0.007 M$ LiCl as background precedes the aluminum wave and is very close to it. The half-wave potential of aluminum depends on the acid concentration [210]:

pH	$E_{1/2}$, V	pH	$E_{1/2}$, V
1.9	-1.88	3.0	-1.76
2.15	-1.86	4.0	-1.66
2.6	-1.80		

At below pH 3 the aluminum wave is masked by the hydrogen wave. The pH can be raised up to a certain limit only, because of the hydrolysis of aluminum salts, which makes the relationship between the diffusion current and the aluminum concentration no longer linear. It follows that the direct polarographic determination of aluminum will be successful only if the pH is rigorously controlled. The optimum pH range is 3.5–4.0. This range is narrow, and it also varies with the concentration of aluminum. Various electrolytes can serve as background: sodium and potassium chlorides [114], calcium chloride [173, 460, 1111], magnesium chloride [1040], barium chloride [1078], and lithium chloride; calcium gluconate [47], tetrabutyl-ammonium chloride [1221], tetraalkylammonium chloride [806, 890], a mixture of dimethyl sulfoxide with acetylacetone [782], sodium salicylate, etc.

The height of the aluminum wave depends on the pH of the solution, and on the concentration of the background electrolyte [173]. The shape of the aluminum wave readily changes with the experimental conditions [1111]. For all these reasons, the potentialities of direct methods are very limited.

Indirect methods

Indirect methods for the determination of aluminum are based on two principles: on the reduction of the wave height of certain organic reagents in the presence of aluminum, when the aluminum complex formed is not reduced, and on the change in the shape of the polarograms of certain dye-stuffs in the presence of aluminum, owing to complex formation [1271]. If small amounts of aluminum are added to a solution containing the dye-stuff in excess, a double wave is obtained. The first wave corresponds to the amount of unreacted dyestuff, and the second is proportional to the concentration of the aluminum complex.

In methods involving the formation of nonreducible aluminum complexes, hydroxyquinoline [619, 678, 1047], quinalizarin [364, 365], Acid Chrome Blue K [68, 413], Erichrome Black T [53], and the tetrasodium salt of EDTA [944] are used.

Parks and Lykken [1047] measured the decrease in the height of the wave of a standard hydroxyquinoline solution after the addition of the aluminum solution to be determined. The aluminum content was read from the calibration curve. The disadvantage of the method is that alumi-num hydroxyquinolate is precipitated to an extent of about 90% only under the conditions employed, even if the solution is held for one hour.

Terent'eva [413] determined aluminum in alumoorganosiloxanes by differential polarography with Acid Chrome Blue. The sample was decom-posed by hydrofluoric and sulfuric acids, the residue was fused with sodium carbonate, and the solution of the melt was polarogrammed between 0 and –0.8 V (with reference to a mercury anode). The height of the second peak (at –0.7 V) was then determined starting from the base of the first peak; the absolute error of the method was $\pm 0.3\%$. The same reagent was also employed to determine aluminum in waste waters [68], with a relative error of $\pm 4\%$.

Skobets et al. [364, 365] determined aluminum in soils, using the forma-tion of a quinalizarin complex, which is not reduced on a dropping mercury electrode. Quinalizarin, with a background of a mixture of $0.1M$ NH_4Cl with $0.1M$ NH_4OH, forms a wave with $E_{1/2} = -0.6$ V (relative to s.c.e.). In the presence of aluminum the wave height varies proportionally to the con-centration of aluminum. Complexone III will mask Mg, Ca, Mn and Zn, while iron is bound as a complex by potassium ferrocyanide.

Aluminum can also be determined by measuring the decrease in the height of the wave given by Eriochrome Black T (at $E_{1/2} = 0.625$ V) in the presence of aluminum (as a result of complex formation).

Lydersen [944] determined aluminum by measuring the decrease in the height of the wave given by the tetrasodium salt of EDTA in the presence of aluminum at pH 4.7.

Methods based on the formation of reducible aluminum complexes are more numerous.

Determination by Solochrome Violet. Willard and Dean [1271] proposed a method based on the formation of the aluminum complex of a di-*o*-hydroxyazo dye, the sodium salt of 2-hydroxynaphthalene-(1-azo-1)-2-hydroxybenzene-5-sulfonic acid (this reagent is also known as Pontachrome Violet SW, Solochrome Violet RS, Superchrome Violet B). The half-wave potential of the dye is $E_{1/2} = -0.3$ V (s.c.e.). In the presence of aluminum the wave height of the dye decreases and a new wave appears, at a potential which is 0.2 V more negative (Figure 27). The height of the second wave is proportional to the concentration of aluminum. Reynolds and Perkins [1057, 1110] studied this method and found that the wave height of the aluminum complex increases with increase in pH. They found that two complexes were formed, with an aluminum-to-dye ratio of 1 : 2 and 1 : 3, respectively, and stable at pH 4.2–5.0 and 5.7–7.0, respectively, as shown by constancy of the wave height.

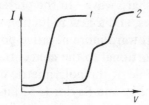

Figure 27. Polarographic wave of Solochrome Violet [1110]:

1) in absence of aluminum; 2) in presence of Al (pH 4.6).

Aluminum is usually determined at pH 4.7 (acetate buffer solution). Thousandfold amounts of Be, B, Ca, Mn, Mg and Zn do not interfere [1056]. Beryllium does not interfere, which is particularly important, as it interferes with the determination of aluminum by many other methods. Aluminum can be determined in beryllium [1056].

The elements Co, Ni, Fe, Ti and V form complexes with Solochrome Violet RS and interfere with the determination of aluminum. Copper forms an insoluble complex. Fluorides, citrates and oxalates reduce the height of the wave. The complex of aluminum with Solochrome Violet RS is formed slowly at room temperature, but within 2 min at 60°C [739]. At a content of 0.01−1.5 mg Al in 50 ml, the calibration curve is rectilinear and passes through the origin of the coordinates [739]. The sensitivity of the method is $2 \cdot 10^{-4}\%$ and the relative error is 3%.

Solochrome Violet is used to determine aluminum in steel [739, 1121], ferrotitanium [778], Fe−V, Fe−Zr and Fe−Ti alloys [251a], Pb−Sn alloys [566], soils [1], ores [257], and zinc coatings [257], etc. Aluminum and zinc can be determined in the presence of one another in magnesium alloys [744], while aluminum and magnesium can be determined in the presence of one another in rocks [708]. Aluminum can be polarographically determined after its complex with Solochrome Violet RS has been oxidized on a rotating pyrolytic graphite electrode [726]. The reagent and aluminum with a background of $0.2 M$ acetate buffer solution at pH 4.7 give anodic waves with $E_{1/2} = +0.53$ V and $+0.87$ V, respectively. The wave of the complex can be used to determine 25 μg Al/ml. At pH 4.7, 20-fold amounts of Ag, As, Be, Bi, Ge, Cd, Ca, Cr, Cu, Hg, Li, Mg, Mo, Ni, Pb, Pr, Sb, Sn, Th, Tl, U, W, Zn, Zr, PO_4^{3-} and dissolved oxygen do not interfere with the determination of aluminum; Fe(III), V(V), Ti(IV), Co, Mn and F^- interfere.

Determination by Superchrome Garnet Y. The sodium salt of 5-sulfo-2,4,2-trihydroazobenzene (Superchrome Garnet Y) in an acetate buffer solution (pH 3−6) gives a sharp wave. In the presence of aluminum the height of the wave decreases, and a new reduction wave of the aluminum complex, which is shifted toward more negative potentials, appears; the height of this wave is proportional to the concentration of aluminum in solution. This is the principle of the polarimetric determination of aluminum [199, 653, 723]. Cooney and Saylor [653] used Superchrome Garnet Y for the determination of aluminum and gallium present together at pH 5.5; Zn, Mn, Ce(IV), Fe(III), V(IV), Ni, Cd and Pb interfere, while Mg, Ca, K, Na, Cs, phosphates, nitrates, sulfates and acetates do not interfere. The relative error of the method is 3−5%. Kostitsyna and Skobets [100] determined aluminum in alloys after iron and copper had been complexed with ferrocyanide. The determination was carried out at pH 4.6; the solutions were heated to 60°C to accelerate the complex formation. The relative error was 5%. Floerence and Izard [723, 725] determined aluminum in thorium compounds and in beryllium by oscillographic polarography using Superchrome Garnet Y. The adsorption peak of the reagent in an acetate buffer solution (pH 3.4−3.5) was observed at −0.3 V (with reference to a mercury cathode), while at −0.51 V a new peak appeared, proportional to the concentration of aluminum (Figures 28 and 29). Beryllium in high concentrations reduces the peak height, so that the concentration

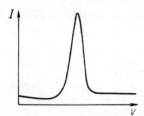

Figure 28. Oscillogram of Superchrome Garnet Y in acetate buffer solution at pH 3.4 [725].

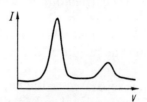

Figure 29. Oscillogram of Superchrome Garnet Y in presence of aluminum in acetate buffer solution at pH3.4 [725].

of beryllium in solution should be about the same (to within ± 3%) during the determination and the plotting of the calibration curve. The optimum pH is 3.4–3.5. Ni, Pb and Mo interfere and are removed by electrolysis on a mercury cathode. The relative error is 2%.

Aluminum is determined in thorium compounds at pH 5.75 [723]. For quantitative complex formation the solutions must be left to stand overnight, which is a serious drawback of the method. The interfering elements are removed by electrolysis on a mercury cathode. For aluminum contents of 0.04% and 0.001% the relative errors are ± 3% and ±7%, respectively.

Determination by hydroxyquinoline. Dehn et al. [678] proposed a method for the simultaneous determination of Al and Fe, and of Al, Fe, and Ti by taking polarograms of solutions of the hydroxyquinolates of these elements in dimethylformamide. By precipitating these metals as hydroxyquinolates, it is possible to separate them from alkali and alkaline-earth metals, and also from ammonium ions which interfere with the determination of aluminum if present at high concentrations. These metals can be determined in the presence of one another, owing to the large differences between their $E_{1/2}$ values. A solution of aluminum hydroxyquinolate in dimethylformamide gives three waves with $E_{1/2}$ at –1.75, –2.10, and –2.37 V, respectively. The first of these waves is the most marked. Water, present in contents of up to 1%, has no effect on the shape or the height of the wave; for this reason the precipitate need not be dehydrated, and drying at 120°C is sufficient. Fifty mg of the dried and weighed precipitate are dissolved in dimethylformamide. The solution is diluted to 50 ml with this solvent, and then polarogrammed. For determining Fe, Ti and Al, the waves at $E_{1/2} = -0.64, -1.37$, and –1.75 V, respectively, are employed.

Determination by Eriochrome Violet BA. This reagent was used by Mikula and Codell [969] to determine aluminum in titanium alloys. The determination is carried out at pH 4.6; the complex formation is accelerated by heating for 5 min on a water bath at $55-70°C$. The polarogram is taken between –0.2 V and –0.8 V with reference to s.c.e. The content of aluminum is found from the height of the second wave, with the aid of a calibration curve.

Titanium is separated as the cupferronate and chromium as chromyl chloride. Copper, nickel and cobalt, as well as the residual chromium, are separated by electrolysis on a mercury cathode. Chlorides are removed by evaporating the solution with perchloric acid. The relative error of the method is 4% for aluminum contents of 1%.

Aluminum can also be determined by the indirect polarographic technique with the aid of Diamond Chrome Red A [906], Alizarin Black SN [1057], and certain other reagents. For a review of the polarographic methods, see [514].

RADIOACTIVATION METHODS

The radioactive method is one of the most sensitive techniques for the determination of aluminum. Its sensitivity is 10^{-5}% [234, 924, 1026]. Radioactive methods for the determination of aluminum are based on the following nuclear reactions:

$$Al^{27} (n, \gamma) Al^{28} \quad (T_{1/2} = 2.3 \text{ min});$$
$$Al^{27} (n, p) Mg^{27} \quad (T_{1/2} = 9.5 \text{ min});$$
$$Al^{27} (n, \alpha) Na^{24} \quad (T_{1/2} = 15.0 \text{ hr});$$
$$Al^{27} (\alpha, n) P^{30}.$$

The first reaction is most often employed. The sample to be analyzed together with the etalons are irradiated by thermal neutrons (flux $2 \cdot 10^{10} - 3 \cdot 10^{11}$ neutrons/$cm^2 \cdot$ sec) in a reactor for $1-5$ min. One or two minutes afterwards, the activity of Al^{28} is determined by the γ-peak ($E_\gamma = 1.78$ MeV) with the aid of a scintillation γ-spectrometer. The sensitivity of the method is 10^{-5}% and the relative error is $6-20$%. This method has been used to determine aluminum in rocks [594, 1112], kaolin clays [235], diamond [1112], graphite [1026], and in solutions [859]. When aluminum is determined in metallic Zr, Fe and Cu by means of this reaction, zirconium is previously separated by precipitation as mandelate, iron in the form of chloride is extracted by ether, and copper is removed by electrolysis. Aluminum and vanadium impurities are coprecipitated with $Fe(OH)_3$, and aluminum is determined after irradiation (without separation) by the γ-spectrometric method [730].

Another source of thermal neutrons is the Po-Be neutron source [594] and the nuclear reaction $Be^9(\gamma, n)Be^8$ [1112]. A method has been described for the simultaneous determination of aluminum and silicon in ores. It is based on the difference in their activation cross sections as regards slow and fast neutrons [171]; the nuclear reaction $Al^{27}(n, p)Mg^{27}$ has also been used. When the nuclear reaction $Al^{27}(n, \alpha)Na^{24}$ is employed, the sample is irradiated for 1½ hr by a flux of $5 \cdot 10^7$ neutrons $\cdot$ cm²/sec. The activity is determined 2 hr after the irradiation, that is, after the short-lived Al^{28} and Mg^{27} isotopes have decomposed. Magnesium and iron do not interfere. The relative error of the method is 5% [1234].

Aluminum can also be determined with the aid of the nuclear reaction $Al^{27}(\alpha, n)P^{30}$ [1022]. The source of the α-particles is a Po^{210} (about 15 mcurie) layer on a platinum target. The time of irradiation is 6 min. The activity of the sample is measured within 20 sec after the termination of the irradiation by a Geiger-Müller counter for 6 min, and a correction is introduced for the decay of Po^{210}. The relationship between the aluminum content and the activity of the sample is linear. The accuracy of the method is affected by the distance between the source of α-particles, the duration of the irradiation, and the dimensions and density of the sample. The error of the method is 5%.

The radioactivation method has been employed for the determination of aluminum in copper ores [236], tungsten [1096], mica [234], and petroleum [234].

SPECTROSCOPIC METHODS

Spectroscopic methods for the determination of aluminum have been widely applied in the analyses of metals, alloys and other materials. The lines of aluminum used in spectroscopic analysis are in the UV part of the spectrum. Table 13 gives the most important sensitive lines of aluminum. The most sensitive lines of aluminum in the arc are $\lambda = 3,961.53$I, 3,944.03I, and 3,082.16I Å. The lines $\lambda = 3,082.16$ and 3,961.53 Å are the ones most frequently used. The most sensitive spark lines are $\lambda = 3,961.53$I, 3,944.03I, 3,092.71I, 3,082.16I, 2816.18II and 2,669.17II Å. Lines $\lambda = 3,082.16$ and 3,092.71 Å are most often employed.

Determination in metals and alloys

Metals and alloys are usually determined directly in the sample in the lump form. If the metal is not homogeneous, it is solubilized, the solution

Table 13. Analytical lines of aluminum and their intensity

Wavelength, Å	Line excitation energy, eV	Line intensities in arc	Line intensities in spark	Wavelength, Å	Line excitation energy, eV	Line intensities in arc	Line intensities in spark
6,243.36II*	15.1	–	80	3,082.16I	4.0	800 R	800 R
6,231.76II	15.1	–	35	2,816.18II	4.6	10	100
3,961.53I	3.1	3,000	2,000	2,669.17II	4.6	3	100
3,944.03I	3.1	2,000	1,000	2,631.55II	15.3	–	60
3,092.71I	4.0	1,000 R	1,000 R				

* I – line emitted by a neutral atom;   II – line emitted by an ionized atom;   R – self-reversal.

is evaporated, and the residue is converted to the oxide by ignition. The oxide powder is ground and mixed with an internal standard (sometimes various carriers are also added). The mixture is then placed in the cavity of a carbon electrode and the spectra are excited in an arc or spark. In most cases briquets are pressed from the prepared powder mixture of the sample, internal standard and carrier. Analyses are carried out in solutions of metals and alloys for the same reason, namely, to avoid errors due to inhomogeneity of the samples. The solutions can be introduced into the discharge gap in various ways: the simplest one is to impregnate the carbon electrode with the sample solution. In this way it is possible to determine aluminum in iron and steel [139, 152, 225, 668, 861, 1285], in nickel and nickel alloys [603, 604, 844] and bronzes [917]. This technique is not very sensitive; a better method is to feed the solution continuously through a hole in the upper carbon electrode [650] with the aid of a fulgurator [20, 60], or by means of the lower carbon electrode which becomes wetted by rotation in the solution [554, 830].

Determination of aluminum in iron, cast iron, steel and ferroalloys

If high sensitivity is not required, aluminum in iron can be determined by the spectroscopic method directly with the metal sample.

To determine aluminum in reduced iron, 7–10 g briquets, 20 mm in diameter and 5–7 mm in height, are prepared from the samples. The spectra are excited for 30 sec in a 5 A, AC arc, without preliminary arcing; capacity 4 μF, copper counter-electrode. The line pair Al 3,092.71 – Fe 3,098.2 Å is employed in the determination [341].

To obtain more uniform samples during the analysis of iron sponge, 0.2g of the sponge is dissolved in $\sim$8 N HNO$_3$, 1–2 ml of sulfuric acid are added, the solution is evaporated to dryness, and the residue ignited at a temperature not exceeding 900°C. The residue is ground, and 5 mg of the finely ground powder are placed in the crater of the anode and mixed with 5 mg of graphite powder. The mixture is covered by a thin layer of graphite and is quantitatively evaporated. The excitation source is a constant temperature arc. The line Al 3,082.16 Å is employed. The mean-square-error of the method is 10–11% [950a].

During the determination of aluminum in iron by the method of the globular arc, the sample is dissolved in HNO$_3$, the hydroxides are precipitated by ammonia and dried, and ignited to the oxides in a muffle furnace for 20 min at 700°C. Briquets are pressed from the ignited sample. The excitation source is a DC arc, and the sample briquet is the cathode. The slit width of the spectrograph is 0.015 mm, the exposure time is 40 sec; current intensity 7 A. The line pair Al 3,082.16 – Fe 3,029.24 Å is employed. The sensitivity of the method is $1 \cdot 10^{-3}$–$5 \cdot 10^{-3}$% [228a].

The sensitivity of the determination of aluminum in iron can be increased by means of preliminary concentration.

A 2g sample of iron or steel is dissolved in acid, and the iron is removed by extraction with methyl isobutyl ketone. The aqueous solution is evaporated to dryness and the residue is dissolved in 10 ml of 6 N HCl containing 10 mg K in 1 ml. The solution is introduced into the crater of a porous electrode with the following dimensions: depth 20 mm, diameter 3 mm, bottom thickness 0.6 mm. The lower electrode, ground to the shape of a truncated cone, is placed 2 mm from the top electrode. The spectrograph slit width is adjusted to 0.010 mm. The spectra are excited for 40 sec after preliminary sparking for 40 sec, with the aid of a unipolar low-voltage spark discharge with adjustable parameters. The highest intensity of the aluminum line corresponds to C = 50 μF, resistance of 2 ohms and self-induction of 30 mhenry, when the porous electrode is connected as cathode. The line pair Al 3,961.53 – K 4,044.14 Å is employed. The relative error of the method is 8% [1285].

A condensed or high-frequency spark has been recommended as the excitation source [565] for the determination of aluminum in alloyed cast iron.

If a high-frequency spark is employed as excitation source, the following procedure is recommended.

Voltage in secondary circuit of transformer 3,000 V, feed current to transformer 0.8 A, discharge gap width 0.9–1 mm. Capacitance of discharge circuit 0.01 μF, capacitance of shunt capacitor 120 pF, coil inductance 0.01 mhenry, analytical gap 1.8 mm, width of spectrograph slit 0.015 mm. An electrolytic copper rod, 5–6 mm in diameter, ground to a cylinder 1.6 mm in diameter in its working part, serves as the supporting electrode. The spectra are taken without a condensing lens; the distance between the spark and the spectrograph slit is 100 mm. Preliminary sparking takes

35–40 sec and the exposure is 25–30 sec. Type 1 spectroscopic grade photographic plates are used; the analytical lines are Al 3,082.15 – Fe 3,083.74 Å. The working range is 0.04–2.0% Al; the relative error is not more than 4.5% [212a].

A study [388] of the different methods for the determination of small amounts of aluminum in steel showed that the most reproducible results are obtained if rods made of the metal being analyzed are used as electrodes, or if the original samples serves as one electrode, and the counter-electrode is made of carbon. The spectra are excited in an AC arc, and a medium dispersion quartz spectrograph is used. The exposure time is 20–40 sec, the preliminary arcing takes 10–20 sec, the current intensity is 3–6 A. The line pair Al 3,082.16 – Fe 3,116.63 Å is employed. The relative error is 3–6%.

Aluminum can be determined in simple carbon steels and low-alloy steels by using any one of the conventional excitation sources. If the determination is carried out in an AC arc, the following experimental conditions are recommended.

Voltage 220 V, arc current intensity 5–6 A, width of spark gap 0.7–0.8 mm, analytical gap 2 mm. The permanent electrode consists of a copper rod, 7–10 mm in diameter, ground to a truncated cone, with a working area 1.5 mm in diameter. Quartz spectrograph, medium dispersion, width of spectrograph slit 0.015–0.02 mm. The spectra are taken with the aid of a three-lens condenser system or without a condensing lens; the distance between the arc and slit is 180–200 mm. Preliminary arcing takes 30 sec. The time of exposure depends on the sensitivity of the photographic plate (type I spectroscopic grade plates). The line pair Al 3,082.16 – Fe 3,083.74 Å is employed. The range of aluminum concentrations is 0.02–1.5%. The relative error is 3.5–5% [212a].

A similar method is described in [1219].

The following method [178a] can be recommended for the determination of aluminum in high-alloy steels, with a condensed spark.

Voltage 220 V, I = 3–3.5 A, C = 0.005–0.01 μF, induction coil switched off. Analytical gap 2.5 mm; permanent electrode consists of an electrolytic copper rod, 3–4 mm in diameter. Quartz spectrograph, medium dispersion, spectrograph slit width 0.025 mm. Preliminary sparking for 60 sec; exposure time varies with the sensitivity of the plates (spectroscopic grade, type I). The line pair Al 3,082.16 – Fe 3,083.74 Å is employed. The relative error is 3.5%.

Chrome-nickel austenite steel is analyzed in an AC arc [54a].

Current intensity in the arc 5 A, analytical gap 2 mm. The permanent electrode consists of a copper rod, 7 mm in diameter, ground to a truncated cone with a working surface 2 mm in diameter. The spectra are taken in a medium dispersion quartz spectrograph, illuminated through a cylindrical condenser, with a nonsharp image of the source in the slit. The width of the spectrograph slit is 0.020 mm. Preliminary arcing for 20–30 sec, exposure time 30–40 sec. Diapositive or spectroscopic grade type I plates are employed. Line pairs Al 3,082.16 – Fe 3,055.27 Å and Al 3,082.16 – Fe 3,116.64 Å are employed. Range of aluminum concentrations 0.015–1.5%; relative error 3.3%.

Kudelya [211] determined aluminum in austenite steels and in welding seams by using a high-frequency spark.

AC arc, condensed spark and high-frequency spark can all be employed as excitation source for the determination of aluminum in magnetic alloys. The following experimental conditions are recommended when an AC arc is employed [99a, 387].

Current intensity in the arc 5 A, slit width of spectrograph 0.025–0.030 mm, analytical gap 1.8–2.0 mm; preliminary arcing for 10 sec, exposure time 20–30 sec. The constant electrode consists of a carbon rod 5 mm in diameter, ground down to a hemisphere. The line pair Al 2,652.48 – Fe 2,656.79 Å is employed. Range of aluminum concentrations 5.0–15.0% Al is suitable. Relative error 4–8%.

Kudela [212] determined aluminum in magnetic alloys by using a high-frequency spark.

A medium dispersion spectrograph is employed. The experimental conditions are as follows: feed current from generator 0.8 A, voltage in secondary circuit of transformer 4,000 V; capacitance oscillation circuit 0.01 μF, capacitance of capacitor of shunt of analytical gap 100 pF, analytical gap 1 mm. Permanent electrode consists of a magnesium rod, ground to a cylinder 1.6–1.8 mm in diameter in its working part. Preliminary sparking 40 sec, exposure time 80 sec. To reduce the errors originating from the inhomogeneous structure of the alloy, spectra are taken at five different sites on the sample surface. The line pairs employed are Al 3,082.16 – Fe 3,047.61 Å or Al 3,082.16 – Mg 3,091.08 Å. Aluminum concentration range is 4.0–16.0% and the relative error is 4.45%.

A number of workers have proposed that aluminum be determined in steels after solubilization of the sample. By these methods total aluminum and acid-soluble aluminum can be separately determined. The total content of aluminum is determined directly on the sample in the lump form. The acid-soluble aluminum is determined by solubilizing the sample in an acid, and then the spectroscopic analysis is carried out on the solution obtained.

According to Levitin and Smirnova [225], total aluminum is determined by exciting the spectra in an AC arc at 5 A, using a copper electrode ground to a cone. Medium dispersion spectrograph is employed; width of spectrograph slit 0.015 mm, analytical gap 1 mm. Preliminary arcing time 10 sec, exposure time 20 sec. Analytical line pair Al 3,082.16 – Fe 3,055.26 Å is employed. To determine the acid-soluble aluminum, 1 g of the sample is dissolved in 10 ml 1 : 1 HCl, the Al_2O_3 residue is filtered off, and the solution is treated with concentrated nitric acid to oxidize the iron. Synthetic etalons are prepared by dissolving Armco iron and adding $AlCl_3$. Both carbon electrodes are impregnated with the sample solution and dried for 30 min at 400°C, and the spectra are excited at 5 A with an arc gap of 1 mm. The ends of the electrodes are ground to hemispheres. The relative error is about 9%.

According to another method [650], the sample solution is continuously fed into the excitation site through a channel in the top electrode, which is connected to a reservoir fitted with a stopcock. Five ml of the sample solution are placed in the reservoir, and the stopcock is adjusted to give a flow rate of 10—20 drops of solution per minute. The duration of the exposure is 80 sec. The excitation source used was a Feussner type high-voltage condensed spark. The lines employed were Al 3,092.71 — Fe 3,116.6 Å for aluminum contents below 0.001% and Al 3,082.16 — Fe 3,116.6 Å for aluminum contents between 0.03 and 0.08%.

This technique of feeding the sample solution into the excitation source makes it possible to determine small amounts of aluminum in steel. A number of workers [139, 668, 861] have recommended that a few drops of the sample solution be placed on the graphite electrode, but with this technique the sensitivity at low aluminum concentrations is insufficient.

Kolbovskii and Krizhanovskaya [175] conducted analyses of steels in solution in the presence of added sodium chloride, which in their view increases the slope of the curve.

In ferroalloys the distribution of the components is typically inhomogeneous; the errors which are caused are minimized by solubilizing the sample, or else the sample is ground, mixed with copper or copper oxide powder, and briquetted.

Aluminum in ferrosilicon can be determined by the method of Veselovskaya [79a].

Ferrosilicon is ground in an iron mortar to a powder passing a 200-mesh sieve, and is thoroughly mixed with copper powder in the ratio of 3 : 7. One gram of this mixture is pressed into briquets 7 mm in diameter. The light source is an IG-2 generator, supplying a 3 A current; capacitance of capacitor 0.01 μF, coil inductance 10 μhenry, discharge gap width 3.7 mm, analytical gap width 2.7 mm. The permanent electrode is a carbon rod 5 mm in diameter, ground to a truncated cone with a working surface 1 mm in diameter. The width of the spectrograph slit is 0.025 mm; duration of preliminary sparking is 60 sec, duration of exposure 30 sec. The photographic plates are spectroscopic grade type I or diapositive, with a sensitivity of 0.5 GOST unit. The line pair employed is Al 3,082.16 — Cu 3,108.60 Å. The suitable range of aluminum concentrations is 1.50—5.0% Al. The relative error is 2.9%.

According to [543], 0.5 g of ferrosilicon are mixed with 0.5 g of graphite powder and 0.5 g CuO. The mixture is pressed into tablets 10 mm in diameter and 10 mm in height. The spectra are excited in a condensed spark produced by a Feussner generator (capacitance 1,650 pF, self-inductance 0.8 mhenry) with a top electrode made of carbon and a 2 mm spark gap. A medium dispersion spectrograph, with a 0.005 mm slit width is employed, using the line pair Al 3,082.16–Cu 3,063.41 Å.

When determining aluminum in ferrotitanium, the ground sample is placed in the form of a strip onto a nickel plate which serves as the bottom electrode, and is moved at a uniform rate during the exposure [71b]. An AC arc serves as the excitation source. The line pair used is Al 2,652.48– Fe 2,679.07 Å; The relative error is 1.35%.

Ferroniobium is analyzed in a similar manner, by placing a mixture of the powdered sample with Fe_2O_3 and graphite onto a slowly rotating (10 rpm) copper or nickel electrode. The lines employed are Al 2,660.39– Fe 2,635.81 Å. The suitable range of concentrations is 0.50–1.2% Al; the relative error is 4–5%.

Determination in nickel and nickel alloys

Nickel can be determined directly in the sample using a low-power spark produced by a DG-1 or DG-2 generator by switching off the arc feed [346a].

Current intensity in primary circuit 0.5 A, spark gap 1.0 mm. Medium dispersion spectrograph, slit width 0.020 mm. The lines employed are Al 3,092.71–Ni 3,097.12 Å.

To obtain more accurate results the nickel is first dissolved in nitric acid, the resulting solution is evaporated, and the residue is ignited to the oxides in a muffle furnace. The oxide (0.2 g) is pressed into briquets 5 mm in diameter. The briquets are placed on cylindrical graphite supports, 10–12 mm in diameter. Spectroscopic-grade carbon serves as the counter-electrode. The excitation source is an activated DC source; a medium dispersion quartz spectrograph, slit width 0.012 mm, is used. The analytical line pair is Al 3,092.71–Ni 3,066.44 Å.

The nickel sample can be solubilized in acid and analyzed [603, 604, 844].

The sample is dissolved in nitric acid and the solution diluted to a concentration of 100 mg Ni in 1 ml of solution. Then 0.025 ml of the solution is placed on the end of a carbon electrode 4.5 mm in diameter, which has previously been impregnated with a 3% solution of polystyrene in benzene and dried in an oven for 3 hr. A necked electrode is employed to facilitate the evaporation of the low-volatile elements. The spectra are excited in a 4.5 A AC arc; the duration of preliminary arcing is 15 sec, duration of exposure 30 sec. The line pair employed is Al 3,092.71–Ni 3,131.8 Å. The standards, which should contain 0.005–0.1% Al, are prepared by dissolving 1 g of 99.95% pure Ni in 5 ml HNO_3 and introducing known amounts of aluminum.

Nichrome is analyzed in AC arc [387].

Arc intensity 3–4 A, width of spectrograph slit 0.02 mm, analytical gap 1.5 mm. The permanent electrode consists of a nickel rod, 8–9 mm in diameter, ground to a hemisphere. Preliminary arcing takes 10 sec. The line pair Al 3,082.16 – Ni 3,097.12 Å is employed. Working range of Al concentrations is 0.1–2.0% Al; the relative error is 3–4%.

According to Kudelya [212a], the accuracy of the determination of aluminum in nichromes can be enhanced by using a high-frequency spark.

Determination in cobalt

Cobalt of low grade purity is determined in cast samples.

The samples are chill-cast into rods, 7–8 mm in diameter and 40–50 mm in length. A medium dispersion spectrograph with a 0.010 mm slit width is used. The spectra are excited in an AC arc, and the counter-electrode is a carbon electrode ground to a cone. The duration of preliminary arcing is 40 sec. Spectroscopic grade type II plates. Line pair Al 3,092.71 – Co 2,975.46 Å; suitable concentration range 0.01–0.15% Al.

If more accurate results are desired (cf. determination in nickel), the sample is dissolved in acid, the salts are dried and are then ignited to oxides.

Determination in chromium

To determine aluminum in chromium [159], a medium dispersion spectrograph is employed; the spectra are excited in AC arc at 6 A; preset gap 0.8 mm, arc gap 3 mm. The counter-electrode consists of a copper rod 6–8 mm in diameter, ground to a truncated cone, 2 mm face diameter. Preliminary arcing 10 sec, exposure time 15 sec. Type I photographic plates; line pair Al 3,082.16 – Cr 3,077.83 Å. Suitable working range 0.11–1.0% Al; mean-square error 3%.

Chwastowska [640] determined aluminum and certain other elements in high-purity chromium, and suggested that the impurities be preliminarily concentrated by separation in the form of N-benzoylphenylhydroxylamine complexes. The concentrate is mixed with graphite powder, with AgCl added as carrier

The mixture to be analyzed is placed in the crater of a carbon electrode (anode), 4 mm in diameter and 4 mm in depth. A graphite electrode with a sharp conical end serves as the counter-electrode. The analytical gap is 2 mm, the sample weight is 35 mg. A medium dispersion spectrograph is used in conjunction with a DC arc; exposure time is 20 sec. The standard error is about 12%.

Determination in manganese

To determine 0.003–0.1% Al in manganese [113], the sample is dissolved in HNO_3, the solution is evaporated, and the residue is ignited at 200°C. The powder thus obtained is placed in the crater of a carbon electrode 2 mm in diameter and 6 mm deep, and is volatilized in a DC arc at 13 A, with an arc gap of 4.5 mm and an exposure time of 90 sec. A medium dispersion spectrograph is employed with a slit width of 0.01 mm; the sensitivity of the photographic plates is 22 GOST units. Standards are prepared from a solution of $Mn(NO_3)_2$. The line used is Al 3,092.71 Å. The relative error of the method is 20%.

Determination in copper alloys

Determination in bronzes. Aluminum in tin bronzes can be spectroscopically determined as follows [48a].

A medium dispersion spectrograph is employed. The excitation source is a condensed spark in a simple circuit (without interruptor). The feed current intensity from the generator is 1.5–2 A, the voltage in the secondary transformer circuit is 12,000 V; the capacitance of the capacitor is 0.01 μF, the coil inductance is 0.1 mhenry. The analytical gap is 3 mm, the width of the spectrograph slit is 0.025 mm. The permanent electrode is a carbon or a graphite rod, ground to a truncated cone with a working area 2–2.5 mm in diameter. Preliminary sparking lasts for 120 sec; spectroscopic grade type I or diapositive plates are employed. The line pair is Al 3,082.16 – Cu 3,073.90 Å; range of aluminum concentrations is 0.01–0.2% Al.

When small (0.005–0.10%) amounts of aluminum are determined, an AC arc is preferably employed. For the special features of the analytical procedures applicable to various brands of tin bronzes, see [212a].

A condensed spark is commonly employed in the analysis of aluminum bronzes [113a, 354, 511]. The following procedure can be recommended [113a, 354].

A medium dispersion spectrograph with a slit width of 0.025 mm is employed; voltage 220 V, feed current from generator 2 A, voltage in secondary transformer circuit 13,000 V; capacitance of capacitor 0.01 μF, coil inductance 0.01–0.05 mhenry, gap in preset discharger 2.5 mm, analytical gap 1.8–2 mm. The permanent electrode is a 5–6 mm carbon rod, ground to a truncated cone with a working area 1.5–2 mm in diameter. Duration of preliminary sparking 60–80 sec. Line pairs Al 3,082.16 – Cu 2,961.16 Å, Al 3,082.16 – Cu 2,824.37 Å, Al 2,816.17 – Cu 2,824.37 Å.

Beryllium bronzes are analyzed by dissolving the sample in HNO_3, evaporating the solution and igniting the residue at 500°C to the oxides. High-purity copper is dissolved together with the sample; thus, ignition yields a mixture of copper oxides with the components of the sample. The oxides are carefully ground in an agate mortar and pressed into briquets 6 mm in diameter and 1 ± 0.05 g in weight. Standards are prepared in the same way from pure metallic Cu, Ni, Fe, Al, Pb and Be.

The excitation source is a 400 V, 5.8 A DC arc, in conjunction with a high-powered high-frequency discharge. Briquets are placed on a carbon cathode, 10 mm in diameter. The electrode has a small cavity for a drop of the oxide melt. The permanent electrode consists of a copper rod 6–8 mm in diameter, ground to a truncated cone with a blunt end. A medium dispersion spectrograph with a 0.02–0.025 mm slit width is employed; the exposure time is 60 sec. For a detailed description of the procedure see [212a]. The line pair Al 3,082.16 – Cu 3,088.13 Å is employed; range of aluminum concentrations 0.002–0.40% Al, mean-square error 5%.

Beryllium bronzes can also be analyzed in the form of solutions [107a].

The solution is fed into the discharge zone by means of a fulgurator. The spectra are excited by an IG-2 or IG-3 generator that forms part of a compound circuit ($C = 0.01 \mu F$, $L = 0.15$ mhenry). The counter-electrode consists of a graphite rod ground to a truncated cone. The line pair used is Al 3,092.71 – Cu 3,093.99 Å; the mean-square error is 7–12%.

Spectroscopic analysis in solution [917] has been proposed for the determination of 0.001–0.1% Al in bronzes, containing not less than 80% Cu and not more than 20% Sn, Zn and Pb.

The solution obtained by dissolving the sample in HF and HNO_3 is applied onto a carbon electrode connected as the anode; the cathode is also a carbon electrode. The spectrum is excited in the spark.

Determination in brasses. Brasses are analyzed with an AC arc as the excitation source.

The determination is carried out under the following conditions: width of spectrograph slit 0.015–0.020 mm, current intensity in arc 6–8 A; the counter-electrode consists of a copper rod 8–10 mm in diameter, ground to a hemisphere. The analytical gap is 2 mm. Preliminary arcing 15 sec; exposure time 20 sec. The line pair used is Al 3,082.16 – Cu 2,882.93 Å [483a].

Brasses can be successfully analyzed in an AC arc [331b]. Between 0.002 and 0.20% Al can be determined in this way; the line pair used is Al 3,082.16 – Cu 3,088.13 Å.

Determination in magnesium alloys

Aluminum can be determined in magnesium alloys with a condensed spark and AC arc as excitation sources. The permanent electrode consists of a high-purity magnesium or spectroscopic grade carbon rod. Twin electrodes made of the sample to be analyzed are also frequently employed.

The following experimental conditions have been recommended for the analysis of MA 4, MA 5, ML 4, and ML 5 alloys, using a spark excitation source [387].

Medium dispersion spectrograph, IG generator connected in simple circuit. Generator feed current intensity 2 A, capacitance of capacitor 0.003–0.005 μF; secondary circuit voltage 12,000 V, slit width 0.025–0.30 mm, analytical gap 2 mm. The permanent electrode consists of a magnesium rod 9 mm in diameter, ground to a hemisphere. Preliminary sparking 30 sec, spectroscopic type I or diapositive plates. Line pairs Al 3,587.06 – Mg 3,329.93 Å or Al 3,082.16 – Mg 3,073.99 Å. Range of working concentrations 3–12% Al, relative error 5–6%.

Very similar experimental conditions have also been employed elsewhere [711].

For the determination of aluminum in an AC arc, a medium dispersion spectrograph with a PS-39, DG-1 or DG-2 generator is employed; arc intensity 2 A, spectrograph slit width 0.03–0.035 mm, analytical gap 2 mm, preliminary arcing 5–10 sec, exposure time 30 sec. Spectroscopic grade type I plates. Line pairs Al 3,082.16 – Mg 2,915.52 Å or Al 3,944.03 – Mg 3,329.93 Å. Working range of aluminum concentrations 2–13%, relative error 8.5% [212a].

Determination in titanium and titanium alloys

Metallic titanium and titanium alloys are usually analyzed with spark excitation sources. Magnesium or carbon electrodes are the most suitable permanent electrodes.

Below is a description of the analysis of a titanium alloy in condensed spark [212a].

An IG-2 generator, connected in a simple circuit, voltage 220 V, current intensity 3 A, capacitance of capacitor 0.02 μF, analytical gap 0.8 mm, width of spectrograph slit 0.03 mm. Preliminary sparking for 10 sec, exposure for 30–40 sec. The permanent electrode consists of a magnesium rod, ground to a cylinder 1.8 mm in diameter. The line pair employed is Al 3,944.03 – Ti 3,987.61 Å; the working range of aluminum concentrations is 0.8–6.5% Al; the maximum deviation of an individual determination from the average aluminum content is 3%.

In a method proposed by Runge and Bryan [1127], the spectra are excited in a spark discharge between a flat-shaped sample of the titanium alloy and a graphite counterelectrode. The working parameters are: capacitance 0.005 μF, inductance 0, resistance 0.5 ohm, current intensity in hf circuit 12 A, analytical gap 2 mm. Duration of preliminary sparking 30 sec; line pair Al 3,044.03 – Ti 3,904.78 Å. Argon is passed through the analytical gap in a horizontal direction to prevent the oxidation of the sample surface in the air. For aluminum contents of 6.6%, the mean-square error is 0.9%.

A high-frequency spark can also be employed in the analysis of titanium alloys [212a].

If aluminum is determined in a powdered titanium sample, one-gram briquets 8 mm in diameter and 5 mm in height are prepared. The spectra are excited in an AC arc (4 A); preliminary arcing 40 sec, exposure time 40 sec. The line pair used is Al 3,092.71 – Ti 3,046.69 Å [514a].

High-purity titanium is previously converted to the dioxide by ignition at 1,000– 1,100°C in a platinum crucible, or else the metal is dissolved in HCl (sp. gr. 1.19), the solution is evaporated with HNO_3 to convert the chlorides to nitrates, and the residue is ignited at 600°C for 30 min. The analysis is carried out in a DC arc (10 A). Twenty mg of TiO_2 are thoroughly mixed with 20 mg of a 4 : 1 mixture of carbon powder with NiO. The mixture is placed in the crater of a carbon electrode connected as the anode; the crater should be 4 mm in diameter and 5 mm deep. The conventional length of the anode is 16 mm, 10 mm of which are ground to a diameter of 4 mm. A carbon rod, 6 mm in diameter and ground to a cone, is used as the permanent electrode. A medium-dispersion quartz spectrograph is employed; the line pair used is Al 3,082.16 – Ni 3,080.76 Å [253a].

Titanium can also be analyzed by the pouring method [310a]. Titanium dioxide powder is poured across an 18 A AC arc with the aid of a special device. The spectrograph employed must have a dispersion of about 4 A/mm (DFS).

Titanium can also be analyzed in the form of a solution in an 4–6 A DC arc [212a].

The bottom electrode (anode) in the form of a graphite disk about 20 mm in diameter and 3 mm thick is dipped into the sample solution of titanium in HCl and rotated at the rate of 6 rpm. A graphite rod is used as the permanent electrode.

Determination in zinc and zinc alloys

Aluminum in zinc can be determined in the arc or in the spark. If the analysis is carried out in the spark, the compound circuit of an IG-2 or IG-3 generator is employed; capacitance of secondary circuit 0.01–0.02 μF, inductance 0.15–0.55 mhenry. Both electrodes are made of the metal sample. Duration of preliminary sparking 10 sec, duration of exposure 60 sec; type I photographic plates. The line pair employed is Al 3,092.71– Zn 3,075.90 Å, both in the arc and in the spark.

Aluminum in TsAM-4-3 alloy is determined by the spectroscopic method, using a medium-dispersion quartz spectrograph with a three-lens system for slit illumination; slit width 0.025 mm, excitation source IG-2 (or IG-3) generator connected to a complex circuit ($C = 0.01 \mu$F, $L = 0.01$ mhenry, one train in half-period of feed current). The counter-electrode is made of carbon, ground to a truncated cone, or a copper rod which has been ground flat. The interelectrode gap is 2.0 mm; preliminary sparking takes 60 sec. The line pair Al 2,567.99 – Zn 2,525.81 Å is employed.

Zinc-plating baths containing 0.5% Al and 0.5% Pb can be analyzed as follows for control purposes [1075].

A flat-shaped zinc sample serves as the cathode, while a graphite support electrode serves as the anode. The arc spectra are excited by a generator (3 A; 55 μF, 360 mhenry) and a 50 ohm quenching resistance. Duration of preliminary sparking 20 sec; duration of exposure 10 sec. The spark gap width is 3 mm. The aluminum lines employed in the determination are 3,961.53 Å for aluminum contents of 0.015–0.06% and 3,082.16 Å for aluminum contents of 0.06%–0.30%. A step reducer is mounted in the slit. The analytical error is 2.7%.

Determination in uranium and uranium alloys

Metallic uranium is first converted to the oxide U_3O_8 and analyzed by fractional distillation using Ga_2O_3 [152] or $BaCO_3$ [79] as carriers.

In the former case U_3O_8 and Ga_2O_3 (1 : 49) are placed in a deep channel of a graphite electrode and the spectrum is excited in a 10 A DC arc. Exposure time 30 sec; up to 0.002% Al can be determined in this manner. The line pair employed is Al 3,082.16 – Cr 3,040.8 Å; the relative error is 5–10%.

If the last technique is employed, 10% $BaCO_3$ and 5% of carbon powder [79] are mixed with the U_3O_8 sample, 25 mg of the resulting mixture are placed in a carbon electrode crater 3.5 mm in diameter and 6 mm deep. The diameter of the electrode tapers off to 2.5 mm at a distance of 5 mm from the working end, over a distance of 4 mm. The spectra are excited in an 18 A AC arc, with an upper carbon electrode, which has been ground to a truncated cone with a working surface 2 mm in diameter, and the spectrum is photographed for 30 sec using an ISP-51 spectrograph with a UF-85A camera. The slit width is 0.015 mm, and the slit is illuminated with the aid of a one-lens condenser. The line pair employed is Al 3,944.03 – Ba 3,995.6 Å. The sensitivity of the method is 10^{-3}%; if the spectra are photographed in triplicate, the mean-square error is 9–11%. The $BaCO_3$ carrier has no effect on the course of the evaporation, but enhances the excitation of the aluminum atoms; the carbon powder attenuates the spectrum of uranium and intensifies the spectrum of aluminum.

To determine aluminum in uranium alloys, uranium is first separated by ion-exchange chromatography [1092].

After dissolution, the sample* is passed through a column 1.5 cm in diameter and 33 cm high, containing 7 g of Deacidite FF (60–120 mesh), at the rate of 1 ml per minute.

* The publication describes the methods for the solubilization of uranium alloyed with Zn, with Mo and Si, with Mo, Si and Nb, and with Fe, V, Ge and Cr.

The resin is previously washed with 150 ml of 1.86 : 1 HCl. The aluminum is eluted with 90 ml of this acid; to the eluate 1 ml of a solution containing 20 mg Fe_2O_3 and 0.5 mm Sc in 1 ml is added. The solution is made alkaline to methyl red with ammonia and is then boiled for 3–5 min. The precipitate is separated by centrifugation, washed with ammonium chloride solution, and dissolved in 0.5 ml HCl (sp.gr. 1.19). The resulting solution is diluted to 5 ml.

The spectra are excited in a spark discharge ($L = 0.13$ mhenry, $C = 0.005 \mu F$); auxiliary gap 1.8 mm, analytical gap 1.5 mm. The solution is fed into the discharge gap by means of a graphite disk rotating at 4 rpm, 12.8 mm in diameter and 3.2 mm thick, which dips into the solution to a depth of 1 mm. The width of the spectrograph slit is 0.015 mm; the exposure time is 2 min. If the aluminum content is higher than 0.02%, the sample is sparked for 30 sec and the exposure time is reduced to 30 sec. The line pairs used are Al 3,944.03 – Sc 3,907.48 Å and Al 3,961.53 – Sc 3,911.81 Å.

Between 0.005% and 0.02% Al can thus be determined in solutions containing not more than 10% Fe, Mo, Nb and Ti, 5% Cr, 1% Ge, 0.6% Si, and 0.05% V. If the aluminum content is 0.01%, the square error of a single determination is 7%.

Determination of aluminum in other metals and alloys

Aluminum in tin can be determined by the spectroscopic method in the spark [483a]. To determine aluminum in cadmium, the sample is dissolved in HNO_3 (sp. gr. 1.4), the solution is evaporated, the salts are ignited to the oxides, and the spectrum is excited in a DC arc with promoter ignition [483a]. For determining aluminum in crystalline silicon, the sample is ground in an agate mortar to a fine powder, which is then mixed with pure CuO in the ratio 1 : 20, and 0.25 g samples of the mixture are briquetted. For details the reader is referred to [483a].

During the spectroscopic determination of aluminum and other impurities, metallic zirconium and zirconium compounds are converted to ZrO_2 [313].

The spectra are excited in a DC arc (10 A) between carbon electrodes. Zirconia samples, which have been mixed with graphite in the ratio of 1 : 1, are vaporized from an electrode channel, 3 mm deep and 3 mm in diameter. The spectrograph slit width is 0.015 mm; the exposure time is 20 sec. The line pair employed is Al 2,660.39 – Co 3,147.06 Å (with cobalt as internal standard).

Metallic calcium is also first converted to the oxide [1138].

To determine aluminum in gold alloys, see [830]; for its determination in platinum metals, see [20, 60].

Spectroscopic determination of aluminum in other materials

In the analysis of other materials the samples are usually introduced into the discharge zone in the form of powder or briquets. In the first case the powdered sample is introduced into the crater of a carbon electrode, which should be 3−4 mm in diameter and 4−6 mm deep. To improve the heatup in the carbon electrode, a constriction is made below the bottom of the channel. The powdered sample is usually mixed with the internal standard and with carbon. The introduction of carriers may improve the vaporization of aluminum in some cases. Thus, in the analysis of Nb_2O_5 the sample is mixed with carrier AgCl to ensure that aluminum is preferentially vaporized at the first moment the AC arc is struck [279].

Both electrodes may be made of carbon, or else the top electrode can be of copper. It has been suggested [233] that the samples be poured onto a moving copper electrode, and in this case it is best to coat the samples with a bakelite lacquer [346]. It has been proposed [353] that aluminum be determined by the powder-pouring method, when a dust-collecting fan is employed to produce a vertical air stream in the arc zone. If the sample is mixed with $NaPO_3−Na_2B_4O_7$ [1116], $Li_2CO_3−H_3BO_3$ [774, 935] and $Na_2CO_3−Na_2B_4O_7$ [304] fluxes, the error due to the inhomogeneity of the sample is reduced.

The preparation of briquets from the powdered sample is widely applied. The briquets are prepared by pressing mixtures of the sample with powdered SiO_2−carbon or nickel−carbon mixtures, or with powdered copper or powdered carbon. Briquetting ensures more uniform vaporization of the sample.

Aluminum can be spectroscopically determined in solutions as well. The analysis of solutions is more time-consuming than that of powders and briquets, and special devices are required for feeding the solution into the discharge zone. This technique should be employed if other techniques give poor reproducibility owing to the inhomogeneity of the materials. The samples are solubilized by fusing with various fluxes. Slags are solubilized by fusing with sodium hydroxide in a silver dish, and the flux is leached with water containing some HCl [748]. Slags can also be fused with NaOH and Na_2O_2 [748], or with a 2 : 3 mixture of Na_2CO_3 with $Na_2B_4O_7$ [134]. Silicates are dissolved in HF and HNO_3 and then in HCl [297, 298], or are fused with NaOH in a nickel crucible; the melt is leached with water containing HCl [803].

The solution can be introduced into the discharge zone by the following techniques: introduction into the crater of a carbon electrode [47], application onto the tip of a carbon electrode impregnated with polystyrene [410], introduction through a channel which is coaxial with the lower electrode [748], introduction with the aid of a fulgurator [134], with the aid of a rotating graphite disk [297, 298], or as an aerosol with the aid of an atomizer. Another technique is to introduce an ashless filter paper impregnated with the sample solution into the discharge zone [100].

Table 14 lists the accuracy and sensitivity parameters of spectroscopic methods for the determination of aluminum in certain nonmetallic materials.

Table 14. Analysis of different materials by the spectroscopic method

Substance	Al concentration range	Relative error, %	References
Silicates	$5 \cdot 10^{-1}$–4.0	2–3	[197]
Quartz sand	$1 \cdot 10^{-3}$–1.0	2	[799a]
Glass powder	$1 \cdot 10^{-3}$–0.5		[789]
Iron ore	$>5 \cdot 10^{-2}$	8–10	[179]
Martensite slags		3–5	[134]
” ”	2–25		[748]
Blast furnace slag		2.6–6.4	[303a]
” ”		1–3	[304]
Agglomerates		3.7–16.7	[303a]
”		1–3	[304]
Limestone	$1 \cdot 10^{-1}$–1	7.8	[778a]
Magnesite			[517a]
Glass	1.3–8	2.3	[923a]
”	$1 \cdot 10^{-1}$–1.0	1.7	[796a]
Zirconium concentrate	$1 \cdot 10^{-2}$–$1 \cdot 10^{-1}$	5	[176]
Tungstic anhydride	$3 \cdot 10^{-3}$–$1 \cdot 10^{-1}$	10	[973]
Niobium pentoxide	$5 \cdot 10^{-3}$–1.10^{-1}	10–12	[279]
Chromic oxide			[124]
Cement			[233]
Industrial nickel slags	$5 \cdot 10^{-1}$–8.0		[489a]
Zinc-plating and nickel-plating electrolytes		6–7	[100]
Binary mixture Nb-Al		7	[353]
Titanium tetrachloride	$4 \cdot 10^{-3}$–$2.5 \cdot 10^{-1}$	2–20	[48]
Commercial-grade boron	$1 \cdot 10^{-2}$–$2 \cdot 10^{-1}$		[122]
Tellurium		5.4–12.1	[510a]
High-purity silicon	$>0.6 \cdot 10^{-5}$	7–10	[489]
Mica		3–9	[333a]
PCl_3 and PCl_5	$>10^{-5}$	30	[207a]
$SiCl_4$ and $SiHCl_3$	10^{-8}–10^{-3}		[1240a]
Polyethylene	$>5 \cdot 10^{-2}$	5.5	[454]
Polypropylene	$4 \cdot 10^{-3}$–$4 \cdot 10^{-1}$	8–15	[303]

Table 15. Concentration methods used in spectroscopic determinations of small amounts of aluminum

No.	Substance	Concentration technique	Sensitivity, %	Relative error, %
1	Si, SiO_2	Distillation of SiF_4	$6 \cdot 10^{-7}$	36
2	$SiCl_4$, $SiHCl_3$	Distillation of $SiCl_4$ and SiF_4	$6 \cdot 10^{-7}$	36
3	SiO_2	Distillation of SiF_4	$7 \cdot 10^{-7}$	25–30
4	$SiHCl_3$	Distillation with CCl_4	$1 \cdot 10^{-7}$	15–20
5	Ge, GeO_2, $GeCl_4$	Removal of Ge as $GeCl_4$	10^{-7}–$2 \cdot 10^{-8}$	13
6	Ga, $GaCl_3$	Extraction of Ga by ether	$1 \cdot 10^{-5}$	
7	Gallium arsenide	Distillation of As as $AsCl_3$, $AsBr_3$, extraction of Ga by ether	$1 \cdot 10^{-5}$	30–40
8	In	Extraction of indium bromide by ether	$5 \cdot 10^{-5}$	30
9	Indium antimonide	Extraction of In and Sb as bromides	$5 \cdot 10^{-5}$	30
10	Indium phosphide	Removal of P as PH_3, extraction of indium bromide by ether	$3 \cdot 10^{-5}$	15
11	Tl	Extraction of thallium chloride by ether	$5 \cdot 10^{-5}$	30
12	As	Distillation of As as chloride	$1 \cdot 10^{-5}$	
13	Sb	Extraction of Sb by butyl acetate	$3 \cdot 10^{-7}$	20
14	P	Extraction of aluminum hydroxyquinolate	$1 \cdot 10^{-6}$	15
15	Pb	Precipitation of $PbSO_4$	$1 \cdot 10^{-5}$	20
16	Bi	Precipitation of BiI_3 or of the basic nitrate	$1 \cdot 10^{-5}$	20
17	Sn	Distillation of $SnCl_4$	$3 \cdot 10^{-4}$	15
18	B, B_2O_3, H_3BO_3	Distillation of BF_3	$1 \cdot 10^{-5}$	
19	B	Removal of B as methyl borate	$2 \cdot 10^{-5}$	30
20	S	Combustion of sulfur	$5 \cdot 10^{-6}$	10–30
21	Se	Distillation after conversion to SeO_2	$1 \cdot 10^{-5}$	20–35
22	Iodine	Sublimation of iodine mixed with carbon powder	$1 \cdot 10^{-6}$	10–25
23	HF, HCl, HNO_3	Volatilization of acid with carbon powder	$4 \cdot 10^{-8}$	
24	HCl, HF, H_2SO_4, CH_3COOH	The same	$3 \cdot 10^{-8}$	15

* Excitation sources: Nos. 1, 2, 4–18, and 23 in DC arc; Nos. 3 and 24, gas discharge tube with a hollow cathode; Nos. 19–22, in AC arc.

Very small amounts of aluminum in metals and other materials are determined after preliminary concentration by chemical methods. Numerous concentration techniques are quoted in [254]; brief descriptions are given in Table 15.

Photoelectric methods for the determination of aluminum are promising [323, 387a].

Flame photometry

Aluminum can be determined by emission photometry in the flame with the aid of lines at 394.4 and 396.15 mμ, and the molecular band of aluminum at 484 mμ. Aluminum is difficult to excite in aqueous solutions, and for this reason the sensitivity of such direct determination methods is low.

An indirect method, based on the decrease in the absorption by calcium, has been proposed by a number of authors. This method is more sensitive than the direct method of determination in aqueous solutions; the radiation intensity given by calcium decreases almost linearly with increase in the amount of aluminum from 20 to 250 μg/ml [1002]. However, the intensity of the radiation emitted by calcium is also weakened by many other ions, such as NO_3^-, Fe(III), Ti(IV), Cr(III), and others.

Hegemann and Osterried [799] showed that the intensity of the molecular band of aluminum at 484 mμ increases considerably if large amounts (up to 20%) of ammonium chloride are introduced into the solution before its atomization; the introduction of HCl enhances the radiation intensity even further. The atomization is improved if up to 4% of acetic acid is added. According to these authors, the introduction of methanol also increases the radiation intensity. Butanol is introduced for the same purpose by others [797, 798, 893]. All these methods are nonselective; Na, K, Ca, Fe and other elements interfere, which means that their contents must first be determined and corresponding amounts introduced into the standard solutions.

Hydroxyquinoline has also been introduced [676, 677] to increase the sensitivity of the method, but this technique is also of little interest.

Methods in which aluminum is first converted into a complex with an organic compound, and a solution of this complex in an organic solvent is then introduced into the flame, are the most important. Aluminum can be introduced into the flame as solutions of the cupferronate in methyl isobutyl ketone [703, 729, 1073], of the 2-thenoyltrifluoroacetonate in methyl isobutyl ketone [703], of the acetylacetonate in chloroform [1149], or in methyl isobutyl ketone [729].

Eshelman et al. [703] found that the intensity of aluminum radiation at 396.2 and 484 mμ increases by a factor of 100 if the aluminum is introduced into an oxygen-acetylene flame in the form of a solution of the cupferronate or the 2-thenoyltrifluoroacetonate in methyl isobutyl ketone, rather than in the form of an aqueous solution of a salt. If aluminum is isolated in the form of complexes, the selectivity of the determination also increases considerably. Extraction as the thenoyltrifluoroacetonate is carried out at pH 5.5–6.0. Washing the extract with $0.1M$ HNO_3 eliminates small amounts of alkali and alkaline-earth metals, which might be partially extracted if present in high concentrations. The elements Zr, Ti, Th, U(VI), Fe(III), Ce(IV), and Cu are preliminarily separated as thenoyltrifluoroacetonates from solutions at pH 1. Up to 10 mg Zn, Ni, Fe(III) and Cu can be extracted as diethyldithiocarbamates by chloroform.

In many cases it is preferable to remove the interfering elements by N-nitrosophenylhydroxylamine, when aluminum is extracted over a wide pH range. The radiation intensity is only slightly weaker than it is when the thenoyltrifluoroacetonate is extracted. During the analysis of steel, most of the iron should first be removed by electrolysis on a mercury cathode, or by extraction as ferric chloride from $5-7M$ HCl by methyl isobutyl ketone. The small amounts of iron which remain in the solution after the extraction can conveniently be removed by extraction with N-nitrosophenylhydroxylamine from 1 : 9 HCl.

Aluminum is selectively extracted from magnesium alloys as the complex with N-nitrosophenylhydroxylamine from methyl isobutyl ketone at pH 2.4–4.5. Magnesium is not extracted, and the contents of the other elements are within the permissible limits.

The method has been employed for the determination of aluminum in steels, bronzes, magnesium and zinc alloys, and also in minerals [703].

Atomic absorption spectroscopy

In the method of atomic absorption spectroscopy, aluminum is vaporized in a graphite cell at 2,400°C [305–307], and in a flame: oxygen-acetylene flame [626, 686], oxyhydrogen flame [626, 686], and acetylene–N_2O flame [506a, 514b, 525, 615].

Flame methods (absorption photometry in the flame). The lines at 309.27 mμ (a practically unresolved doublet at 309.27 and 309.28 mμ) and at 396.2 mμ can be employed. The sensitivity found for the 309.27 mμ line, using C_2H_2–N_2O and (50% O_2 + 50% N_2)–C_2H_2 flames was $1 \cdot 10^{-4}\%$ [526, 1274]. The sensitivity of the 396.2 mμ line in an oxygen-acetylene flame is $6 \cdot 10^{-4}$–$8 \cdot 10^{-4}\%$ [686, 1174]. Ramakrishna et al. [1099] quoted the following sensitivity values for the various resonance lines of aluminum, obtained in a C_2H_2–N_2O flame:

λ, mμ	Sensitivity, ppm/1% absorption	λ, mμ	Sensitivity, ppm/1% absorption
309.27	1.2	237.3	3.5
396.1	1.0	257.5	7.0
398.2	1.8	265.2	8.5
394.0	2.2		

According to these data, the lines at 309.27 and 396.1 mμ have equal intensities.

Amos and Thomas [525] determined aluminum by means of an acetylene-(50% oxygen + 50% nitrogen) flame with a slit (0.45 × 30 mm) burner. The minimum detectable amount of aluminum by the 309.27 mμ line was 1.7 μg/ml. Up to 20 mg of Cu, Zn, Pb, Mg, Na, PO_4^{3-} and SO_4^{2-} per ml do not interfere. In the presence of 10–20 mg Fe(III) per ml and 2–5% HCl the results are too low, and the error must be compensated by introducing these substances into the standard solutions.

Ramakrishna et al. [1099] studied the effect of a large number of ions on the determination of aluminum by the 309.27 mμ line, using a C_2H_2–N_2O flame. Twenty mg of Al per liter can be determined in the presence of not more than 200 mg/liter of NH_4^+, Li, Na, K, Rb, Cs, Cu(II), Ag, Be, Mg, Ca, Sr, Ba, Zn, Cd, Hg(I), Hg(II), Zr, Ce(IV), Sn(II), Sn(IV), Pb, Sb(V), Cr(III), Bi, Fe(III), Co, Ni, Pd(II), F^-, Cl^-, Br^-, I^-, NO_3^-, SO_4^{2-}, HPO_4^{2-}, SO_3^{2-}, $B_4O_7^{2-}$, HCO_3^-, CO_3^{2-}, SiO_3^{2-}, VO_3^-, $HAsO_4^{2-}$, SeO_3^{2-}, MoO_4^{2-}, TeO_3^{2-}, WO_4^{2-}, UO_4^{2-}, citrates, tartrates, oxalates, EDTA, HCl, H_2SO_4, HNO_3, and $HClO_4$. Acetic acid, if present in amounts of 5%, causes a 10% increase in the absorption. Titanium also enhances the absorption, and for this reason its contents should be the same in the sample and standard solutions. If the amount of aluminum being determined is 20 ppm, the greatest effect is produced by Ti(IV) (80 ppm); the absorption is not appreciably increased if the concentration of titanium is further increased to 300 ppm. The calibration curve is rectilinear between 5 and 60 mg Al per liter; the sensitivity of the method is 1 ppm. The relative error is about 2%.

A similar method has been described for the determination of aluminum in cement [615]. Aluminum has been determined [626, 686] by the atomic absorption method, using oxygen-acetylene and hydrogen-acetylene flames. Aluminum was introduced in the form of aluminum cupferronate in methyl isobutyl ketone; the extraction was carried out at pH 3.5. The absorption of the extract was measured at 394.4–396.2 mμ, with a spectral width of the instrument of 6.2 mμ. The calibration curve was practically linear for aluminum contents between 50 and 1,500 μg Al/ml for an acetylene-oxygen flame, and between 500 and 2,500 μg Al/ml for an oxyhydrogen flame. In the former case the sensitivity is 9 times as high. The location of the flame employed significantly affects the sensitivity: the maximum absorption was noted 19 mm from the cone of the flame, while at distances of 15 and 24 mm the absorption decreased to one half. If the amount of acetylene is decreased, while that of oxygen remains constant, the aluminum absorption decreases. If the oxygen : acetylene ratio is decreased from 2.7 to 4.5, the absorption decreases by one half.

The standard deviation of the method is 53 μg/ml for an aluminum concentration of 750 μg Al per ml.

Aluminum has been determined in steel [514b] by the atomic absorption method in an acetylene-nitrous oxide flame; aluminum was introduced into the flame in the form of cupferronate dissolved in an organic solvent (methyl isobutyl ketone, cyclohexanone, n-butyl acetate or isopropanol).

Atomic absorption method with a graphite cell. Vaporization in a graphite cell has significant advantages over vaporization in the flame. In the first case the vaporization process takes 1,000 times longer than in the flame. When a graphite cell is used, most compounds highly dissociate and the effect of third components is less marked. Low-volatile foreign substances do not usually affect the rate of vaporization of aluminum in a graphite cell. It has been shown [306] that even 100,000-fold amounts of Fe, Ni, Co, Cr, Ti and Cu do not interfere.

Nikolaev and Aleskovskii [305–307] used a graphite cell to determine between 0.00002 and 38% aluminum in Fe, Cr, Ni, W, Nb, Ta, Mo, Ti, Cu, Cd, and Sb in steels, bronzes, brass, Silumin, in nickel alloys, ferrovanadium, silicocalcium, and graphite. The relative error is about 5% for aluminum contents between 0.5 and 1.5%, and is smaller for higher aluminum contents. Two determinations can be completed in 35 min.

If a graphite cell is employed, the sensitivity of the method is much higher than in the flame. Nikolaev and Aleskovskii succeeded in determining aluminum in double-distilled water at a content of $8 \cdot 10^{-7}$%; the optical density of the solution was 0.12, that is it was sufficient (sample volume 0.020 ml). The absolute sensitivity of the method was $1.5 \cdot 10^{-11}$ g, and the relative sensitivity was $2.5 \cdot 10^{-5}$% for a $6 \cdot 10^{-5}$ g sample size.

When aluminum is determined by the atomic absorption method using a graphite cell, the 309.27 mμ line is employed. The cell is heated to 2,400°C and is placed in an argon-filled chamber with quartz-glass walls under a pressure of 3 atm. For details, see [307].

The use of a graphite cell is described in [125, 240].

X-RAY SPECTROSCOPY

The determination of aluminum by X-ray spectroscopy usually includes the use of fluorescence spectra which are produced when the sample is exposed to X-rays. Less frequently, primary X-ray spectra are employed; these are obtained by direct excitation of the sample with high-energy electrons.

The K-radiation of aluminum is employed in the fluorimetric determination. Sagrera [1129] pointed out that Si, Ca and Fe do not affect the determination of aluminum. However, Savelli [1143] reported that the $Al-K_\alpha$ line intensities increase in the presence of silicon (4% SiO_2 are equivalent to 0.1% Al_2O_3). A correction for SiO_2 must accordingly be introduced.

The $Al-K_\alpha$ line intensities are measured by an X-ray fluorescence spectrometer (Model XRD-3 or XRD-5, manufactured by Philips Co.), with chromium and tungsten tubes. A chromium anode tube is preferable, since then the intensity of the fluorescent radiation emitted by aluminum is four times higher than when a tungsten anode tube is employed [544, 620, 1178]. The tubes operate under an applied voltage of 40–50 kV at a current intensity of 20–50 mA. Pentaerythritol and ethylenediamine ditartrate are used as crystals in the analyzers, to decompose light beams into spectra. The detector used to measure the intensity of the spectral lines consists of a gas-flow proportional counter with an amplitude analyzer (a mixture of 90% of argon and 10% of methane). It is recommended that the windows of the proportional counter be made of especially thin films.

To prepare the windows, 1 g of polyvinyl acetate is dissolved in 400 g of CH_2Cl_2, a microscope slide is dipped into the solution, taken out again, and dried. The film is cut at the edges and is cautiously removed from the glass. An aluminum layer (about 0.1 mμ) is applied to the side of the film facing the inside of the counter.

A more accurate method for determining aluminum involves the preparation of beads [1129]. When determining aluminum in minerals, the granulometric conditions are very important.

The sensitivity of the determination of aluminum by the X-ray fluorescence method is usually reported to be of the order of a few thousandths of one percent; according to [620], it is 10^{-5}%. For aluminum contents of 10–40%, the relative error is 1.5–2%.

The method has been employed for the determination of aluminum in clays, kaolin [143, 1012, 1018, 1243], various minerals [1018], flotation products of clays and bauxites [1161], and aluminum-iron alloys [462].

OTHER METHODS

These methods include: radiometric determination of aluminum in sillimanite ores and ore dressing products, with Fe^{55} and Co^{60} [107]; analysis of a mixture of Al, Ga and In hydroxyquinolates by IR spectrometry [794]; determination of aluminum in iron alloys by measuring its thermoelectric potential [901]; sedimentometric determination of aluminum [1035], and thermometric determination (measuring the change in temperature of the solution being analyzed after the addition of titrant) [1137]. These methods are not very frequently employed.

CHAPTER III

Separation of Aluminum from Accompanying Elements

PRECIPITATION BY ORGANIC AND INORGANIC REAGENTS

METHODS BASED ON THE PRECIPITATION OF ALUMINUM

Aluminum can be separated by two precipitation methods that have already been described under "Gravimetric Methods." Only one other method can be employed for this purpose, namely, separation of aluminum as chloride. This technique is mainly employed to eliminate the bulk of the aluminum before the determination of other elements, for example, in the determination of impurities in high-purity aluminum. The method is unsuitable for the determination of aluminum, since the precipitation of aluminum is not quantitative.

The isolation of aluminum as chloride was first proposed by Gooch and Havens [766], and was later modified by other workers [715, 716, 1157]. It is based on the low solubility of aluminum chloride in a mixture of hydrochloric acid and HCl-saturated ether. The optimum conditions for the separation are: 50% concentration of ether in the mixture and a temperature of 0°C. According to Seidel and Fischer [1157], the solubility of $AlCl_3$ is as low as 0.8 mg Al in 100 ml of 44.3% HCl at 0°C. Its solubility in a 1 : 1 mixture of HCl with ether is even lower, 0.15 mg Al in 100 ml [715]. Aluminum can be separated from Be, Fe(III), Co, Zn, Mn, Ca, SO_4^{2-}, PO_4^{3-} by this method. Owing to the low solubilities of Ni, Mg, K, NH_4^+ and Na^+ chlorides, these metals can be present in very small concentrations only.

The separation from chromium is unsatisfactory. In the presence of large amounts of titanium or vanadium, aluminum chloride must be reprecipitated. For details of the separation technique, see [89].

METHODS BASED ON THE PRECIPITATION
OF INTERFERING ELEMENTS

Separation by alkali. This method is the one most frequently used to separate interfering elements from aluminum. Aluminum hydroxide dissolves in excess of sodium hydroxide to form an aluminate (Chapter I), whereas the hydroxides of other elements usually present do not. The disadvantage of the method is that aluminum is initially precipitated as hydroxide together with the hydroxides of other elements. Hydroxide precipitates are usually bulky (this applies particularly to Fe(III) and Cr), so that some of the aluminum hydroxide is occluded and is not completely dissolved even if the concentration of sodium hydroxide is increased. As a result, the analytical data obtained for aluminum are too low.

The alkali separation method is successful only if calibration curves are plotted, or titers of the working solutions are determined, under exactly the same conditions as those prevailing during the analysis. In this case, the method proves to be more rapid and convenient than, for example, the separation of accompanying elements from aluminum by electrolysis on a mercury cathode, or by chromatography.

Better separation is obtained if a weakly acid boiling solution of the sample is slowly introduced into a hot, concentrated solution of sodium hydroxide, with constant stirring.

The sample solution can also be introduced from a funnel with a very fine stem bore (1 mm [1115], or even 0.4 mm [1069]. Richter [1115] states that better separation is obtained at room temperature. According to the precipitation theory of Tananaev [404], amorphous precipitates should be precipitated from hot concentrated solutions, since then less coprecipitation occurs. Therefore, precipitation by alkali is preferably carried out from hot solutions. With increase in the concentration of NaOH, the adsorption of aluminum is lower. A suitable concentration of the alkali is 6 N NaOH. Small amounts of iron remain in the filtrate (the higher the NaOH concentration, the more iron remains in the filtrate). Therefore, if small amounts of iron interfere, very high NaOH concentrations should not be employed. Ferrous iron strongly sorbs aluminum and should, therefore, be first oxidized to Fe(III). As the amount of aluminum increases, its sorption on $Fe(OH)_3$ becomes proportionally less [1024]:

Amount of Al, μg	Al sorbed, %	Amount of Al, μg	Al sorbed, %
10	11	70	3.4
30	4.7	90	3.3
50	4.4	100	3.4

The sorption of aluminum increases with increase in the amount of iron. Werz and Neuberger [1258], and also other workers, carried out a preliminary enrichment by the magnetic method to reduce the sorption of aluminum, but this complicates the analytical procedure and cannot be recommended. It was shown by Seuthe [1160] that aluminum is more readily separated from iron if the $Fe(OH)_3$ precipitate is homogenized by mechanical stirring. The method was checked by Hill [809], who found that the separation of aluminum was incomplete. To decrease the sorption of aluminum on $Fe(OH)_3$, this author suggested that a large quantity of an aluminum-like element be introduced, which chemically resembles aluminum but does not interfere with the subsequent determination of the aluminum. If aluminum is determined photometrically by Eriochrome Cyanine R, Li, B, Zn, and particularly a Zn−B mixture are best for the above purpose. During separation by sodium hydroxide, vanadium accompanies aluminum. Chromium is an amphoteric element, and so is also found in the filtrate, but is completely precipitated in the presence of iron. In the absence of iron some titanium passes into the filtrate; in the presence of iron, titanium is quantitatively precipitated.

Korenman et al. [185, 186, 189, 190] studied the possibility of separating a number of elements from aluminum by sodium hydroxide. Magnesium hydroxide sorbs aluminum very strongly; an ignited MgO precipitate may contain as much as $30-40\%$ Al_2O_3 [185]. If the precipitation is carried out at room temperature, $1-7\%$ Al is lost with $Y(OH)_3$ [190] and $3-20\%$ Al is lost with $La(OH)_3$ [189]. The separation is practically quantitative if the precipitation is carried out in a boiling solution [189, 437]. Only about 1.5% Al is lost with $Cd(OH)_2$ [186].

Aluminum is sometimes separated, together with iron, chromium, and certain other elements, as the hydroxide by ammonia or urotropin to separate it from divalent metals. The precipitate is dissolved in acid, and the separation is carried out by NaOH [524, 777, 1003]. This method is recommended especially during the analysis of samples containing calcium and magnesium, which strongly sorb aluminum. Přibil and Veselý [1083,

1085] separate titanium from aluminum by precipitation with sodium hydroxide in the presence of triethanolamine, which retains aluminum in solution.

The sample solution is diluted to 100 ml, 20–50 ml of a 20% solution of triethanolamine are added, and then 2 ml of $2M$ NaOH (until the solution is decolorized). The mixture is boiled for 1 min and the precipitate is filtered and washed, first with 5 portions of a hot 1% solution of triethanolamine, then twice with water. Aluminum in the filtrate is determined by the complexometric method.

Pritchard [1087] modified the sodium hydroxide separation by adding diaminocyclohexanetetraacetic acid (DCTA) to bind aluminum into a complex [1087]. In this case, aluminum is more readily separated from magnesium, and in the presence of DCTA aluminum is not coprecipitated with $Mg(OH)_2$.

The solutions of sodium hydroxide should be freshly prepared, if possible, and should be stored in polyethylene vessels. Old solutions, which contain CO_2 sorbed from the atmosphere, give unsatisfactory results.

Aluminum can be separated from interfering elements by fusion with NaOH. Fusion with Na_2CO_3 also gives an effective separation, and aluminum is separated from the same elements as in the case of precipitation by a solution of NaOH. Sometimes fusion with sodium carbonate is preferable, since alkaline-earth elements are separated from aluminum as well. Vanadium, Mo, Sn and Zn pass into the filtrate together with aluminum. If Fe, Co and Ni are present in amounts more than 100 times the amount of aluminum, fusion with sodium carbonate will not give complete separation.

Separation by other inorganic precipitants. Taimni and Tandon [1218] separate As(V), Sb(III), Sb(V), Te(IV), Se(IV), Mo(VI), Hg(II), Au(III), Pt(IV), Re(VII) from aluminum in the form of sulfides. The solution is treated with $1M$ Na_2S in excess, and is then acidified with HCl to a concentration of about $6N$ for As, Pt and Re, to a concentration of $2N$ for Se, and to a concentration of $1N$ in the presence of other elements. The relative error in the determination of aluminum is 0.7%.

Kunenkova and Ostroumov [220] suggested that indium be separated from aluminum as sulfide from a solution containing monochloroacetic acid at pH 2.5. The separation of zinc as sulfide results in a better separation than precipitation by ammonia [611]. Pilz [1063] separated iron as FeS after the reduction of Fe(III) by sodium trithiocarbonate or ammonium trithiocarbonate. The reagent is in short supply and has no advantage over other separation methods. Iron can be separated from aluminum as sulfide from solutions containing tartaric acid [1155].

Burke and Davis [600] removed manganese by boiling perchloric acid solutions with sodium chlorate. Manganese dioxide tends to entrain certain metals, but according to these workers, aluminum is not coprecipitated.

Gordon et al. [756, 768] separated iron from aluminum by homogeneous precipitation as periodate. The precipitation is best effected from nitrate solutions at pH 1.1–1.2. If iron is fully precipitated, aluminum is precipitated as well, and for this reason reprecipitation is recommended.

Zinc chloride is introduced during the separation of iron from aluminum by ferrocyanide, to reduce the coprecipitation of aluminum [33].

Tanaka [393] described a method for the separation of titanium by hydrolytic precipitation as $Ti(OH)_4$, which takes place when a solution containing sulfosalicylic acid is boiled at pH above 9. The phosphate ions are separated by precipitation as Li_3PO_4 [523] from a strongly alkaline solution; the method is not very efficient. Tin can be separated from aluminum as metastannic acid [820]. The separation of chromium by precipitation as $PbCrO_4$ after preliminary oxidation [1141] is of little interest.

Separation by organic precipitants. The following can be added to supplement the list of precipitants given under "Gravimetric Methods." Phenylarsonic acid will separate titanium and iron from aluminum at pH 1.5–1.7 [105, 106]. All of the aluminum, 2–5 μg TiO_2, and less than 1μg Fe remain in the filtrate; such small amounts of iron and titanium do not interfere with the determination of aluminum by Aluminon. Titanium can be separated by *p*-hydroxyphenylarsonic acid [845]. The conditions of precipitation of several metals by different isomers of phenylenediarsonic acid have been studied [1064], and this study may serve as a starting point for the development of methods for the separation of a large number of metals from aluminum. Titanium, iron and certain other elements [453, 1051] can be separated from aluminum by precipitation as cupferronates.

Banks and Edwards [547] separated thorium from aluminum by precipitation as oxalate in the presence of excess dimethyl oxalate and oxalic acid. With potassium ethylxanthate, $KSCSOC_2H_5$, it is possible to separate zinc from aluminum [1254]. The precipitation of zinc is most effective at pH 5–5.5. Aluminum can be separated from indium by double precipitation of the latter by hexamethylenediamine in the presence of ammonium salts [195].

SEPARATIONS BY EXTRACTION

Extraction of chloride complexes

Many elements can be separated from aluminum as chloride complexes by extraction with organic solvents. The optimum acidity values for the extraction are quoted in the monograph by Morrison and Freiser [280].

Extraction by diethyl ether. Diethyl ether is the most frequently used organic solvent, not necessarily because it is the best, but because it is the most readily available of all organic solvents. The optimum HCl concentration during the extraction is 6.2 N. One extraction of the aqueous layer by an equal volume of the organic layer results in the extraction of about 99% iron [659]. Some of the Fe(III) is reduced to Fe(II) by ether and is not extracted [1058, 1258]. Werz and Neuberger [1258] suggested that the reduced iron be oxidized by potassium chlorate. All of the aluminum remains in the aqueous layer.

Gallium [653] and certain other metals are also extracted by ether from 6 N HCl.

Extraction by isopropyl ether. According to Craft and Makepeace [659], isopropyl ether is a more efficient extractant than diethyl ether. This solvent extracts 99% Fe at acidities of 6.5−8 N, while with diethyl ether the limiting acidity is 6.2 N. Moreover, isopropyl ether is less soluble in hydrochloric acid solutions. Phosphate, molybdate and vanadate ions are partially extracted by isopropyl ether, if iron is present in the ferric form; aluminum is not extracted.

Extraction by amyl acetate. It was shown by Wells and Hunter [1253] that amyl acetate extracts iron more effectively than diethyl and isopropyl ethers. If the volumes of the aqueous layer and of the solvent are equal, 99.95% Fe is extracted; after three isopropyl ether extractions 99.9% of the iron was extracted when 1 g Fe was taken. Under the same conditions ∼83% Mo(VI), ∼68% Sn(IV), ∼14% V(V), 0.2% Ti and 0.5% Co are extracted, while Ni, Cr, Ca, Mg, and Cu(II) are not. Aluminum is extracted in an amount of 0.1%, while as mentioned above, it is not extracted by diethyl and isopropyl ethers. If small amounts of aluminum are determined, the 0.1% of the extracted aluminum are insignificant. When mixtures of iron with aluminum were extracted (in ratios varying from 100,000: :1 to 5,000:1), between 99 and 104% of the amount of aluminum introduced were recovered [1253], while the percent of iron extracted by two extractions was 99.88−99.97%. Another advantage of amyl acetate is that the concentration of HCl need not be strictly controlled.

Extraction by methyl isobutyl ketone. This reagent is effectively employed to separate Fe(III), Mo(VI), Ga, Te(IV) from aluminum [145, 156, 592,

772, 820, 857, 1216]. The partition coefficients of certain metals for extraction from $6N$ HCl have the following values [145]:

Element	K_p	Element	K_p
Te(IV)	360	Cr(III)	$25 \cdot 10^{-3}$
Bi	$3.1 \cdot 10^{-4}$	Al, Ni	$2.5 \cdot 10^{-4}$
Cr	$5.15 \cdot 10^{-2}$	Co	$<2.5 \cdot 10^{-2}$

Iron and molybdenum are extracted from $8N$ HCl [592, 857], while tellurium is extracted from $6N$ HCl [145]. Two extractions by 20 ml of methyl isobutyl ketone each (with an equal volume of the aqueous solution) will remove 0.5 g Fe almost quantitatively. If 1 g Fe is to be extracted, three extractions by 30 ml of solvent each time are required.

Extraction by other solvents. Elliot and Robinson [700] extracted the chloride complexes by dichlorodiethyl ether; the optimum HCl concentration is $9N$. This solvent does not extract iron as effectively as those just described. Dichlorodiethyl ether has the advantage that it is somewhat heavier than water, which is convenient during repeated extractions. Other extractants sometimes used are isoamyl alcohol, methyl amyl ketone [213], butyl acetate [1216], and a mixture of diethyl ether with tetrahydrofuran [1186].

Extraction of cupferronates

The extraction of cupferronates of interfering elements is widely employed during the determination of aluminum. Figures 30–32 show the percent of cupferronate extracted as a function of the pH. In a number of cases the separation may be effected by extracting aluminum cupferronate [1162]. The extraction of aluminum cupferronate is practically complete at pH 3.5 or above [555, 703, 1162, 1193]. Even at pH 0.5 some aluminum is extracted, and for this reason the separation of Fe, Ti, Zr, V and other elements is preferably conducted from more acid solutions. The usual practice is to extract from $1N$ HCl or H_2SO_4. In this case, insignificant amounts of aluminum are lost, and these can be neglected. Miller and Chalmers [970] treated a standard solution of aluminum with Cupferron, and extracted this by dichlorobenzene. The aluminum content in the aqueous phase was then determined; these authors report that their absolute experimental error was $1-5\ \mu g$.

Erroneous data on the optimum pH value for the extraction of the cupferronates of interfering elements are found in the literature. For example, it has been reported [550a] that aluminum can be accurately determined

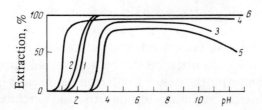

Figure 30. Extraction of cupferronates as a function of pH [1193]:

1) aluminum; 2) beryllium; 3) lanthanum; 4) scandium; 5) yttrium; 6) antimony.

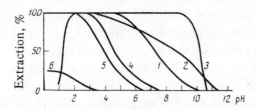

Figure 31. Extraction of cupferronates as a function of pH [1193]:

1) titanium; 2) zirconium; 3) thorium; 4) vanadium (V); 5) molybdenum (VI); 6) tungsten (VI).

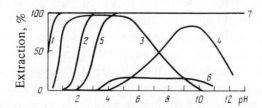

Figure 32. Extraction of cupferronates as a function of pH [1193]:

1) copper; 2) lead; 3) mercury; 4) zinc; 5) cobalt; 6) manganese; 7) iron.

after the extraction of cupferronates at pH 2; in fact, aluminum is partially extracted from such solutions as its cupferronate, and the results must necessarily be too low.

Cupferronates are not formed instantaneously, and for this reason the solutions must be left to stand for a few minutes after the addition of Cupferron before they can be extracted. According to Short [1162], iron cupferronate is more readily extracted in the presence of 2–3 g of ammonium chloride. If the iron is present in the ferrous form, it should first be oxidized After the cupferronates have been removed by extraction from strongly acid medium, aluminum in certain materials can be determined directly, without any further separation (after the excess Cupferron has been decomposed by evaporation with nitric and sulfuric acids, or nitric and perchloric acids, to the evolution of SO_3 or $HClO_4$ fumes). If interfering elements are still present after the cupferronates have been removed from the acid medium, aluminum can be isolated from the solution by extraction as its cupferronate at pH 3.5. This is, in fact, the technique adopted by Corbett [655] during the determination of aluminum in titanium by Aluminon, by Short [1162] during the determination of aluminum in iron, by Rooney [1121], and others. Attempts to reextract aluminum from a chloroform solution of the cupferronate did not give satisfactory results [787, 1218]. Therefore, the residue obtained after evaporation of the chloroform and decomposition of excess Cupferron should be dissolved in acid, or fused with sodium carbonate, and then the melt leached with water. When large amounts of aluminum are determined, satisfactory results can be obtained without decomposition of the Cupferron residue, if the extraction is effected with small amounts of Cupferron and if small aliquots of the aqueous layer are taken for the determination [417, 418, 420].

When highly sensitive photometric methods are employed, the concentration of Cupferron in the final solution used for the photometric determination is found to be negligible. A freshly prepared 6% aqueous solution of Cupferron is usually employed; the solution of Cupferron should be cooled to prevent decomposition. The following organic solvents have been proposed for the extraction of cupferronates: chloroform, diethyl ether, benzene, carbon tetrachloride, o-dichlorobenzene, 1 : 1 mixtures of chloroform with benzene and diethyl ether with benzene, and a 1 : 1 : 1 mixture of benzene, diethyl ether, and acetone. It has been reported [970] that o-dichlorobenzene gave the best results, but nevertheless chloroform must be considered as the most efficient extractant. o-Dichlorobenzene is inconvenient in use, because it boils at a much higher temperature (180–183°C) than chloroform (58–61.5°C). The following procedure is recommended for the extraction of cupferronates.

Five to ten ml of the sample solution, 1 N in HCl, are placed in a separatory funnel, 5–7 ml of a 6% aqueous solution of Cupferron are added (if the concentrations of metals to be extracted are high, a larger amount is taken) and the mixture is shaken for 0.5–1 min to coagulate the cupferronates. After 4–5 min, 15 ml of chloroform are added (more if the precipitate is voluminous), and the mixture is shaken for one minute.

The extract is discarded, and the completeness of the precipitation is checked by adding a few drops of the Cupferron solution to the aqueous layer. If a white precipitate appears, the precipitation is complete; if not, more Cupferron is added and the extraction is repeated until the extract is colorless. The aqueous layer is shaken with two 10-ml portions of chloroform to remove excess Cupferron.

Then, 2 ml $HClO_4$ (sp. gr. 1.54) and 2–3 ml HNO_3 (sp. gr. 1.4) are added to the aqueous layer, or to a part of it, in a beaker, and the mixture is evaporated to the appearance of $HClO_4$ fumes. If the solution is not colorless, the decomposition of the organic residues is repeated. Aluminum is determined by a suitable method in the colorless solution.

Extraction of complexes with N-benzoylphenylhydroxylamine

N-Benzoylphenylhydroxylamine is an analog of Cupferron (for more details see p. 60). The completeness of the extraction of certain elements at optimum pH values can be described by means of the following data [640]:

Element	pH	Extraction, %	Element	pH	Extraction, %
Fe	2–3	100	Hg	7.5–9	88
Bi	2–13	100	Ni	9–11	97
Cr	3–4	23	Zn	9–11	100
Cu	4–11	100	Mn	10	100
Al	6	100	Co, Cd	10–11	100
Pb	7–11	97			

According to Chwastowska and Minczewski [640, 641], the maximum extraction of aluminum is observed at pH 6.

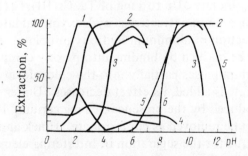

Figure 33. Extraction of the complexes of a number of metals with N-benzoylphenylhydroxylamine as a function of the pH [640]:

1) iron; 2) copper; 3) aluminum; 4) chromium; 5) zinc; 6) manganese.

Figure 33 represents the percent extraction of complexes of a number of metals as a function of the pH. The extraction of complexes with N-benzoylphenylhydroxylamine has been used to separate aluminum from Ti, Zr and V [263, 776], and alkali and alkaline-earth metals [703], and concentrate aluminum and other metals appearing as impurities during their determination in silver and in high-purity chromium [640].

Extraction of hydroxyquinolates

Elements which interfere with the determination of aluminum are usually not removed by extraction as hydroxyquinolates. On the contrary, attempts are made to choose the experimental conditions so that aluminum hydroxyquinolate is extracted, whereas the interfering elements remain in the aqueous layer (cf. p. 83).

Extraction of hydroxyquinaldates

8-Hydroxyquinaldine (2-methyl-8-hydroxyquinoline) forms sparingly soluble compounds with several metals. The reagent does not react with aluminum under ordinary conditions [811, 815]. The precipitates of metal hydroxyquinaldates can be separated by filtration, but extraction with chloroform is preferable. Hinek and Wrangell [811] studied the conditions for the extraction of the hydroxyquinaldates of a large number of metals. Complexes of Zn, Ni, Co, Pb, Sn(IV), Sb(III), Mn(II), Cd, Cu, Fe(II), Fe(III) can be extracted at pH 9.2. Titanium and Ce(III) are not readily extracted; Ta, Mg, Cr, Mo, W, V, Nb, Th, U, and Zr are not extracted. These authors used hydroxyquinaldate extraction during a photometric determination of aluminum as hydroxyquinolate. Up to 1 mg of Ta, Cr(III), Cr(VI), Mo(VI), W(VI) and V(V), which are not extracted as hydroxyquinaldates, do not interfere with the extraction of aluminum hydroxyquinolate. The metals Nb, Th, U, and Ce can be masked by binding into peroxy complexes. The peroxy complex of cerium passes partially into the chloroform layer, but if anhydrous sodium sulfate is added, its effect decreases. Under such conditions, a 1% error is produced by the presence of 1 mg cerium. It is recommended that zirconium be removed as cupferronate. Hinek and Wrangell [811] used this technique for the separation of interfering elements during the determination of aluminum in aluminum bronzes and stainless steels. The error of the determination is 0.028 and 0.14% at aluminum contents of 0.5–1.5 and 1.5–7.0%, respectively.

Ashbrook and Ritchey [534] consider hydroxyquinaldine as the best reagent for the elimination of the interfering elements during the determination of aluminum by hydroxyquinoline. These authors extract the hydroxyquinaldates at pH 9.5–10.0. Unlike Hinek and Wrangell, they found that manganese cannot be extracted as hydroxyquinaldate; however, the formation of the manganese complex has been confirmed by other workers [993].

Extraction of thiocyanates

Certain elements can be separated from aluminum by extraction as thiocyanates. The method is employed mainly for the separation of iron. It should be noted that aluminum is also extracted under certain conditions. According to Bock [577], the percent of aluminum thiocyanate extracted increases with decrease in temperature and increase in the concentration of thiocyanate.

To reduce the losses of aluminum to a minimum, the concentration of the thiocyanate introduced should not exceed $1\,M$. Under these conditions, that is in $1\,M$ NH$_4$SCN and $0.5\,M$ HCl, the conditions produced are the most favorable for the extraction of ferric thiocyanate. According to Bock [577], about 89% of the iron is then extracted. Ferric thiocyanate can be extracted by diethyl ether [577, 706, 1235], amyl alcohol [1001], a mixture of amyl alcohol with ether [409, 568] and butyl alcohol with ether [19]. It was shown by Specker et al. [1186, 1187] that the distribution coefficient of ferric thiocyanate between diethyl ether and the hydrochloric acid solution increases greatly on addition of tetrahydrofuran, whereas the distribution coefficient of ammonium thiocyanate remains unchanged.

Thirty ml of the solution, $1.2\,N$ in HCl, are placed in a separatory funnel, and 12 ml of HCl (sp. gr. 1.19) and 15 ml of NH$_4$SCN solution (500 g/liter) are added. Ferric thiocyanate is extracted by a mixture of 30 ml of diethyl ether and 30 ml of tetrahydrofuran. The extraction is repeated, using 20 ml of ether and 20 ml of tetrahydrofuran, after the addition of 10 ml of ammonium thiocyanate solution. The mixture is then shaken again with 20 ml of ether and 20 ml of tetrahydrofuran [1186, 1187].

By using this method, the authors obtained satisfactory results in the determination of aluminum, even at NH$_4$SCN concentrations much higher than the optimum range (see above) of $1-4\,M$ NH$_4$SCN, when 2–3 extractions were made from solutions $0.5-3\,N$ in HCl and at Fe : Al ratios ranging from 1 : 30 to 30 : 1 [1186]. The separation process takes 10 min.

Zhivopistsev and Minin [119] employed the extraction of thiocyanate to separate titanium from aluminum.

Extraction of diethyldithiocarbamates

The most detailed study of extraction with sodium diethyldithiocarbamate was carried out by Bode [579]. Aluminum is not extracted at any pH value; the optimum pH values for the extraction of the diethyldithiocarbamates of numerous metals are given in the monograph by Morrison and Freiser [280]. It should also be pointed out that manganese diethyldithiocarbamate is extracted quite satisfactorily by carbon tetrachloride from solutions at pH 5.0–5.5 [424, 428, 432]. At pH 4–6 aluminum can be separated from Fe, Mn, Zn, Ni, Co, Cu, Cd, Bi, Se, Ag, As(III), Sb(III), Sn(IV), Pb, Mo, V, In, Ga, and Tl. Eckert [697] determined aluminum in nickel alloys, and successfully removed nickel and admixtures of Co, Fe, Mn, and Cu as diethyl-dithiocarbamates. According to this author, in this case diethyldithiocarbamate gives better results than any other carbamate. For example, if piperi-dinedithiocarbamate is employed, the results are 10–20% too low. When large amounts of nickel are to be separated, chloro derivatives of hydro-carbons are the best solvents. Ethers, higher alcohols, and esters are poor solvents of nickel carbamate. Carbon tetrachloride, higher alcohols and esters can also be employed to separate other elements.

Extraction of acetylacetonates

Acetylacetone reacts with many metals to form complexes, which can be extracted by acetylacetone and by carbon tetrachloride, chloroform, benzene and xylene.

Acetylacetonates of different elements are extracted at the following pH values [280]:

Element	pH	Element	pH	Element	pH
Al, Mn	4	Ga, U(IV)	2.5	Th	5.8
Be	2	In	2.8	Ti	1.6
Ce	>4	Fe	1.5	U(VI)	4–6
Cr(III)	0–2	Mo(VI)	-0.8–0.0	Zn	5.5–7
Co(III)	-0.3–2	Pu(IV)	2–10	Zr	2–3
Cu, V(IV)	2.0				

The maximum extraction of aluminum acetylacetonate, $Al(C_5H_7O_2)_3$, is noted at pH > 4 (Figure 34) [1198]. As in the extraction of cupferronates, many elements can be separated from aluminum as acetylacetonates by acetylacetone at pH 0; aluminum itself can be separated as acetylacetonate from Ca, Mn, Mg, and other elements.

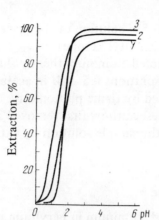

Figure 34. Extraction of aluminum (1), gallium (2), and indium (3) acetylacetonates as a function of the pH [1198].

Miller and Chalmers [970] used acetylacetone to separate aluminum from other elements during the microanalysis of silicates (10 mg samples). At pH 6–7, aluminum acetylacetonate is fully extracted by diethyl ether. After the reextraction of aluminum with 6 N HCl, the element was determined by the gravimetric hydroxyquinoline method, with a relative error of 0.4%. At pH 6–7, many components of silicates are not extracted. About 3% of Co and Cr are extracted, but these metals are present in silicates in small concentrations only, and do not interfere. Beryllium accompanies aluminum. Alimarin and Gibalo [14] succeeded in separating beryllium from aluminum by extracting beryllium acetylacetonate at pH 9 with chloroform from solutions containing Complexone III. Only the beryllium complex is extracted under these conditions, while complexes of aluminum and many other metals are not. The extraction of beryllium acetylacetonate is independent of the concentration of Complexone III and sulfate, chloride, nitrate and ammonium ions.

Extraction of thenoyltrifluoroacetonates

Thenoyltrifluoroacetone was used by Goldstein et al. [763] to separate Th, Zr, Fe and certain other elements from aluminum during the determination of aluminum in thoria. Methyl isobutyl ketone is used as organic solvent. Interfering elements are best separated from solutions 2 M in acetate ions at pH 1.5. Under these conditions Fe, Th, Zr and Cu are very well extracted, while Ti, U, Mo and V are only partially extracted. Aluminum is extracted at pH 5–5.5 [584, 763] at 1 M acetate concentration.

The optimum amount of the $0.5\,M$ solution of the reagent is 10 ml (the reagent is dissolved in methyl isobutyl ketone).

Eshelman et al. [703] extracted aluminum thenoylfluoroacetonate by methyl isobutyl ketone from solutions $0.5-1\,M$ in acetate, at pH 5.5–6.0. Aluminum was then determined by flame photometry. In the analysis of multicomponent materials, these authors first removed Zr, Ti, Th, U(VI), Fe(III), Ce(IV) and Cu from the sample solution by extraction as thenoyl-fluoroacetonates at pH 1.

Extraction by triisooctylamine

Floerence [724] determined aluminum in beryllium materials by 5-sulfo-4-diethylamino-2,2-dihydroxyazobenzene, and proposed that the interfering elements, Fe, Co, Cu, Ti, V, W, U and Th, be previously separated by extraction with triisooctylamine from $8\,M$ HCl.

The solution is evaporated almost to dryness and the residue is dissolved in 5 ml of $8\,M$ HCl. The solution is transferred to a 50-ml separatory funnel by means of 5 ml of $8\,M$ HCl. Ten ml of a 5% solution of triisooctylamine in toluene (which has previously been saturated with $8\,M$ HCl) are added, and the solution is extracted for one minute. The aqueous layer is decanted into a 150-ml conical flask, the organic phase is washed with 5 ml of $8\,M$ HCl, and the washings are transferred to the same conical flask. One ml of $15\,M$ HNO_3 and 3 ml of 72% $HClO_4$ are added, and the solution is evaporated almost to dryness. It is then diluted to a suitable volume in a volumetric flask, and the determination of aluminum is terminated by the spectrophotometric method, using 5-sulfo-4-diethylamino-2,2-dihydroxyazobenzene.

Extraction by tri-*n*-octylphosphine oxide

Extraction with tri-*n*-octylphosphine oxide makes it possible to separate aluminum from a large number of elements. According to White [1263], a $0.1\,M$ solution of tri-*n*-octylphosphine oxide in cyclohexane will quantitatively extract Cr(VI), Au(I), Hf, Fe(III), Mo(VI), Sn(IV), U(VI) and Zr from $1\,M$ HCl; Sb(III), Bi, Cd, In, Hg(II), Pt(II) and Zn are partially extracted. The following are quantitatively extracted from $7\,M$ HCl: Sb(III), Cr(VI), Ga, Au(I), Hf, Fe(III), Mo(VI), Sn(IV), Ti(IV), U(VI), V(IV), Zr.

Ishimori [837] studied the extraction of 60 ions by a toluene solution of tri-*n*-octylphosphine oxide. Burke and Davis [600] used this separation method for their determinations of aluminum in multicomponent refractory alloys, together with a preliminary separation of most of the interfering elements by electrolysis on a mercury cathode. Burke and Davis state that the elements Ti, Zr, Hf, Au, Ga, U(VI), Nb, Sn, In, Sc, Th, Os, Mo, As(III), W, Fe(III), Hg, and V are quantitatively extracted from $6M$ HCl by a solution of tri-n-octylphosphine oxide in cyclohexane. Aluminum is not extracted under these conditions; its partition coefficient is 10^{-3}. Rare earth elements, Y, B, Mg, Be, alkali and alkaline-earth elements are also not extracted.

Extraction by other solvents

Babko and Mikhal'chikhin [32] separate Fe(III) from Al by extracting the complex of iron with α-nitroso-β-naphthol by chloroform at a pH of about 1.5. Zhivopistsev and Minin [119] used chloroform to extract the complex of Fe(III) with diantipyrylmethane from $5-6N$ HCl. Iron may be separated from aluminum as tributylammonium hexathiocyanoferrate from solutions at pH $1-5$ [1291], or by extracting its complex with pyridylazonaphthol at pH $1-2$ [1216].

Banks and Edwards [547] separated thorium from aluminum by extraction with mesityl oxide from a solution saturated with lithium nitrate. The acidity of the solution being extracted should be $1M$ in HNO_3.

Ziegler [1290] pointed out that indium can be separated from aluminum (In : Al $= 15 : 1$) by extracting dibutylammonium indium sulfide, formed by the reaction between indium and dibutylamine in the presence of polysulfide, by dichloromethane.

Banarjee et al. [546] separated aluminum from beryllium by extracting butyrates by chloroform or ethyl acetate from solutions containing Complexone III. Aluminum in the form of butyrate is extracted at pH > 3.40, but in the presence of Complexone III, which must be added before the introduction of butyric acid, it is not extracted and remains in the aqueous phase. Chlorides, nitrates, sulfates, silicates, Mg, Zr, Ti(IV) and small amounts of fluorides do not interfere; phosphates interfere. The effect of up to 4 mg P_2O_5 can be eliminated by adding $ZrOCl_2$ and Complexone III to the mixture to bind the excess zirconium. The method has been successfully adapted to the analysis of beryllium ores. Beryllium can be separated from aluminum by extraction with a solution of di-(2-ethylhexyl)-phosphoric acid in kerosene from sulfate solutions [621]. The beryllium complex is extracted from more acid solutions than the aluminum complex. At pH 2.2, 70% Be and 12% Al are extracted. However, the curves of the percent extraction plotted against the pH for beryllium and aluminum submitted by this author very nearly coincide, so that the effectiveness of the extraction by this method is doubtful.

Alekperov et al. [10] studied the separation of aluminum from gallium by extraction with a solution of naphthenic acids in a 1 : 1 mixture of kerosene and diethyl ether. The optimum pH value is 3.5–3.6. The amount of gallium extracted in a single pass is 80%.

CHROMATOGRAPHIC SEPARATION METHODS

SEPARATION ON CATION EXCHANGERS

Aluminum can be separated from other elements on cation exchangers because of the differences in sorption at a given acidity of the solution, their behavior toward complexing agents, and the amphoteric nature of aluminum.

Separation by controlling the acidity of the solutions. These methods are not as important as those based on complex formation, but may still be useful on a number of occasions. They are mainly employed to separate Be, Zr and U from aluminum.

Gordeeva and Prosviryakov [94] studied the conditions for the separation of beryllium from aluminum and iron on KU-2 cation exchanger. All three metals are sorbed in the acidity range between $2N$ HCl and pH 4. Beryllium is eluted by 150–200 ml of $0.5N$ HCl (or HNO_3); aluminum begins to be eluted after about 300 ml of $0.5N$ HCl have been passed through. The total amount of the ions to be separated must not exceed 5–6 mg per 1 g of cation exchanger, or aluminum will be partially eluted together with the last portions of beryllium. Titanium is desorbed together with beryllium. Aluminum can then be extracted by passing 20–25 ml of $4N$ HCl.

In a similar study, Strelow [1205] eluted beryllium from AG 50 W×8 cation exchanger by passing through 375 ml of $1N$ HCl at the rate of 3–4 ml/minute. In the presence of more than 60 mg of iron beryllium is preferably eluted with 425 ml of $1.2N$ HNO_3. All the aluminum remains on the column, from which it can be eluted by 500 ml of $2N$ HCl.

Beryllium can also be separated from aluminum by passing the solution, which should be $0.5–0.7N$ in sulfuric acid, through a sulfonic acid type cation exchanger FN. Aluminum is then retained on the column, while beryllium is eluted. Aluminum can then be extracted from the column by $2–4N$ hydrochloric or sulfuric acid [1241]. If the Be : Al ratio is between 1 : 8 and 4 : 1, the relative error is not greater than 3%.

Strelow [1204] separates zirconium from aluminum on Biorad AG 50 W×8 cation exchanger. Aluminum is then eluted by 400 ml of $2N$ HCl passed at the rate of 2–2.5 ml per minute. Zirconium is desorbed by passing 400 ml of $5N$ HCl or 250 ml of $3N$ sulfuric acid. Anions which form stable complexes with zirconium (fluorides and oxalates) interfere.

Shikhova and Katsarova [498] also separate zirconium from aluminum by working at an acidity of $2N$ in HCl, with Wolfatite KPS-200 cation exchanger. Usatenko and Gureeva [445] carried out a comparative study of a number of cation exchangers for the separation of these elements. Of the cation exchangers studied by these authors, KU-1 cation exchanger (at a maximum acidity of $1.5N$) was the best for zirconium sorption, the next best resin was SBS-R at $1.25N$, and the poorest was Wolfatite P at an acidity of $0.75N$. The largest difference in the acidity for the sorption of zirconium and aluminum was observed with sulfonated coals KU-1 and MOF-1. Aluminum is not sorbed on sulfonated coal or on KU-1 cation exchanger from solutions at an acidity of $1.5N$. A number of workers [49, 177] pointed out that it was possible to separate zirconium from aluminum by passing the solution, which should be $1N$ in HCl, through KU-2 cation exchanger.

Strelow [1207] separated uranium (VI) from aluminum by utilizing the difference in their partition coefficients between Biorad Ag 50W×8 cation exchanger and $1N$ sulfuric acid (9.6 and 126, respectively).

Uranium is desorbed by 350 ml of $1N$ sulfuric acid passed at 3–4 ml/min, while aluminum is desorbed by 400 ml of $3N$ HCl.

In a separation on Amberlite IR-120 cation exchanger [433], uranium is eluted by 150 ml of $1N$ sulfuric acid, while aluminum is eluted by 150 ml of $2N$ HCl.

Acidity-adjusting cation exchangers are employed to separate aluminum from alkaline-earth elements and magnesium [394], lithium [584] and phosphoric acid [158].

A cellulose column has been employed [565a, 566] to separate aluminum from interfering elements during the analysis of iron, steel and Sn-Pb alloys. By using cellulose which has been treated with a $0.1M$ solution of 2-(ethylhexyl)phosphoric acid in cyclohexane, it is possible to separate aluminum from Ga, In and Tl by selective elution of the sorbed elements with hydrochloric acid of varying concentrations. Thallium, gallium, indium, and aluminum are eluted by $0.3M$, $0.7M$, $3M$, and $8M$ HCl, respectively [624].

A number of workers use silica gel to separate aluminum from the interfering elements [448, 450, 883–885]. Kohlschütter et al. [885] use silica gel to separate aluminum from calcium and zinc. All three metals are sorbed on silica gel from solutions at pH 4, and then calcium can be extracted by passing acetate buffer solution at pH 5.75, while zinc is eluted by acetate buffer solution at pH 5.2. Aluminum remains on the column and can be desorbed by passing 50 ml of 1 : 1 HCl and then washing the column with water. Fedorov and Sokolova [448, 450] determined aluminum in alloy steels, and separated cobalt with the aid of silica gel. Morachevskii and Zaitsev [272] employed natural and artefact phosphorites to separate thallium and gallium from aluminum.

Separation by complexing. Aluminum is separated from a number of metals by taking advantage of the difference in the stabilities of their complexonates.

To determine aluminum in iron ores, clayey carbonate spars and in fuel, Complexone III in excess is added to the sample solution to bind aluminum and iron, the pH is adjusted to 5.3, the solution is boiled for 2 min, and then left to cool. The cooled solution is run through a column of Wolfatite KPS-200. Calcium and magnesium are sorbed on the cation exchanger, while aluminum and iron pass through the column. The overall content of the two metals is complexometrically determined in the filtrate [737].

Iron is determined on an aliquot, and then aluminum is found by difference.

According to Babacěv [539], the determination of aluminum in a mixture with Fe, Cr, Ca and Mg is carried out as follows. Iron is first titrated in the sample solution at pH 1.5–2 by Complexone III with sulfosalicylic acid. To the titrated solution Complexone III is added in excess to bind aluminum and chromium, the solution is brought to the boil, the pH is adjusted to 5, and the boiling is continued for 10 min more. The cooled solution is passed through a column with KU-2 or Wolfatite KPS-200 in the ammonium form, when calcium and magnesium are sorbed. The column is washed with water; NaF is added to the filtrate to decompose aluminum complexonate, and the liberated Complexone III, in an amount equivalent to that of aluminum, is titrated by a solution of zinc acetate in the presence of ferricyanide and ferrocyanide.

In a second study by the same worker [538], aluminum is separated from iron and titanium as follows. The sample solution at pH 1.5–2 is heated to 50–60°C, a few drops of H_2O_2 are added, and the solution is titrated by Complexone III with sulfosalicylic acid as indicator. The titrated solution is passed through a KU-2 or Wolfatite KPS-200 cation exchanger. Titanium and iron are eluted by water, aluminum is desorbed with 3 N HCl, and is then determined in the eluate by titrating the excess Complexone III by a solution of zinc in the presence of Xylenol Orange as indicator at pH 4.6. The method has been used for the analysis of cement, clays, and slags.

A similar procedure was used [22] to separate Fe, Al and Ti from Mn. The only difference was that after the titration of iron and titanium, Complexone III in excess was added to the solution, the pH was adjusted to 3.3–4.4, the solution was boiled for 1–2 min, and then passed through a column with the above cation exchangers. Under these conditions, manganese does not form a complexonate and is sorbed by the cation exchanger, while aluminum passes into the filtrate as complexonate and is determined as described above. The method has been used for the analyses of slags and manganese ores.

Giuffré and Capizzi [757] proposed that titanium be separated from aluminum by passing the mixture of the two metals (pH 1–2) containing Complexone III and H_2O_2, through a column with Dowex W50×8 cation exchanger. Titanium passes into the filtrate as complexonate, while aluminum is sorbed and is then eluted by 150 ml of 4 N HCl.

A similar technique has been proposed [617] for the separation of aluminum from cobalt.

The sample solution is treated with Complexone III and acidified to a pH of about 2. Then 5–10 ml of 6% H_2O_2 are added, and the solution is boiled for 2 min. The solution is cooled and is then passed through a column with Amberlite IR-120 cation exchanger; aluminum complexonate is decomposed and the aluminum is sorbed on the resin, while cobalt passes into the filtrate. Aluminum is then desorbed by $4 N$ HCl.

Aluminum can be separated from beryllium as complexonate by passing the solution at pH 2.5–4.5, containing Complexone III, through Amberlite IRA-120 cation exchanger in the Na^+-form [1019].

Aluminum is separated from magnesium and calcium as the citrate complex by passing the solution at pH 4 through Dowex 50W×8 cation exchanger in the NH_4^+-form [528]. Aluminum passes into the filtrate, while calcium and magnesium are retained on the column.

The method has been used for the analysis of limestones and dolomites.

In the analysis of chromium-iron ores [528] the chromium is oxidized to Cr(VI) by fusing the sample with Na_2O_2. The melt is dissolved and the solution, which should be 0.1–0.2 N in H_2SO_4, is passed through the column. Chromium (VI) passes through the column, while all the other metals are retained. The residual Cr(VI) is eluted by 140 ml of 0.1 N sulfuric acid and 50 ml of water. Magnesium and nickel are then eluted by 200 ml of 1 M acetate buffer solution at pH 6; iron is eluted by 180 ml of 5% citric acid. The aluminum remaining on the column is eluted by 150 ml of 4 N HCl.

During the analysis of bronzes [355], aluminum is separated from copper as the citrate complex from ammoniacal solutions, using SBS cation exchanger.

Iron is separated from Al, Co and Ni in a mixture of 80–90% tetrahydrofuran and 0.6 M HCl with the aid of Dowex 50×8 cation exchanger [897].

Titanium in the form of a peroxide complex can be separated from aluminum by using a highly acidic cation exchanger [473, 1206].

Tsyvina and Kon'kova [473] pass the sample solution, 0.75 N in HCl and containing a few drops of H_2O_2, through KU-2 cation exchanger in the H^+-form. Titanium passes into the filtrate, while aluminum is sorbed on the resin. It is desorbed by 50 ml of 3 N HCl.

Strelow [1206] separates titanium from aluminum and certain other elements by passing the sample solution, 0.3 N in sulfuric acid, through a column of Biorad AG 50W× ×8 cation exchanger in the H^+-form. Titanium is selectively extracted by 350 ml of 1 N sulfuric acid in the presence of 1% H_2O_2. The column is washed with 100 ml of 0.1 N HCl and Fe(III), Be, Mg, Cu, Mn, Zn, Ni, and Co are desorbed by passing 400 ml of 2 N HCl, while aluminum is desorbed by passing 400 ml of 3 N HCl.

Chromium can be separated from aluminum [1055] on Wolfatite KPS-200 cation exchanger in the H^+-form as the thiocyanate complexes $[Cr(SCN)_4]^-$, $[Cr(SCN)_5]^{2-}$, and $[Cr(SCN)_6]^{3-}$.

Aluminum has been separated from Be, Ca, Mg, Mn, Zn, and Cu by taking advantage of the formation of a stable complex between aluminum and Tiron [90, 91]. Aluminum can be separated from copper as the sulfosalicylate complex from solutions at pH 10 [355, 952]; Ni, Zn and Mn [952] can also be separated. Copper is separated from aluminum as the thiosulfate complex [356] on KU-2 in the Na^+-form, while iron can be separated as the pyrophosphate complex [255].

Methods based on the amphoteric nature of aluminum. Aluminum is an amphoteric metal and can be separated from both cations and anions by sorption on cation exchangers in the NH_4^+-form from solutions at pH 2.5 – 3.0. It can then be desorbed by solutions of alkali [222, 238, 239, 356, 357]. Lazarev [222] determined aluminum in Al-Ni-Co alloys and in bronzes by passing the solution through a column with SBS in the H^+-form. Aluminum is then extracted by 300 ml of 1 N NaOH and 50 ml water at the rate of 3.5 ml/min.

Ryabchikov and Osipova [356] proposed that before the separation of aluminum from copper, the copper be converted to the ammoniate complex, and the aluminum to the aluminate. When the solution is passed through KU-2 cation exchanger in the Na^+-form, the copper is retained on the column, while aluminum passes into the filtrate.

SEPARATION ON ANION EXCHANGERS

Methods based on the formation of chloride complexes are especially important. Kraus and Nelson [207] carried out the most detailed study of the sorption of chloride complexes on anion exchangers. According to these workers, Dowex-1 anion exchanger strongly sorbs the chloride complexes of the following metals: Cu(I), Ag(I), Au(III), Zn, Cd, Hg(II), Ga, Tl(III), Ge(IV), Zr, Sn(II), Sn(IV), Hf, V(V), Sb(III), Sb(V), Bi, Pa(V), Mo(VI), W(VI), Te(IV), Po(IV), U(IV), U(VI), Tc(VII), Re(VII), Fe(III), Co(II), Ru(IV), Rh(IV), Os(III), Pd(II), Ir(IV), Pt(IV). The following are weakly sorbed: Cu(II), Sc, In, Ti(III), Ti(IV), Pb(II), V(IV), As(III), As(V), Nb(V), Ta(V), Cr(III), Se(IV), Mn(II), Fe(II), Rh(III), Ir(III). Alkali and alkaline-earth metals, Be, Mg, Al, Y, REE, Tl(I), Ac, Th, and Ni are not sorbed.

Horton and Thomason [819] quote the following values for the partition coefficients between 9 M HCl and Dowex-1 for metals usually encountered in alloys of Fe, Cu, U, Sn, and Pb:

Metal	K_p	Metal	K_p
Fe(II)	4	Zr	100
Cu(II)	8	U(VI)	500
Cd, Co(II), W(VI)	40	Zn(IV)	$5 \cdot 10^3$
Sb(III), Sn(II)	50	Fe(III)	$3 \cdot 10^4$
Zn	60	Sb(V)	$2 \cdot 10^5$
Mo(VI)	90	Cr(VI), Mn(VII), Tc(VII), V(V)	High

According to these workers, the following are not sorbed from $9\,M$ HCl $(K_p < 2)$: Al, Mn(II), Cr(III), Ni(II), V(IV), Ti(III), Ti(IV), Th, Mg, REE, Be, and Pb. Thus, sorption from $9\,M$ HCl can be used to separate aluminum from almost all the metals found in Fe, Cu, U, Sn and Pb alloys, which interfere with the determination of aluminum by Aluminon. Additional separations become possible if the sample solution is passed through the anion exchanger twice, at different acidities. Thus, lead is not sorbed from $9\,M$ HCl, but is strongly sorbed from $2\,M$ HCl. Therefore, Horton and Thomason [819] suggested that the solution be made $9\,M$ in HCl and passed through the anion exchanger column. The acidity of the eluate is then reduced to $2\,M$ HCl, and the eluate passed through a second anion exchanger column.

The variation in the sorption of the chloride complex of Fe(III) on strongly basic Amberlite IRA-400 with the acidity can be illustrated by the following data [590]:

HCl, M	Sorption, %	HCl, M	Sorption, %
10.4	99.8	6.2	99.6
9.4	99.7	5.2	99.5
8.3	99.6	4.2	89.6–93.0
6.7	99.6		

Thus, the separation from iron is satisfactory from $5-10\,M$ HCl. The anion exchangers OAL [908], Amberlite LA-1 [386], and AV-17 [8, 378, 895] have also been employed to separate aluminum from iron. The separation of Cu [278, 1023], Zn [172, 317], Pb [317], and Co [843] is usually carried out in $8\,M$, $2-2.5\,M$, $2.5\,M$, and $9\,M$ HCl, respectively. Gallium is separated on strongly basic anion exchangers OAL and L from $7\,M$ or $6\,M$ HCl [899]. Denisova and Tsvetkova [104] described the analysis of Al-Sb-Ga alloys, with separation as chloride complexes.

Titanium is very weakly sorbed on anion exchangers as the chloride complex; the addition of H_2O_2 does not increase the sorption [580]. Thus, titanium cannot be separated from aluminum as the chloride complex. Plutonium is separated on Dowex-1 anion exchanger from $8\,M$ HCl [979] or $12M$ HCl [705]. The partition coefficient of plutonium from $8M$ HCl is 1,400, while from $12M$ HCl it is 8,000 [979]. Uranium can be separated on the strongly basic anion exchanger Deacidite-FF from $8M$ HCl [1237] or from $10-11M$ HCl in the presence of HI [572]. Tellurium can be separated from aluminum on EDE-10P anion exchanger; the sorption takes place from $6M$ HCl [382].

Kanzantsev et al. [155] carried out a comparative study of various anion exchangers (AV-17, AMP, EDE-10P, AN-2F, AN-22, AN-31) with a view to their potential use in separating various metals as their chloride complexes. Aluminum is not sorbed by any anion exchangers at acidities between 0.1 and $11N$ HCl; Ni and Ce(III) are sorbed to an approximately equal extent $(1-15\%)$ by all anion exchangers. Copper is practically not sorbed from HCl at concentrations up to $3N$; its maximum (about 50%) adsorption on EDE-10P, AN-31, and AN-2F is noted in $9-11N$ HCl. The maximum sorption of Fe(III) by all anion exchangers takes place from $9-11N$ HCl; the iron is best sorbed on AV-17 and AMP. Zinc is sorbed by all anion exchangers even from $0.1N$ HCl (maximum sorption from $2-5N$ HCl).

Nelson et al. [1011], and Pakholkov and Rylov [331a], studied the possible separation of aluminum from other metals as the fluoride complex. Aluminum dissolved in HF-HCl and HF-H_2SO_4 mixtures is not sorbed on strongly basic anion exchangers, but is sorbed to a considerable extent on weakly basic anion exchangers (EDE-10P, AN-2F), and can be separated from a large number of metals in this way. The highest sorption is observed from $0.1N$ HCl and H_2SO_4 solutions which are $2M$ in HF. The sorption becomes much weaker if the concentration of HCl or H_2SO_4 is increased, while that of HF is decreased. Beryllium is sorbed by strongly basic anion exchangers from a solution which is $0.01N$ in HCl and $1M$ in HF, and can thus be separated from aluminum.

According to [733a], the greatest difference between the partition coefficients of zirconium and aluminum is observed in solutions which are $0.06M$ in HCl and $0.8M$ in HF. The following partition coefficients were found for zirconium and for a number of other metals:

Metal	K_p	Metal	K_p	Metal	K_p
Fe	0	Ti	9.1	Zr	18.4
V	1.2	Nb	10.5	Mo	20
Sn	2.4	Hf	11	Ta	22

Iron passes completely through the column and interferes with the determination of aluminum; it is accordingly separated as the cupferronate.

Aluminum can be separated from beryllium as the oxalate complex [384, 385]. Beryllium is not sorbed by AV-16 anion exchanger from oxalate solutions at pH 4–4.5, while aluminum is readily sorbed under these conditions ($K_p = 545$). If iron is present, it is sorbed as well.

Lusker and Sebba [943] described the separation of aluminum from beryllium in the form of oxalate complexes with the aid of ion flotation at pH 4 and 23°C; the flotation agent was tetradecylamine hydrochloride, $C_{14}H_{29}NH_2 \cdot HCl$, dissolved in methanol, and the molar ratio Al : Be was 2 : 1. The oxalate complex of beryllium dissociates on dilution, so that beryllium passes through the column and is thus separated from aluminum.

Aluminum can be separated from a number of metals, including calcium, as a citrate complex on Dowex-2 or Dowex-50 anion exchanger in the citrate form; calcium passes into the filtrate, while aluminum and iron remain on the column. Aluminum can then be eluted by conc. HCl, and iron by $1\,N$ HCl [1139, 1140].

It has been suggested [571, 1224, 1225] that Fe(III) and V(IV) be separated from aluminum as thiocyanate complexes.

Iron is separated from $0.68\,M$ thiocyanate solution (pH 2) on Amberlite IRA-400A. The sample solution is passed through the column, and the column is washed with fifteen 10-ml portions of $0.68\,M$ KSCN, acidified to pH 2 with HCl. Aluminum passes through the column almost quantitatively (about 99%), whereas iron in the form of a complex anion is sorbed on the anion exchanger. The relative error of the method is 1%.

During the isolation of V(IV) as the thiocyanate complex on Amberlite IRA-400 anion exchanger, the partition coefficients from solutions $0.1\,M$ in HCl have the following values (after Bok and Schuler [580]):

NH$_4$SCN, M	K_p	NH$_4$SCN, M	K_p
0.1	220	1.00	3,220
0.25	3,655	2.00	2,875
0.50	4,335		

If $0.5-2\,M$ thiocyanate concentrations are employed, the sorption of vanadium is equally satisfactory from $0.1\,M$ and $1.0\,M$ HCl (99.7–99.8% and 99.6–99.7%, respectively). Pentavalent vanadium is sorbed as the complex ion $[VO(SCN)_4]^{2-}$. The relative error involved in the separation of 10–20 mg V(IV) from 20–50 mg Al is not more than 0.8%.

Vanadium can be separated from aluminum as the peroxy complex on Wolfatite L-150 anion exchanger [1000].

To 5 ml of the sample solution, about $1\,N$ in nitric acid, 3 g of swollen Wolfatite L-150 and 10 ml of water are added, and the mixture is stirred for 10 min. During this period the vanadium complex is sorbed on the resin. The resin is filtered and washed with 20–30 ml of water. The filtrate is boiled to expel H_2O_2, and then aluminum is determined by Aluminon.

Zirconium can be separated from aluminum as the sulfate complex. Korkisch and Farag [898] found that the sorption of zirconium decreases greatly with increase in the concentration of sulfuric acid; complete sorption is observed from 0.05–0.2 N sulfuric acid. Up to 10 g Na_2SO_4 and $(NH_4)_2SO_4$ in 100 ml of solution have no effect on the sorption of zirconium, but with more than 10 g the sorption of zirconium decreases. Sodium chloride greatly reduces the sorption of zirconium; the sorption is quantitative in the presence of up to 0.25 g NaCl in 100 ml. Nitrates produce an even larger decrease in the adsorption of zirconium; up to 0.1 g $NaNO_3$ in 100 ml are still permissible. If the experimental conditions are optimum for zirconium, Sn(II), V(V), Mo(VI), W(VI), and U(VI) are sorbed on the anion exchanger, while Mg, Ca, Cu(II), Zn, Cd, Ti(IV), Th, Cr(III), Mn, Fe(III), Co, Ni, and REE pass into the filtrate. Phosphates interfere with the separation of zirconium.

Molybdenum [271], uranium [1021, 1238], and plutonium [536, 850] can be separated on anion exchangers from nitric acid solutions.

According to Urubay et al. [1238], uranium is separated from aluminum by passing the ethereal sample solution, 0.03 M in HNO_3, through a column of Dowex-1 anion exchanger, which has been previously washed with 50 ml of 1 M HNO_3, at the rate of 0.5 ml/min. Then 50 ml of 0.03 M HNO_3 in diethyl ether are passed through. Uranium, with a low partition coefficient under these conditions (about 4.5), passes into the filtrate, while aluminum is completely sorbed on the anion exchanger. Thorium and iron are sorbed together with aluminum. The column is washed with a 0.03 M solution of HNO_3 in ether. Aluminum is eluted by 1 M HNO_3. The relative error is 1%.

To separate plutonium from aluminum, the sample, 7–8 M in HNO_3, is passed through a column of Deacidite-FF, which has been previously treated with 7–8 M HNO_3. The column is then washed with 60 ml of 7 M HNO_3 [536, 850].

Nazarenko et al. [291] increased the selectivity of the separation of thorium from aluminum and other elements by using AV-17 anion exchanger, previously treated with Thoron. This treatment has the practical effect of converting the anion exchanger to a cation exchanger on which thorium cations are sorbed. The optimum acidity is 0.05 N in HCl; the solution should be passed at the rate of 0.5 ml per minute.

Gelis [749] separated the various components of steel by using cation exchanger and anion exchanger columns.

OTHER CHROMATOGRAPHIC SEPARATION METHODS

Babko and Gridchina [29] proposed that aluminum be separated from zirconium by electrodialysis. The separation is based on the different tendencies of these metals to form polyions in solution. Nakagawa [296] separated Al from Fe and Ti by the selective extraction of Ti(IV) and Fe(III) from HCl solutions with N-dodecenyltrialkylmethylamine dissolved in kerosene; Ti and Fe are extracted from solutions of HCl at concentrations above $8N$ and $6N$, respectively. Aluminum is not extracted from sulfuric or hydrochloric acid solutions.

The separation of aluminum from interfering elements by partition chromatography has been repeatedly studied.

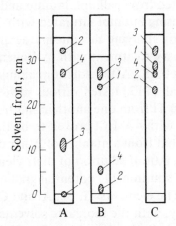

Figure 35. Separation of aluminum (1), gallium (2), indium (3), and thallium (4) by paper chromatography [624]:

A and B) paper treated with di-(2-ethylhexyl)phosphoric acid; developing solvent $1M$ HCl (A) and $8M$ HCl (B and C).

Magee and Scott [946] described a method for the separation and determination of Al, Ga, In and Tl by chromatography on Whatman No.1 paper in a solvent mixture (phenol : methanol : conc. HCl = 50 : 30 : 30 pts. by wt.). The R_f-values found for Al, In, Tl and Ga were 0.10, 0.24, 0.50 and 0.76, respectively. The chromatogram zones are revealed by spraying with a solution of hydroxyquinoline in a 48 : 48 : 4 mixture of methanol, chloroform, and water. The paper is then held in an atmosphere of ammonia at

80°C for 10 min, washed with hot water, and dried at 120°C for 2 hours. Yellow zones, fluorescing in UV light, are observed. The optimum amounts for the separation are 20–500 μg. The zones are eluted by chloroform, and then aluminum can be determined by spectrophotometric methods. The R_f-values found for Cu, Fe(III), Cd, Ni, Cr and Zn are 0.25, 0.40, 0.33, 0.15 and 0.35, respectively, which means that these metals interfere.

The same elements were also separated [624] on Whatman No.1 filter paper impregnated with a 0.1 M solution of di-2-ethylhexyl)-phosphoric acid in cyclohexane. The mobile phases employed were 1 M and 8 M HCl; the R_f-values of Al, In, Tl and Ga in 1 M HCl were 0.00, 0.32, 0.78 and 0.92, and in 8 M HCl 0.78, 0.89, 0.17 and 0.04, respectively. The R_f-values obtained on paper which had not been impregnated with the reagent were similar for all four elements, and the separation was not possible, whereas it was easily performed on impregnated paper (Figure 35).

Aluminum was separated from gallium, indium and thallium [97] by paper chromatography, using butanol saturated with 2 N HCl as solvent. Aluminum was revealed by spraying the chromatogram with a solution of Aluminon, after preliminary treatment with hydrogen chloride (red spot); the R_f for aluminum was 0.05. After isolation, aluminum was determined gravimetrically. Fisel et al. [713] tested various solvent mixtures in order to separate Al, Ga, In and Tl from one another and from other elements. A mixture of pentanol-1 with 4 N HCl proved to be the most satisfactory. Aluminum can be separated from a number of other cations by paper electrophoresis [731]. A 0.06 M solution of diethylenetriamine was used as the electrolyte; pH of solution 11, potential gradient 10 V/cm, time of migration one hour. The elements Ti, Be, Al and Ca can be separated by paper chromatography, with the inorganic solvents NaOH and NH₄OH [1165]. The chromatogram is developed best by a 0.5 N solution of ammonia. The R_f-values of Ti, Be, Al and Ca are 0.00, 0.67, 0.87 and 0.96, respectively. The elements are revealed by spraying with a saturated solution of alizarin in 96% ethanol and holding in NH_3 and CH_3COOH vapors.

Alimarin et al. [12] separated gallium from aluminum in the system HCl-tri-n-butyl phosphate by column chromatography. Polyfluoroethylene resin was used as carrier; tributyl phosphate served as the stationary phase, and 0.1 N HCl as eluent.

Aluminum was determined in aluminum bronzes by paper chromatography [997]. A 75:15:10 mixture of n-butanol-conc. HCl-water was used as solvent. The section of the chromatogram which contained aluminum was cut out and treated with boiling water acidified with hydrochloric acid. Aluminum was photometrically determined in the eluate by Chrome Azurol S.

The separation of Al from Be, Ga, In, Tl, Fe and Cu by gas-liquid chromatography has been repeatedly described [987, 988, 996, 1153]. The metals are first converted to trifluoroacetonate complexes, which are then extracted by benzene.

DETERMINATION OF INTERFERING ELEMENTS BY ELECTROLYSIS ON A MERCURY CATHODE

Electrolysis on a mercury cathode is one of the most important methods for the separation of interfering elements from aluminum. This technique is especially valuable in the analysis of multicomponent materials, since a large number of metals can be eliminated at the same time by electrolysis.

The following metals are separated by electrolysis on a mercury cathode: Fe, Cr, Co, Ni, Cu, Zn, Mo, Cd, Sn, Pb, Bi, Hg, Tl, In, Ga, Ge, Ag, Au, Pt, Pd, Rh, Ir, Re. The metals Al, Ti, Zr, V, U, Th, Be, Nb, Ta, W, P, As, Sc, Y, REE, Mg, alkali and alkaline-earth metals are not separated. Manganese is not completely separated; some of it becomes oxidized to MnO_2 and separates out at the anode, or is oxidized to MnO_4^- and colors the solution raspberry red. According to Dübel and Flurschütz [689], the separation of manganese is quantitative if a few drops of 30% hydrogen peroxide are added to the electrolyte during the electrolysis. Chromium is slowly removed by electrolysis. Therefore, in the analysis of steels containing more than 5% of chromium, it is recommended that the bulk of the chromium be removed before hydrolysis as chromyl chloride [555]. A small proportion of the iron always remains in the electrolyte, but this does not interfere with the determination of aluminum by many photometric methods, provided that the iron is reduced to Fe(II) by ascorbic acid. Traces of chromium and molybdenum may also remain in the electrolyte.

Electrolysis on a mercury cathode is mainly employed for the preliminary separation of the bulk of the interfering elements. As a rule, the elements which remain in the electrolyte after electrolysis are removed by some other method.

Traces of Fe and Cu and also Ti, Zr and V, are removed by chloroform extraction of their cupferronates. Traces of certain metals are removed by extracting their hydroxyquinolates and hydroxyquinaldates.

The metals which remain in the electrolyte are separated from aluminum by fusion with sodium carbonate [1047], precipitation by NaOH in the presence of H_2O_2 [1156], or by double precipitation with ammonia in the presence of NH_4Cl [814]. Aluminum mixed with Ti, Zr, V and residual iron is separated from chromium and phosphates by extracting the cupferronates of these metals at pH 3.5. Aluminum is then separated from Ti, Zr, V and Fe by extracting the cupferronates of the latter metals from $4N$ HCl [555].

To cool the electrolyte, the electrolyzer is equipped with a water jacket [1038, 1264] or with a spiral condenser [568, 776]. The cooling is necessary to produce the optimum temperature of about $40°C$, since above $40-50°C$ the electrolysis slows down. If the electrolyte is cooled, it is permissible to work at relatively high current intensities (5-8 A) so that the interfering elements are rapidly removed and the dissolution of mercury is prevented. The solution can be stirred, either mechanically with a glass stirrer [776, 1264] or else magnetically [568, 656, 689]. According to a number of workers, stirring the upper mercury layer maintains the purity of the mercury surface and favors rapid precipitation [1264]. The advantage of magnetic stirring is that the ferromagnetic metals which separate out at the mercury surface and do not form amalgams (Fe, Co, Ni, Cr) penetrate inside the mercury layer, and the mercury surface remains clean throughout. As a result, the duration of the electrolysis is somewhat shorter [689].

The electrolysis is carried out from solutions of sulfuric, perchloric or acetic acid, but not from solutions of hydrochloric acid. Chlorides must be absent from the electrolyte. If the sample solutions contain chlorides, these must be removed by evaporating with sulfuric acid or perchloric acid to the appearance of SO_3 or $HClO_4$ fumes; the evaporation is best repeated twice.

Hinek and Wrangell [811] prefer to perform the electrolysis from solutions of perchloric acid. The volatile acids are readily removed before electrolysis and vaporization during electrolysis is insignificant. Moreover, sparingly soluble salts are not formed during electrolysis, as is sometimes noted in electrolysis from sulfuric acid solutions.

Holler and Yeager [814] used acetic acid solutions for the electrolysis. Acetic acid foams strongly during electrolysis, but this can be prevented by adding 2-3 drops of octanol or Antifoam LFX. The disadvantage of the method is that many samples cannot be solubilized in acetic acid.

The concentration of sulfuric acid usually recommended is $0.1-0.2 N$. At lower acidities the electrolysis is slower; if the acid concentration is too high, the amalgam formed may dissolve. The optimum amount of mercury is 20-30 ml. Current intensity 4-15 A, voltage 4-12 V. Under optimum conditions the duration of the electrolysis is 40–60 min. The end of the electrolysis is indicated by the absence of certain ions in the electrolyte, indicated by spot tests. For the purification of the mercury, see [343].

Due to the toxic nature of mercury, electrolysis on a mercury cathode is seldom employed in laboratories. It may be employed in exceptional cases, but if at all possible, other methods should be used.

OTHER SEPARATION METHODS

Other methods include the elimination of the interfering elements by evaporation, sublimation, and distillation.

The elimination of chromium as chromyl chloride is widely employed. Tin, arsenic and antimony are often evolved as bromides by treating the sample with an $HBr-Br_2$ mixture [260, 938]; tin can also be expelled as chloride by repeated evaporations with HCl until thick, white fumes are no longer evolved [378, 488]. During the determination of aluminum in boron, the boron is evolved by treatment with a mixture of concentrated nitric and hydrofluoric acids [284].

Singh et al. [1166] separated uranium from aluminum by the photolytic method. The bulk of the aluminum was separated as fluoride and the precipitate was filtered off. Then 10 ml of ethanol were added to the filtrate and the solution was allowed to stand in sunlight. There followed an almost quantitative (99%) photolytic precipitation of uranium as $NH_4F \cdot$ $\cdot UF_4 \cdot H_2O$; it could thus be separated from the residual aluminum.

CHAPTER IV

Determination of Aluminum in Natural and Industrial Substances

DETERMINATION OF ALUMINUM IN MINERALS, ORES AND INDUSTRIAL CONCENTRATES

The following methods are employed to solubilize minerals, ores and industrial concentrates: treatment with $HF-H_2SO_4$ or $HF-HClO_4$ mixtures, followed by fusion of the dry residue with $K_2S_2O_7$; fusion with $NaKCO_3$, Na_2CO_3, $Na_2B_4O_7$ or with a mixture of Na_2CO_3 with $Na_2B_4O_7$; sintering with Na_2CO_3; fusion with $NaOH$, Na_2O_2, or with a mixture of $NaOH$ and Na_2O_2; fusion with $KHSO_4$ or $K_2S_2O_7$; dissolution in a mixture of HCl, HNO_3 and $HClO_4$, evaporation to the appearance of $HClO_4$ fumes and fusing the residue with $K_2S_2O_7$.

Determination of aluminum in ores

Bauxite, nepheline concentrates, nepheline-apatite ores, cryolite and aluminum fluoride are decomposed by fusion with sodium hydroxide in silver, nickel or iron crucibles. Aluminum in these materials is determined mainly by the complexometric method. The following procedure was proposed by Artem'eva [16] to determine aluminum in nepheline-apatite ores and nepheline concentrates.

Sodium hydroxide (3–4 g) is fused in an iron crucible and cooled, and 0.25 g of the sample is introduced. The mixture is wetted with a few drops of an ethanol solution of NaOH, and is fused at 600–700°C for 15–20 min. The melt is dissolved in hot water, 15 ml of a 20% solution of NaOH are added, and the mixture is boiled for 20 min. When cool, the mixture is diluted to 250 ml in a volumetric flask, and filtered through a double layer of dry filter paper.

The filtrate (100 ml) is neutralized with HCl to Congo Red paper, 5–20 ml of 0.05 M Complexone III are added, and the mixture is heated to 50°C. It is then made alkaline with ammonia until the indicator turns pink, 20 ml of acetate buffer solution (pH 5.2–5.8) are added, and the mixture is boiled for 1–2 min. When cool, 5–7 drops of a 0.1% solution of Xylenol Orange and Complexone in excess are added, and the mixture is titrated by 0.05 M $ZnSO_4$.

The relative error of the method is less than 2%.

Busev et al. [71] suggested that aluminum be determined in bauxites, chamotte, clinker, and other materials by a more selective method involving the use of sodium fluoride.

One gram of the sample is decomposed by any suitable method. Silica is separated and the filtrate is diluted to 250 ml with water. A 50-ml aliquot is diluted to about 200 ml with water, 10–60 ml of 0.05 M Complexone III are added (60 ml for bauxite, 40 ml for chamotte, 15 ml for clinker, 10 ml for crude cement mixtures). The mixture is heated to about 50°C, the solution is neutralized with 1 : 1 ammonia to phenolphthalein, and the color is discharged by adding 1 : 1 HCl drop by drop. Twenty ml of acetate buffer solution (500 g of ammonium acetate and 20 ml of 80% acetic acid are diluted to 1 liter with water). The mixture is boiled for 2–3 min and rapidly cooled under the tap. Then 3–4 drops of a 0.1% solution of Xylenol Orange are added, and the excess Complexone III is titrated by 0.05 M $Zn(NO_3)_2$ until the yellow solution turns violet. To the titrated solution are added 40 ml of a 4% solution of NaF, and the solution is boiled for 2–3 min. The solution is rapidly cooled and the liberated Complexone III is titrated by 0.05 M $Zn(NO_3)_2$. The consumption of zinc nitrate during the second titration is equivalent to the content of aluminum.

The determination of aluminum in bauxite can be carried out by a similar method, except that $CuSO_4$ is employed as titrant [877]; at the end point the color changes from green to blue-violet. The color change is less sharp than when zinc is employed as titrant. Aluminum in nepheline concentrates can also be determined by back-titration against iron sulfosalicylate after the sample has been decomposed by fusing with sodium hydroxide [138].

Aluminum in cryolite and in aluminum fluoride can be determined by the gravimetric hydroxyquinoline method [123].

The sample (0.2 g) is fused in a silver crucible with 3 g of NaOH at 700°C for 10 min. The melt is leached with hot water, and then the solution is cooled and diluted to 200 ml in a volumetric flask. The solution is filtered through a dense filter paper into a dry beaker; the first portion of the filtrate is discarded. A 100-ml aliquot (for cryolite) or a 50-ml aliquot (for AlF_3) is withdrawn for the determination of aluminum by the gravimetric hydroxyquinoline method.

The gravimetric phosphate method has been proposed [1102] for the determination of aluminum in iron and manganese ores. However, since the aluminum content is low, photometric methods are preferable: with hydroxyquinoline [144, 864], Eriochrome Cyanine R [463, 808, 855], or Chrome Azurol S [596].

Kassner and Ozier [864] proposed the following method for the determination of aluminum in iron ores.

Two grams of the ore are dissolved by heating in 20 ml HCl (sp. gr. 1.19). Between 5 and 10 ml of HNO_3 (sp. gr. 1.42) and 12–15 ml of 60–70% $HClO_4$ are added, and the mixture is heated to the evolution of thick $HClO_4$ fumes. The fumes are allowed to escape for 5 min to dehydrate the silica. The residue is cooled, and then diluted to 75–100 ml and filtered. The filter is washed with 1% HCl and then with hot water. The residue on the filter is fused with pyrosulfate, and the solution of the melt is combined with the main filtrate and diluted to 1 liter. To 100 ml of the resulting solution are added 15 g NH_4Cl and 5–6 drops of Bromocresol Purple indicator. The solution is heated almost to boiling. Ammonia (1 : 1) is added until the solution turns purple and the solution is boiled for 1 min. It is then filtered and the precipitate is washed with a hot 2% solution of NH_4Cl. The precipitated sesquioxides are dissolved in 20 ml of hot HCl (1 : 1), the filter is washed with hot 5% HCl, and then with hot water until the acid has been removed.

The solution and the washings are diluted to 250 ml in a volumetric flask. To a 25-ml aliquot are added 0.2 g of tartaric acid, 1 ml NH_4OH (sp. gr. 0.9) and 1 ml of 3% H_2O_2. Five ml of a saturated solution of sodium sulfite are added after 5 min; after 3 more minutes a solution of 1.3 g KCN in 5 ml of water is added, the solution is heated to 70–80°C, cooled, and 2 g of ammonium nitrate are added. The pH of the solution is adjusted to 8.9 ±0.3 by the addition of ammonia or hydrochloric acid. Five ml of a 2% solution of hydroxyquinoline in chloroform are added and the mixture is shaken for 2 min. Two minutes later the chloroform layer is decanted into a 50-ml volumetric flask. The extraction is repeated until the separation of aluminum is complete. The extract is diluted to 50 ml with chloroform. The optical density of the extract is measured at 389 mμ and the content of aluminum is read off a calibration curve.

The relative error is 1–7% for aluminum contents between 0.03 and 2%.

Itikuni [144] described a simple photometric determination of aluminum in iron ores by hydroxyquinoline.

Iron is separated from aluminum as a chloride complex by sorption on an anion exchanger from 9 N HCl. To the eluate are added 1 ml of 3% H_2O_2 and ammonia to neutralize the acid, and aluminum hydroxyquinolate is extracted. The photometric determination is performed at 420 mμ.

Up to 2 mg V and 2 mg Fe do not interfere with the determination of 3.3–23.3 μg Al. The drawback of the method is that it is applicable to limonite only, which does not contain Ni, Cr, or Mn.

The photometric determination by Chrome Azurol S is a very satisfactory method for the determination of aluminum in iron ores [596].

The ore (2.5 g) is dissolved in 2 : 1 HCl, the solution is evaporated to dryness, and the residue is dissolved in 1 : 1 HCl and filtered. The residue is treated with HF to remove SiO_2; it is then fused with pyrosulfate, the solution of the melt is combined with the main filtrate, and the solution is made up to 500 ml. A 20-ml aliquot of this solution is diluted to 500 ml and 20 ml of the diluted solution are taken for the determination. Ten ml of a 2% solution of ascorbic acid are added and the mixture is allowed to stand for 10 min. Fifty ml of acetate buffer solution at pH 5.6 (50 g $CH_3COONa \cdot 3H_2O$ and 3 ml of glacial acetic acid in 1 liter) are introduced, 10 ml of a 0.1% solution of Chrome Azurol S in 50% ethanol are added, and the solution is allowed to stand for 10 min. The photometric determination is performed in 1-cm cells at 545 mμ against a blank solution.

To obtain more accurate results, equivalent amounts of iron should be introduced into standard solutions used to plot the calibration curve.

A number of methods for the photometric determination of aluminum in iron ores by Eriochrome Cyanine R have been described [463, 808, 855]. The determination is conducted at pH 6 and many interfering elements are masked with thioglycolic acid [808, 855]. A correction is introduced for vanadium; to do this, a solution is prepared as for the determination of aluminum except that 2 ml of a 2.4% solution of NaF are added, the solution is made up to the mark with water, and its optical density is determined against a blank solution. The value of this optical density is subtracted from the optical density of the sample solution.

Aluminum can be determined in manganese ores [508], and in chromium ores and refractories [507] by a complexometric method with diaminocyclohexanetetraacetic acid (DCTA).

To determine aluminum in manganese ores, 0.2 g of the sample are fused with $NaKCO_3$ and the melt is leached with 200 ml of water. The solution is acidified with hydrochloric acid, a little 3% H_2O_2 is added, and the solution is boiled to decompose the excess of hydrogen peroxide. When cool, 10 ml of a 50% solution of ammonium acetate and ammonia are added to pH 2.2, and Fe(III) is titrated by Complexone III with salicylic acid. To the titrated solution are added 2 ml of a 0.05 M solution of Cu(II) and 2–3 drops of a 0.1% solution of PAN. The solution is heated to boiling and aluminum is titrated by a solution of DCTA.

A similar method has been described for the determination of aluminum in chromium ores and refractories after the sample has been fused with

$KHSO_4$ [507]. The pH of the melt solution is adjusted to 4–6 and the solution is boiled so that Fe(III), Al, and Cr(III) complexonates are formed. Hydroxyquinoline is introduced, the solution is made alkaline with ammonia and is heated to 80°C to precipitate Fe(III) and Al hydroxyquinolates. The solution is filtered after 10 min and the precipitates are dissolved in HCl. Iron and aluminum are determined in the solution, as in the analysis of manganese ores, except that it is the hydroxyquinoline liberated from the solution which serves as the indicator for Fe(III). The presence of hydroxyquinoline does not interfere with the titration of aluminum; Cr(III) is masked by Complexone III and does not interfere.

Determination of aluminum in silicate and carbonate rocks

The determination of aluminum in silicate and carbonate rocks by complexometric methods with Xylenol Orange as indicator is worth noting. The different variants of the method proposed by Babachev et al. [23] can be recommended on account of their simplicity.

Analysis of clinker, cement, carbonates, and other
materials which contain iron, aluminum, and small
amounts of titanium

The sample (0.5 g) is fused with sodium carbonate or with a mixture of Na_2CO_3 with $Na_2B_4O_7$, and then the melt is dissolved in HCl. The silicic acid is separated by evaporating twice with HCl. After the removal of silica, 5–6 drops of HNO_3 (sp. gr. 1.4) are added. Then the solution is heated to boiling and kept boiling for 2–3 min. Salicylic acid (8–10 drops of a 10% alcoholic solution or aqueous solution) is added, and the solution is neutralized by a 10% solution of ammonia until it turns blue-violet (pH 1– 1.5). The solution is heated to 50–60°C and titrated by 0.025 M Complexone III until the blue-violet color has disappeared (determination of iron).

A solution of Complexone III is added in an amount required to bind the expected content of aluminum, and then 10 ml in excess. The solution is heated to boiling, neutralized to pH 5 with a 10% solution of ammonia, 10 ml of acetate buffer solution at pH 5 (500 g CH_3COONH_4 and 20 ml of glacial CH_3COOH in 1 liter) are added, and the solution is again brought to the boil and boiled for 2–3 min. When cool, the excess Complexone III is titrated by a solution of zinc acetate in the presence of Xylenol Orange.

If silica need not be determined, aluminum and iron are precipitated by ammonia and filtered. The precipitate is dissolved in HCl, and the two elements are precipitated in solution as described above.

The error of the method is 1–2% in the presence of 5–6% Al_2O_3.

Analysis of clays, ceramic products, and other
materials containing Fe, Al, and Ti

After the separation of silicic acid, the filtrate is diluted to 250 ml in a volumetric
flask. Iron is determined complexometrically by salicylic acid in 100 ml of the filtrate.
The sum of Fe, Ti and Al is determined on another 100 ml of the solution as follows.
The pH of the solution is adjusted to 1–1.5, a few drops of 3% H_2O_2 and salicylic acid
are added, and the sum of Fe and Ti is titrated by Complexone III. The solution is then
adjusted to pH 5 and aluminum is determined complexometrically in the presence of
Xylenol Orange (in the same solution). If the sample solution contains more than 10 mg
Ti, the results are too high, since the equivalence point in the titration of the sum Fe+Ti
is not sharp. The aliquots taken should be small and the titration must be carried out
very carefully.

Analyses of Portland Cement, metallurgical slags,
and other materials containing Fe, Al, and Mn

After silica has been separated, Fe, Al and Mn in the solution are oxidized by ammonium persulfate and precipitated by ammonia. The precipitate is immediately dissolved
in 1 : 1 HCl and a few drops of H_2O_2. The resulting solution (about 100 ml) is neutralized
with ammonia to the appearance of a faint turbidity, which is redissolved by adding
2–3 drops of 1 : 1 HCl. Ammonium chloride (2–3 g) is added, the solution is brought
to the boil, and 10–15 ml of a 5% solution of ammonium cinnamate are added drop by
drop, with constant stirring. The solution with the precipitate is boiled for 2–3 min, and
is left to stand on a water bath for one hour, with occasional stirring. The precipitate is
filtered, and washed with a hot solution of 10 g NH_4NO_3 and 10 ml of a 5% solution of
ammonium cinnamate in 1 liter. The precipitate on the filter is washed with cold water and
dissolved in HCl. The solution is then diluted to 250 ml in a volumetric flask. A 100-ml aliquot of this solution is used for the complexometric determination of iron and aluminum,
as described on p. 209.

Analysis of materials containing Fe, Al, and Cr. After the separation of silica, Fe, Al
and Cr are precipitated by ammonium cinnamate from hydrochloric acid solution, as
described above. The precipitate is dissolved in 1 : 2 sulfuric acid. One aliquot is taken
for the complexometric determination of iron. In another part of the solution, Cr(III)
is oxidized to Cr(VI) by ammonium persulfate and aluminum is determined by the
complexometric method, as described on p. 212, with allowance for the iron content.
In a third part of the solution, chromium is oxidized to Cr(VI) and then is titrimetrically
determined.

Pritchard [1087] determined aluminum in silicates by the complexometric
method, using DCTA.

The silicate (0.2 g) is fused at 650–750°C in a silver crucible with 2.5 g NaOH for
10 min. The melt is leached with hot water and a few drops of 1 : 1 HCl are added to
the solution toward the end. The solution is transferred to a 250-ml volumetric flask,
containing 100 ml of hot 2 N HCl. The cooled solution is made up to the mark with

water. Fifty ml of the resulting solution are transferred to a polyethylene (or polypropylene) beaker, and 60 ml of a 5% solution of NaOH are added. Ten ml of $0.08\,M$ DCTA (to a suspension of 14 g DCTA in water, $5\,N$ NaOH is added drop by drop until the suspension has dissolved and the solution is diluted to 500 ml) are immediately added, then some macerated paper. The mixture is heated for one hour on a steam bath, and is then allowed to cool slowly and filtered. The precipitate is washed with six 6-ml portions of a cold 2% solution of NaOH. The filtrate is acidified with 1 : 1 HCl until acid to Bromocresol Green. Sodium hydroxide $(5\,N)$ is then added drop by drop until the solution turns blue.

Buffer solution is added (25 ml; pH 5.5; 200 g urotropin, and 40 ml HCl, sp. gr. 1.19 in 1 liter), and then 5 drops of a 0.1% solution of Xylenol Orange. The excess DCTA is titrated by $0.025\,M$ $ZnCl_2$ until the orange-colored solution turns raspberry-red.

The method can be used to determine 10 mg Al in the presence of 20 mg Si, 4 mg Fe, 2 mg Ca and Mg, 0.6 mg Ti, and 0.7 mg Mn.

Evans [704] also used DCTA in the analysis of silicates and other rocks.

The sum Al+Fe+Ti is first determined as follows. Water is added to a part of the sample solution, which should contain 0.05 g of the sample, to a final volume of about 80 ml, after which 2 ml of 20% H_2O_2 are added. Twenty ml of $0.015\,M$ DCTA (or a sufficient excess for 5−10 ml of the titrant to be consumed in back-titration) are added, and the pH is adjusted to 3.5−3.7 with the aid of 4−5 ml of a 10% solution of sodium acetate. The mixture is heated to $50°C$, cooled, 50 mg of indicator mixture (a mixture of o-dianisidine-N,N,N′,N′-tetraacetic acid with NaCl in the ratio of 1 : 200) are added. The excess DCTA is titrated by a $0.01\,M$ solution of a copper salt until the fluorescence, observed when the solution is viewed in UV light in a dark place, is quenched. Another part of the solution is diluted to about 90 ml, and 2 ml of 20% H_2O_2 and 5 ml of a 4% NH_4F solution are added. The mixture is thoroughly stirred and 10 ml of $0.015\,M$ DCTA are introduced. The pH is adjusted to 3.5−3.7 with a 10% solution of CH_3COONa, 50 mg of indicator are added, and the mixture is titrated in a similar manner. In this titration the consumption of DCTA, corrected for the indicator (10 mg of indicator mixture are equivalent to 0.01 ml of $0.01\,M$ Cu) corresponds to the content of iron. Aluminum is found by difference.

A number of methods have been proposed for the complexometric determination of aluminum in silicates by titration of excess Complexone III by a solution of zinc in the presence of dithizone. In one such method iron and titanium are preliminarily precipitated as cupferronates [1244].

The precipitate of cupferronates is filtered and washed 4−5 times with cold washing liquor containing 13 ml HCl (sp. gr. 1.19) and 2 ml of a 6% solution of Cupferron in 100 ml. The excess of Cupferron is extracted from the filtrate with 5−8 ml of chloroform. The chloroform layer is not separated, the solution is neutralized with ammonia to methyl orange and is again acidified with a few drops of HCl (sp. gr. 1.19). Depending

on the content of aluminum, 10–25 ml of 0.05 M Complexone III are added and sufficient acetate buffer solution (80 g CH_3COONH_4 and 60 ml glacial acetic acid in 1 liter)
for the solution to turn yellow, then 10 ml are added in excess. The solution is boiled
for 10 min, cooled under the tap, and an equal volume of ethanol is added, followed by
a few drops of a 0.025% solution of dithizone in ethanol until the solution turns green.
If the solution is brown, more ethanol is added. The Complexone III in excess is then
titrated by 0.05 M $ZnCl_2$ until the solution turns red.

Bennet et al. [557] determined aluminum in alumosilicates by titration by
zinc in the presence of dithizone, and separated iron and titanium by extracting as cupferronates with chloroform. This method is more accurate
than filtering through paper.

Wiebel [1265] used a similar method in the analysis of silicates and separated Fe, Ti, Zn, Ga, V, and some other elements with the aid of nitrosophenylhydroxylamine. Aluminum is separated in this way from all the
interfering elements usually encountered in rocks with a low content of
manganese. Manganese interferes with the determination of aluminum if
present in large amounts, since it is then titrated, albeit not quantitatively,
together with the aluminum [424, 428].

Falchi and Tonani [707] proposed that iron and titanium be extracted
with chloroform as the N-benzoylphenylhydroxylamine.

Titration with dithizone was used during analyses of cements [623]. The
overall iron-plus-aluminum content is determined complexometrically, and
iron is determined by a chromatographic-volumetric method. In a similar
method, developed for kaolin [751], the sum is determined in the same
manner, but if titanium is present, 2 ml of a 10% solution of tartaric acid
are added, and iron is determined by titration with Complexone III at pH 2
in the presence of sulfosalicylic acid in another part of the solution. The
aluminum content is found by difference.

The method proposed by Zalessky and Voinovitch [1288] is worth noting.

To an aliquot part of the solution is added a known volume of Complexone III to
bind Al, Fe and Ti. The pH is adjusted to 5–6 with the aid of acetate buffer solution,
and tartaric acid and a solution of $(NH_4)_2HPO_4$ are added to decompose titanium complexonate. The excess Complexone III is titrated by a solution of $ZnCl_2$ in the presence
of dithizone. Aluminum complexonate is then decomposed by introducing saturated
NaF solution, and the liberated Complexone III is titrated in the same manner. The consumption of $ZnCl_2$ in the second titration is equivalent to the content of aluminum.

The relative error is 0.5%.

To determine aluminum in silicate and carbonate rocks, the excess Complexone III can be titrated by iron sulfosalicylate [161, 373, 975, 1074].
Milner and Woodhead [975] separate Ti, Fe, Zr and certain other metals as

cupferronates and then precipitate aluminum as benzoate. The benzoate precipitate is dissolved in HCl, and then aluminum is determined complexometrically. Sochevanova and Sochevanov [373] determine aluminum in ferruginous carbonate rocks by the following method.

The sample (0.5 g) is dissolved in dilute HCl. The insoluble residue is filtered off, the Fe(II) in the filtrate is oxidized by 10–15 drops of 30% H_2O_2, which is then decomposed by boiling. The cold solution is neutralized with ammonia to first turbidity, which is dissolved in 2–3 drops of HCl. Urotropin (15 ml of a 25% solution) is then added, and the mixture is heated to 80–90°C (boiling should be avoided). The precipitate is filtered off, washed 8–10 times with a warm 0.5% solution of urotropin, and washed back into the beaker used for the precipitation. The residue on the filter is dissolved by 6–8 treatments with hot 5% sulfuric acid. Five ml of sulfuric acid (sp. gr. 1.84) are added, and the mixture is heated until the hydroxides have completely dissolved, and then diluted to 250 ml in a volumetric flask.

To determine iron, 25 ml of this solution are mixed with 75 ml of water, ammonia or a 50% solution of CH_3COONH_4, the pH is adjusted to 1.0–2.0, the solution is heated to 40–50°C, and then 20 drops (0.7–0.8 ml) of a 25% solution of sulfosalicylic acid are added. The solution is titrated by Complexone III until it turns lemon-yellow. To the titrated solution are added 20 drops (0.7–0.8 ml) of HCl (sp. gr. 1.19) and Complexone III in excess. The solution is brought to the boil, neutralized with ammonia to phenolphthalein, and the pink color is discharged by adding 2–3 drops of HCl. Twenty ml of acetate buffer solution (500 g CH_3COONH_4 and 20 ml of glacial CH_3COOH in 1 liter) are added, and the solution is boiled for 3 min. When cool, the solution is titrated by 0.1 M $FeCl_3$ until the lemon-yellow solution turns golden-yellow.

For the complexometric determination of aluminum in micro samples of minerals, see [166].

Photometric methods for the determination of aluminum in silicate and carbonate rocks by Aluminon [1074] and Eriochrome Cyanine R [241, 1116, 1247] have been described.

The method for the determination of aluminum in quartz sand by Aluminon [1074] is given below.

The sand (0.1 g) is dissolved in 5 ml of 40% HF and 5 drops of H_2SO_4 (sp. gr. 1.84) in a platinum crucible. The residue from the evaporation is ignited and fused with 0.5 g of Na_2CO_3. The melt is leached with hot water and the solution is transferred to a beaker. To the crucible, 5 ml HCl (sp. gr. 1.19) are added and the contents of the crucible are cautiously heated for 5 min. Hot water is added and the solution is left on the hotplate for another 5 min, and is then added to the beaker. The resulting solution is boiled for 5 min, cooled, and diluted to 100 ml in a volumetric flask. To determine aluminum, a 50-ml aliquot (for contents not exceeding 0.1% Al), a 25-ml aliquot or a 10-ml aliquot (for contents of 0.1–0.5% Al and 0.5–2% Al, respectively) is taken for the determination of aluminum. If the aluminum content exceeds 2% Al, the solution is diluted to 250 ml and a suitable aliquot is taken.

The aliquot is placed in a 100-ml volumetric flask, 1 drop of a 1% solution of p-nitrophenol is added, and ammonia is then introduced drop by drop until the solution turns yellow. Hydrochloric acid (1 : 1) is then added drop by drop until the yellow coloration disappears. Sixteen drops of 1% solution of thioglycolic acid are added, followed by 16 ml of a solution of Aluminon (0.25 g of Aluminon is dissolved in 250 ml of water; 5 g of gum Arabic are suspended in the solution, 87 g of ammonium acetate and 126 ml of 1 : 1 HCl are added, and the mixture is diluted with water to 500 ml). The volume of solution in the flask is adjusted to about 5 ml below the mark, and the flask is placed in a boiling water bath for 10 min and then cooled. The solution is made up to mark and the optical density is determined.

Determination of aluminum in phosphate rocks

Aluminon in phosphate rocks can be determined photometrically by Alizarin S. The interfering elements are eliminated by cupferronate extraction with chloroform [840].

To a part of the aqueous layer is added 0.5 ml of 1 : 1 H_2SO_4 and the mixture is evaporated to SO_3 fumes. The solution is cooled, and the contents of the beaker are transferred to a 100-ml volumetric flask with the aid of 40 ml of water. Three ml of a 0.1% solution of Alizarin S, 10 ml of 3 N NH_4OH, and 5 ml of 5 N acetic acid are added, the solution is made up to mark, and its optical density is determined at 370 mμ against a blank solution.

At aluminum contents of 0.9–1.8%, the absolute error is 0.02–0.10%.

Determination of aluminum in slags

Aluminum can be determined in slags by complexometric methods; Xylenol Orange is best used as indicator.

Fedorov and Ozerskaya [447] suggested the following method.

The slag (0.25–0.5 g) is placed in an iron crucible, 8–10 g Na_2O_2 are added, the contents of the crucible are mixed and are fused at 80–850°C, with occasional stirring. The crucible is cooled, placed in a beaker and covered with a watch glass, and the melt is leached with 80–100 ml of water. The solution of the melt is placed in a 500-ml volumetric flask, cooled, made up to mark with water, and left to stand for 12 hr to allow the precipitate to settle. It is then filtered through dry paper into a dry flask and the first portions of the filtrate are discarded. One hundred ml of the filtrate are boiled for 8–10 min to decompose the excess H_2O_2. When cool, 25 ml of 0.05 M Complexone III are introduced, the solution is neutralized with HCl (sp. gr. 1.19) to phenolphthalein, 15 ml of buffer solution (500 g CH_3COONH_4 and 30 ml of glacial CH_3COOH in 1 liter) are added, and the solution is boiled for 3–4 min. When cool, the excess Complexone III is titrated by 0.05 M $ZnCl_2$ in the presence of 0.2–0.3 g of indicator (mixture of Xylenol Orange with NaCl in the ratio of 1 : 100).

Lenskaya and Verzhbovskaya [228] decomposed slags in a similar manner.

Complexometric determination of aluminum is effected in the presence of a redox indicator: 1 ml of mixture (10 ml of a 10% solution of $K_3Fe(CN)_6$, 10 ml of a 1% solution of $K_4Fe(CN)_6$, and 30 ml of water) and 1 ml of a 1% solution of benzidine. The solution is titrated by zinc until a bluish-green coloration appears. Fifty ml of a 3% solution of NaF are added, the solution is boiled for 3 min, 1 ml of ferrocyanide-ferricyanide mixture and 0.5 ml of benzidine solution are added, and the solution is again titrated.

Aluminum in titanium slags and concentrates can be determined by complexometric back-titration (titration of excess Complexone III by a solution of thorium with Alizarin S as indicator) [479]. Stilbazo is employed in the analysis of phosphoric furnace slags [111], and slags of copper and lead melts [446]. Unjustifiably complex methods of analysis [363] occasionally appear in the literature.

Determination of aluminum in refractories

The following method [135] can be employed in the determination of aluminum in chromomagnesite refractories.

The sample (0.5 g) is sintered at 850–950°C with sodium carbonate. The cake is dissolved in a mixture of HCl (sp. gr. 1.19) with $HClO_4$, and is heated on a sand bath to expel CrO_2Cl_2. The residue is treated with a few drops of HCl (sp. gr. 1.19) and 50 ml of hot water. The SiO_2 is filtered and the filtrate is diluted to 250 ml in a volumetric flask. Fifty ml of the resulting solution are withdrawn, the pH is adjusted to 5.5, and aluminum is determined by a complexometric method, using Xylenol Orange as indicator and lead acetate as titrant, in the presence of NaF to improve the selectivity of the method.

A similar method has also been described elsewhere [71].

A method employed for the decomposition of chamotte materials is given below [664].

In a nickel crucible are fused 10 g NaOH; 1 g of the sample is added to the cooled melt, the mixture is moistened with ethanol, covered with another 2 g of NaOH, and fused over a moderate flame until blistering ceases; the heating is then continued at red heat. The melt is leached with hot water, HCl (1 : 1) is added until the precipitate is completely dissolved, and the solution is diluted to 250 ml in a volumetric flask. Iron is determined in 20 ml of the filtrate by a complexometric method; the titration is carried out at pH 2.0 by salicylic acid.

Aluminum is determined complexometrically on the same solution by titration with zinc solution in the presence of Xylenol Orange. The absolute error is less than 0.2%.

Aluminum in clays and refractories can also be determined photometrically by Eriochrome Cyanine R at pH 3.8; Fe(III) is masked by the addition of thioglycolic acid [597, 759].

Determination of aluminum in products of the titanium industry

Aluminum in intermediate products of the titanium industry (melts of chlorides, condensates, etc.) can be determined by a complexometric method, in which direct titration is performed in the presence of copper complexonate, with PAN as indicator; the interfering elements are eliminated by extracting their cupferronates with chloroform [430].

The sample (1–2 g) is treated with 30 ml of hot 1 : 1 HCl, and the solution is filtered through a "white ribbon" paper with some paper pulp. The filter with the precipitate is washed with hot water and ashed in a platinum dish. The residue is fused with 3 g Na_2CO_3 and 1.5 g $Na_2B_4O_7$. The melt is leached with water containing 30 ml of 1 : 1 HCl. The solution of the melt is combined with the original filtrate and made up to 250 ml in a volumetric flask.

Twenty-five ml of the solution are withdrawn into a separatory funnel, 10 ml of a 6% solution of Cupferron and 15 ml of chloroform are added, and the mixture is shaken for 30 sec. The chloroform layer is decanted, Cupferron is added to the aqueous layer until iron and titanium cupferronates no longer separate out (as shown by the white color of the precipitate being formed), chloroform is added, and the solution is again extracted. The extraction is repeated until the chloroform layer is colorless.

The aqueous layer is decanted into a 250-ml conical flask and neutralized with a 10% solution of sodium acetate to pH about 3 (Congo Red paper). Ten ml of acetate buffer solution at pH 3 (mixture of 9 ml of 2 M ammonium acetate with 491 ml of 2 M acetic acid), 2 ml of copper complexonate solution (about 0.005 M), and 5–6 drops of a 0.1% solution of PAN in ethanol are added, then water to a final volume of 80–100 ml and the solution is heated to boiling.

The hot solution is titrated by 0.05 M Complexone III to a pure yellow color. It is again boiled and the titration is completed.

The relative error of the method is 1%. It is not applicable to materials containing appreciable amounts of chromium.

The photometric determination of aluminum in the products of titanium industry can be effected with Chrome Azurol S [417], Xylenol Orange [418], and Methylthymol Blue [420].

Determination of aluminum in other materials

A photometric determination of aluminum in beryllium-containing products with 5-sulfo-4'-diethylamino-2',2-dihydroxyazobenzene was proposed by Floerence [724].

An aliquot of the weakly acid sample solution, containing not less than 15 μg Al and not more than 10 mg Be, is placed in a 25-ml volumetric flask. Two ml of a 0.1% solution of potassium ferrocyanide, 2.5 ml of 1 M acetate buffer solution at pH 4.7 (sodium acetate and acetic acid in equal amounts), and 3 ml of a 0.03% solution of 5-sulfo-4'-diethylamino-2',2-dihydroxyazobenzene solution are added. The solution is diluted to 20 ml with water and the flask is placed for 15 min in a water bath at 40–55°C. The solution is cooled to room temperature, and the optical density is determined after 15 min at 540 mμ in a 1-cm cell against a blank solution.

If the solution contains more than 200 μg of beryllium, the absorption and color attenuation by beryllium must be corrected for (p. 135).

Floerence [724] also proposed another method, in which the determination is preceded by separation of the interfering elements through extraction with triisooctylamine. For details, see p. 192.

Aluminum in thorium compounds can be determined by the method of Goldstein et al. [763].

The sample (0.1 g), containing 0.0005–0.15% of aluminum, is dissolved by gently boiling in 20 ml of 1 : 1 nitric acid containing a few drops of HF. When the dissolution is terminated, 3 ml of 72% $HClO_4$ are added, and the mixture is evaporated to the appearance of $HClO_4$ fumes. The mixture is cooled to room temperature, 5 ml of water and 8 ml of a 50% solution of ammonium acetate are added, and the pH is adjusted to 1.5 with perchloric acid. The solution is transferred to a separatory funnel and made up to 25 ml with water. Ten ml of a 0.5 M solution of thenoyltrifluoroacetone in methyl isobutyl ketone are added, and the solution is shaken for 5 min.

The aqueous layer is decanted into another separatory funnel and the extraction is repeated with an equal amount of the reagent.

The aqueous layer is transferred to another separatory funnel and washed by shaking for 3 min with 10 ml of the solvent. During the extraction, Fe, Zr, Cu, and most of the Th are separated.

If Ti, V, Mo, Mn and Co are present, the solution is heated, 2 ml of a 2% solution of sodium diethyldithiocarbamate are added, the pH is adjusted to 9, and the solution is extracted by 20 ml of chloroform for 2 min. The aqueous layer is transferred to a beaker and diluted to 75 ml. Five ml of 0.2 M nitrilotriacetic acid, 2 ml of 3% H_2O_2, 2 ml of acetate buffer solution at pH 8 (200 g of ammonium acetate and 70 ml NH_4OH (sp. gr. 0.9) in 1 liter), and 2 ml of a 2% solution of hydroxyquinoline in chloroform are added, and the pH of the solution is adjusted to 8 with the aid of 1 : 1 ammonium hydroxide.

The solution is transferred to a separatory funnel and extracted with two 20-ml portions of chloroform for two minutes each time.

If Ni, Zn and Cd are also present in the samples, the extract is shaken with 100 ml of alkaline cyanide solution (40 g of ammonium nitrate and 20 g KCN are dissolved in water, 10 ml of ammonium hydroxide (sp. gr. 0.9) are added, and the mixture is diluted to 1 liter with water) for 2 min. The extract is transferred to a 50-ml volumetric flask containing 1 g of anhydrous sodium sulfate and the solution is made up to mark with chloroform. The optical density is measured at 390 mμ in a 1-cm cell against a blank solution.

The determination can also be terminated by the fluorimetric technique, in which case the sample solution should contain 1–10 μg Al.

A photometric determination of aluminum in tungsten compounds by Arsenazo has been described [503]; tungsten interferes with the determination of aluminum if present in amounts of more than 40 μg. It is therefore previously separated by β-naphthoquinoline, and the excess of this reagent is separated by adding alkali to a strongly alkaline reaction, since it interferes with the determination of aluminum.

One gram of the sample is fused with 1 g Na_2CO_3. The melt is leached with 1 : 1 HCl, 3 ml of a 2% solution of β-naphthoquinoline are added, and the solution is evaporated to 3–5 ml. Ten ml of water are added and the precipitate formed is filtered. The filtrate is neutralized to phenolphthalein with a 20% solution of NaOH, and 2 ml of the alkali are introduced in excess; β-naphthoquinoline is precipitated. The solution with the precipitate is transferred to a 50-ml volumetric flask and made up to the mark with water. A part of the solution is filtered, an aliquot of the filtrate is neutralized with 1 : 10 HCl to phenolphthalein, 1 drop of a 10% solution of ascorbic acid is added, followed by 1 ml of a 0.05% solution of Arsenazo I and 0.2 ml of a 25% solution of urotropin.

The solutions are transferred to a 50-ml volumetric flask and made up to mark with water. The optical density is measured, using a green filter and a 5-cm cell. The sensitivity of the method is 2 μg Al in 50 ml.

The method can be used to determine $2 \cdot 10^{-4}$–$2 \cdot 10^{-2}$% Al in WO_3, tungstic acid, metallic tungsten, and other tungsten preparations. The relative error may be as high as 12%.

Aluminum in glass can be determined by a complexometric method [54].

The glass (0.5–1.0 g) is decomposed by an HF-H_2SO_4 mixture. The salt residue, which has been evaporated twice, is dissolved by heating with 3–5 ml HCl (sp. gr. 1.19) and the solution is made up to 250 ml in a volumetric flask. To a 50–100 ml aliquot of this solution, 0.025 M Complexone III is added in excess. The solution is heated to boiling and the hot solution is neutralized with ammonia to Congo Red paper, then 10 ml of acetate buffer solution at pH 5 (27.22 g $CH_3COONa \cdot 3H_2O$ and 4.9 ml HCl (sp. gr. 1.19) in 1 liter) are added, and the mixture is cooled to room temperature.

Six or seven drops of a ferricyanide-ferrocyanide mixture (0.2 g $K_3Fe(CN)_6$ with 0.1 g $K_4Fe(CN)_6$ in 30 ml of water) and 2–3 drops of a 1% solution of dimethylnaphthidine in glacial CH_3COOH are added, and the excess Complexone III is titrated by a 0.025 M solution of zinc acetate to the first appearance of a violet-pink coloration. Forty ml of a saturated solution of NaF are added, the solution is cooled to room temperature and boiled for 3 min. The solution is left to stand for 5–7 min and cooled to room temperature. The same amounts of ferri- and ferrocyanide and dimethylnaphthidine are again introduced, and the titration is repeated in the same manner. The amount of zinc consumed in the second titration is equivalent to the content of aluminum.

A determination of aluminum in glasses with Stilbazo has been described [245].

DETERMINATION OF ALUMINUM IN SOILS, MATERIALS OF ORGANIC ORIGIN, AND WATER

Determination of aluminum in soils

The correct choice of the extractant is important in the determination of mobile aluminum in soils. Many studies on this subject are available. The proposed extractants include 1 M ammonium acetate at pH 4.8 and 1 N KCl, NaCl, CaCl$_2$ and BaCl$_2$. Little [933], who carried out a critical review of the published data in addition to his own experimental work, concluded that mobile aluminum is extracted in a satisfactory manner by 1 N KCl. But 1 M ammonium acetate was not recommended, since the results are too low (at pH 4.8) or too high (at pH 4.0).

Aluminum in soils can be determined by photometric methods, with Aluminon, hydroxyquinoline, and Chrome Azurol S. If the soil extract is colorless, aluminum can be determined directly. But if it is colored due to the presence of organic compounds, it is evaporated to dryness with 10 ml of 30% H$_2$O$_2$. If fluorides are present, it is evaporated to dryness with 2 ml of sulfuric acid (sp. gr. 1.84). The extract thus prepared can be used for the preparation of colored solutions for the photometric determination of aluminum. The aluminum in the extract can be determined as described under "Photometric Methods," and also according to [933].

Five grams of air-dry soil are treated with 50 ml of 1 N KCl. On the following day the suspension is filtered through filter paper, and the residue on the filter is washed with five 5-ml portions of the extractant. The filtrate and the washings are collected for analysis. Aluminum is determined by the photometric hydroxyquinoline method [581]. Aluminum hydroxyquinolate is extracted by trichloroethylene from alkaline solutions containing Complexone III, thioglycolic acid and KCN to eliminate the effect of iron and other metals.

The optical density of the extract is determined at 410 mμ [933].

Determination of aluminum in organic materials

Aluminum in vegetal materials can be determined according to Middleton [967].

A sample of dry, finely ground leaves, containing 0.008–0.032 mg Al and 0.02–0.08 mg Fe, is placed in a 100-ml Kjeldahl flask. Five ml HNO_3 (sp. gr. 1.42) are added for each 0.5 g of leaves; the leaves are wetted with acid and 0.5 ml $HClO_4$ (sp. gr. 1.70) and 0.5 ml H_2SO_4 (sp. gr. 1.84) are added. The contents of the flask are cautiously heated until brown fumes are no longer evolved, and then the heating is intensified until the reaction becomes violent. When the reaction has ended, most of the excess $HClO_4$ is eliminated by heating for 5 min.

When cool, 25 ml of water are added and the solution is evaporated to a residual volume of 5–6 ml. The cooled solution is transferred with the aid of 10 ml of water to a separatory funnel. One ml of a 0.6% solution of hydroxyquinoline in 2% acetic acid is added and the solution is slowly neutralized with 5 M ammonia, then 5 ml of acetate buffer solution (0.1 N ammonium acetate in 5 N CH_3COOH adjusted to pH 2.85) are added, and the solution is stirred. Ten ml of a 0.3% solution of hydroxyquinoline in chloroform are added and the mixture is shaken.

The optical density of the extract is determined at 470 mμ, and the iron content is read off the calibration curve.

To determine aluminum, 4.5 ml of 5 M ammonium hydroxide are slowly added to the aqueous layer with stirring, 10 ml of the solution of hydroxyquinoline in chloroform are added, and the mixture is shaken. Its optical density is measured at 385 mμ in a 1-cm cell against a blank solution.

The elements Ca, Mg, K, P and Mn, usually present in leaves, do not interfere. The sensitivity of the method is 1.5 ppm; the relative error is 1.4%.

Cholak et al. [639] determined aluminum in biological materials with Alizarin S after ashing the sample (for details, see [360]).

Stafford and Wyatt [1203] decompose the organic matter in the Kjeldahl flask by treating with sulfuric, nitric and perchloric acids. They then eliminate iron and certain other metals by extracting their cupferronates with chloroform. Aluminum is photometrically determined with Aluminon. Tahler and Mühlberger [1217] determine aluminum in biological materials with Eriochrome Cyanine R. The sample is decomposed by ashing. The many metals that interfere are preliminarily removed by extracting their cupferronates from solution at pH 0.4. Aluminum cupferronate is then extracted at pH 4.8. The chloroform is expelled, the residue is fused with Na_2CO_3, and aluminum is determined in the solution of the melt.

Aluminum in rubber can be determined complexometrically by back-titrating excess Complexone III by a solution of iron with sulfosalicylic acid [55]. The rubber sample is ignited in a porcelain dish and the ash is dissolved in 1 : 1 HCl. Aluminum and iron are precipitated as hydroxides and dissolved in HCl, and then aluminum is determined complexometrically. In another variant, aluminum and iron are separated as hydroxides; if the contents of these metals are high, a preliminary separation is carried out on EDE-10P anion exchanger [82].

Aluminum in polyethylene can be determined according to Bolleter [581].

The sample is ignited or decomposed by acid. In the latter case 2 g of the sample are boiled with 20 ml of sulfuric acid (sp. gr. 1.84), then 20 ml HNO_3 (sp. gr. 1.4) are added and the boiling is continued. If a black carbon residue remains, more HNO_3 is added. Five ml of 70% $HClO_4$ are added and the mixture is evaporated. The residue is dissolved in water and the solution is diluted to 100 ml.

In another solubilization technique, 5 g of the sample are mixed with 2 g of a K_2SO_4-KNO_3 mixture (3 : 1), and the mixture is ignited in a platinum crucible. The residue is heated with 1 ml of sulfuric acid (sp. gr. 1.84) and dissolved in 10 ml of 10% sulfuric acid. The solution is diluted to 100 ml with water. To 50 ml of the solution thus obtained, 1 ml of a 10% solution of thioglycolic acid is added, the solution is neutralized with ammonia to phenolphthalein, and then 10 ml NH_4OH (sp. gr. 0.9) are added in excess. One ml of 1 M NaCN, 5 ml of a 2% solution of hydroxyquinoline in ethanol and 10 ml of trichloroethylene are added, and the mixture is shaken. The extract is filtered through paper and its optical density is determined at 390 mμ.

The relative error in the determination of 10–100 μg Al is 3%.

Aluminum in polypropylene can be determined by the photometric hydroxyquinoline method. The interfering elements are previously precipitated by hydroxyquinaldine [1017].

A 1 g sample is heated with 15 ml of sulfuric acid (sp. gr. 1.84) until the mixture has blackened. When cool, 5 ml of 30% H_2O_2 are added and the mixture is heated. The addition of H_2O_2 is repeated another 2–3 times. The excess H_2O_2 is decomposed by boiling and the solution is diluted to 100 ml with water. A 20-ml aliquot of this solution, containing 20–100 μg Al, is transferred to a separatory funnel and diluted with water to about 40 ml. Then 1 ml of a 33% solution of tartaric acid, 1 ml of a 33% solution of ammonium acetate, and 5 ml of a 2.5% solution of hydroxyquinaldine in 5% acetic acid are added.

The pH is adjusted to 8–9 by aqueous ammonia (sp. gr. 0.9), and titanium and iron hydroxyquinaldates are extracted by 15 ml of chloroform. The extraction is repeated twice, using 1.5 ml of hydroxyquinaldine solution and 7.5 ml of chloroform each time. To the aqueous, aluminum-containing phase, 1.5 ml of a 2.5% solution of hydroxyquinoline in 5% acetic acid are added, and aluminum hydroxyquinolate is extracted by 7 ml of chloroform. The extract is diluted to 25 ml with chloroform, and the spectrometric determination is carried out at 395 mμ.

Determination of aluminum in water

Giebler [754] carried out a comparative study of four photometric methods for the determination of aluminum in water: with hematoxylin, Eriochrome Cyanine R, Alizarin S, and Aluminon. In his view, preference should be given to the determination by Aluminon.

The first stage is to remove fluoride ions (if present in concentrations higher than 1 mg/ml) and organic matter by treatment with H_2SO_4 (sp. gr. 1.84) to SO_3 fumes. After these substances have been separated, 2 ml of a 1% solution of thioglycolic acid are added to 50 ml of water (neutral reaction) in a 100-ml volumetric flask, and after 2 min 15 ml of Aluminon solution described on p. 97 are added. The flask is placed for 15 min in a boiling water bath and then cooled. The optical density of the solution is determined at 530 mμ; the pH should be 5.3–5.4 [754].

A similar method was proposed by Rolfe et al. [1119].

Motojima [994] proposed a photometric hydroxyquinoline method for the determination of aluminum in pure water.

Five-hundred ml of water are carefully boiled with 5 ml HCl (sp. gr. 1.19). When cool, 2 ml of a 2% solution of hydroxyquinoline in CH_3COOH are added. The pH of the solution is adjusted to 5.5–6 with 1 N NaOH, 1 M NH_4OH or 4 M CH_3COONa. The solution is transferred to a separatory funnel and made up to 525 ml. Ten ml of chloroform are added and the mixture is vigorously shaken for 2 min. After the layers have separated, the extract is transferred to a second separatory funnel containing 50 ml of cyanide-carbonate wash liquor.* The extract is shaken with the wash liquor, the chloroform layer is placed in a conical flask containing 1 g of anhydrous Na_2SO_4 and shaken gently to remove the water drops, and the optical density is measured at 390 mμ against a blank solution.

Aluminum and iron can be determined in the presence of each other in seawater as follows [459].

Seawater (250–330 ml) is freed from suspended impurities by filtration. The filtrate is acidified with 1 ml HCl (sp. gr. 1.19) and boiled for one minute. To the cooled solution are added 7.5 ml of a solution of beryllium nitrate containing 10 mg Be/ml to eliminate the effect of fluorides. Three ml of a 1% solution of hydroxyquinoline in acetic acid are added, and ammonium acetate and ammonium hydroxide are introduced to adjust the pH to 5.0–5.5. The mixture is diluted to 400 ml with water and extracted with 10 ml of chloroform. The extract is washed with 10 ml of 0.1 M KCN to eliminate copper and the optical density is determined in a 1-cm cell at 390 and 470 mμ.

The contents of aluminum and iron are calculated from the formulas:

$$C_{Al} = A_{390} - 0.746 \cdot A_{470}/0.0310;$$

$$C_{Fe} = A_{470}/0.0134,$$

where C_{Al} and C_{Fe} are the contents of aluminum and iron, respectively, in the water sample, μg/liter, and A_{390} and A_{470} are the optical densities at 390 and 470 mμ, respectively.

* To prepare the wash liquor, 50 ml of a 13% solution of KCN are added to 100 ml of a 30% solution of ammonium carbonate, and the mixture is diluted to 400 ml with water; the pH is adjusted to 9–10 with a 20% solution of NH_4Cl, and the interfering metals present as impurities are separated by extracting with a chloroform solution of hydroxyquinoline.

The method can be used to determine $14-252\,\mu g$ Al/liter and $9-156\,\mu g$ Fe/liter.

Solomin and Fesenko [369, 370] proposed a complexometric method for the determination of aluminum in acid waters in coal mines. Aluminum and iron are first precipitated as hydroxides; the precipitate is dissolved in acid, and the sum of these metals is determined by back-titration by zinc in the presence of dithizone on one part of the solution; iron is determined in the other part of the solution by direct titration by Complexone III at pH $1.0-1.5$ in the presence of sulfosalicylic acid as indicator.

DETERMINATION OF ALUMINUM IN METALS AND ALLOYS*

Determination of aluminum in iron, steels, and ferrous alloys

Steels are dissolved in different acids and acid mixtures: HCl (conc., $2:1$ and $1:1$), HNO_3 ($1:1$), $HClO_4$, H_2SO_4 ($1:9$), HCl-HNO_3 mixtures, $HClO_4$-HNO_3 mixtures, H_2SO_4-HNO_3 mixtures. Most steels can be dissolved in $1:1$ HCl, and HNO_3 is added toward the end of the dissolution process to decompose carbides. Certain high-alloy steels are best dissolved in aqua regia. Ferrous alloys are dissolved in $1:1$ HCl, $1:1$ or $1:4$ H_2SO_4, aqua regia, or an HF-HNO_3 mixture, depending on the composition.

Aluminum is present in steels as metal (acid-soluble) or as oxide (acid-insoluble). To determine each of these forms separately, the steel sample is dissolved in $1:1$ HCl or $1:9$ H_2SO_4, and the iron is oxidized by a small amount of HNO_3. The insoluble precipitate is filtered off and washed, the filtrate is made up to the mark in a volumetric flask, and the "soluble" aluminum is determined on an aliquot of this solution. The filter with the residue is ashed, ignited, and fused with a little pyrosulfate. The melt is leached with water, and "insoluble" aluminum is determined in the resulting solution.

Gravimetric methods for the determination of aluminum. If the content of aluminum in the steel is $0.1-7\%$, it can be determined by the gravimetric fluoride (cryolite) method. This method is standard (GOST 12357-66) for alloy and high-alloy steels [377a].

Aluminum is separated from the interfering elements by sodium fluoride from a weakly acid solution, the cryolite precipitate is dissolved in a boric acid-hydrochloric acid mixture, and aluminum is precipitated by hydroxyquinoline. The precipitate is then ignited to the oxide.

---

* For the determination of aluminum in metals and alloys, see also "Spectroscopic Methods," Chapter II.

The steel sample (2, 1, 0.5, and 0.2 g sample for aluminum contents of 0.1–0.5, 0.5–1, 1–2, and 2–7%, respectively) is placed in a 250–300 ml beaker, 25 ml HCl (sp. gr. 1.19) are added, the beaker is covered with a watch glass and kept at moderate heat until the sample has dissolved. The watch glass is then moved slightly and HNO_3 (sp. gr. 1.4) is added drop by drop until frothing ceases, and then another 5–10 ml in excess. The solution is heated to expel nitrogen oxides and cooled, then 10–15 ml H_2SO_4 (sp. gr. 1.84) are added, and the solution is evaporated to the appearance of SO_3 fumes. The contents of the beaker are cooled, 80–100 ml of water are cautiously added, and the mixture is heated to dissolve the salts. The watch glass is rinsed with water and the precipitate (silicic, tungstic, niobic and other acids) is filtered through two "white ribbon" filters. The filter with the precipitate is washed 4–5 times with hot 1 : 100 sulfuric acid and the filtrate and washings are collected in a 400-ml beaker. The filter with the precipitate is placed in a platinum crucible, ashed, and ignited at 900–1,000°C. When cool, the residue is wetted with 2–3 drops of water, 2–3 ml of 1 : 4 sulfuric acid are added (if W, Nb and Ta oxides are present, 5–6 ml of the acid are added), and then 3–5 ml of 40% HF. The mixture is evaporated until white SO_3 fumes are no longer evolved and the residue is ignited at 900–1,000°C, and then fused with 2–3 g $KHSO_4$.

The melt is leached with 20–30 ml of 1 : 9 sulfuric acid, and the solution of the melt is combined with the main filtrate. From the solution thus obtained, aluminum is precipitated in the form of cryolite. The determination is continued as described on p. 57.

Elliot and Robinson [700] determine aluminum by the gravimetric hydroxyquinoline method, after the bulk of the iron has been removed by extraction with dichlorodiethyl ether, and aluminum has been separated from a large number of interfering elements (Cu, Ni, Mn, Co, etc.) by precipitating with ammonia in the presence of ammonium chloride. The precipitate is dissolved in acid, Complexone III is added and the solution is made alkaline. Aluminum is precipitated as hydroxyquinolate.

Dübel and Flurschütz [689] determined aluminum in magnetic alloys by the gravimetric hydroxyquinoline method, after interfering elements had been removed by electrolysis on a mercury cathode. The reproducibility of the results of this method is satisfactory. In the analysis of Alnico alloy, which contained about 8.8% Al, the deviation of individual results from the arithmetic mean was 0.01–0.04%. A similar method was employed in a microgravimetric determination of aluminum (weight of samples 0.02–0.05 g) [878].

Aluminum in steels can be determined by the gravimetric phosphate method [1102].

Two grams of steel (for aluminum contents above 0.04%) or 5 g (for aluminum contents below 0.04%) are dissolved in 30 ml of $12 N$ HCl. To the solution are added 10 ml of a 20% solution of sodium hypophosphite and 20 ml of bromine water, and the solution is evaporated to a small residual volume. If zirconium and titanium are present, they separate as a white, crystalline precipitate. The solution is filtered, the filtrate is evaporated to dryness, the residue is wetted with $12\,N$ HCl, and the evaporation is repeated. Then 30 ml of HCl $(1:1)$ are added to the residue. The mixture is heated to dissolve the salts and filtered. The precipitate, consisting of silica and insoluble aluminum, is washed with hot 5% HCl.

If it is desired to determine total aluminum, silica is removed by treatment with HF, the residue is fused with $KHSO_4$, and the melt is leached with water. The solution thus obtained is combined with the main filtrate.

The solution is evaporated to a small residual volume and is made $6\,N$ in HCl. Thirty ml of the solution are shaken in a separatory funnel with an equal volume of ether saturated with hydrochloric acid. The ether layer is discarded, and ether is expelled from the aqueous layer by heating. The acid in the solution is neutralized with $1:1$ ammonia to a final acid content of 5%. Five ml of 80% thioglycolic acid are added to complex nickel, chromium, and other metals, as well as residual iron. Ammonium thiocyanate is then added (5–8 g or 10 ml of saturated solution). Ammonia $(1:1)$ is cautiously added to pH 0.5. Ten ml of a 10% solution of $(NH_4)_2HPO_4$ are added and the solution is stirred. Ammonium acetate (40% solution) is added drop by drop to adjust the pH to 3.7–3.9. The solution is heated for 5 min at $50-60°C$ with continuous stirring. The aluminum phosphate precipitate is allowed to settle, and then filtered, washed with 5% ammonium nitrate solution, ignited, and weighed as $AlPO_4$. If the precipitate is not perfectly white, it is dissolved in HCl (sp. gr. 1.19) and the precipitation is repeated.

Complexometric methods. Aluminum in steels can be determined [363] by a complexometric method in the presence of Xylenol Orange.

The solution of the steel is neutralized until the hydroxides begin to separate out, and these are then dissolved in a few drops of $1:1$ sulfuric acid. The solution is heated to boiling and poured in a thin jet, with constant shaking, into a 500-ml volumetric flask containing 100 ml of a hot solution of NaOH or KOH (200 gm/liter). The solution is made up to the mark, the precipitate is allowed to settle, and the solution is filtered. Aluminum is determined by a complexometric method on an aliquot of the filtrate by back-titrating the excess Complexone III by a solution of zinc in the presence of Xylenol Orange. The concentration of the zinc solution is determined by means of a standard sample of steel, which has been run through all stages of analysis under exactly the same conditions as the sample.

For an aluminum content of 5%, the relative error is about 2%.

This method, in which the interfering elements are separated by NaOH, is the most rapid, but it must be borne in mind that vanadium and molybdenum pass into the filtrate together with the aluminum; if the titration is conducted slowly, with vigorous stirring, molybdenum does not interfere [712]. The effect of vanadium is eliminated by adding 30% H_2O_2 [712].

After interfering elements have been separated by sodium hydroxide, aluminum can be determined by a direct complexometric method in the presence of copper complexonate and PAN [1184]. Small amounts of vanadium and molybdenum have no significant effect.

Burke and Davis [600] described a complexometric method involving the use of diaminocyclohexanetetraacetic acid (DCTA). A large number of interfering elements are eliminated by electrolysis on a mercury cathode, while other elements are separated by extraction with a solution of tri-*n*-octylphosphine oxide in cyclohexane. The method is more complicated than those described above, but the results obtained are more accurate. For a content of aluminum of about 3%, the error involved is 0.02%.

Complexometric methods are also employed in the analysis of ferrous alloys. It has been proposed [63] that aluminum and titanium be determined in ferrotitanium complexometrically, by conducting successive titrations on the same solution.

The excess Complexone III is titrated at pH about 5 by a solution of zinc in the presence of Xylenol Orange. A solution of tartaric acid (20%, 10–15 ml) is added, the pH is adjusted to the desired value, and the liberated Complexone III, equivalent to the content of titanium, is titrated by a solution of zinc. Fluoride ions are added to the titrated solution, aluminum complexonate is decomposed, and the liberated Complexone III, equivalent to the content of aluminum, is titrated by a solution of zinc.

The relative error in the determination of titanium is up to 1.2%; the relative error in the determination of aluminum is up to 2.8%.

Ferrotitanium can also be analyzed by back-titration by a solution of an iron salt in the presence of sulfosalicylic acid, after the interfering elements have been separated by sodium hydroxide [160]. Aluminum in ferrosilicon is determined after elimination of interfering elements by electrolysis on a mercury cathode. If more than 1% Mn is present, the element is removed by precipitation with sodium hydroxide in the presence of hydrogen peroxide. Iron is first titrated by salicylic acid at pH 2, and then aluminum is determined by back-titration by a solution of iron salt at pH 5.

Photometric Aluminon method. Aluminum in iron can be determined by the method of Short [1162].

Ten g of iron are dissolved in HCl (sp. gr. 1.19), and the iron is oxidized with the minimum amount of HNO_3. The solution is evaporated to dryness, the residue is dissolved in 80 ml HCl and is again evaporated to incipient separation of salts. Fifty ml HCl (sp. gr. 1.12) are added and the solution is transferred to a separatory funnel. Iron is extracted with ethyl ether or isopropyl ether. The aqueous layer is transferred to a beaker and heated on a steam bath to expel the ether. The solution is oxidized with a few drops of HNO_3 and the solution is evaporated almost to dryness. A little HCl is added (sp. gr. 1.12) and the ether extraction is repeated, while keeping the volume of the aqueous layer as small as possible. The solution is heated to remove ether, a few drops of HNO_3 are added, and the solution is again evaporated to a small residual volume.

The solution is diluted to 50 ml and 3 g NH_4Cl are added; the pH of the solution is adjusted to 0.3–0.4 (pH-meter). The solution is then transferred to a separatory funnel, and 0.1 g of Cupferron dissolved in a small volume of water is added. After 5 min the solution is extracted with chloroform; if necessary, the extraction is repeated until a colorless chloroform layer is obtained. The aqueous layer is heated for a few minutes to remove chloroform, and the solution is made up to 50 or 100 ml with water. Aluminum is then photometrically determined by Aluminon on an aliquot of this solution.

The method is suitable for the analysis of high-purity iron.

Craft and Makepeace [659] determined aluminum by Aluminon in low-alloy steel, after iron had been removed by extraction with isopropyl ether and chromium as chromyl chloride.

According to GOST-11658-65, aluminum is determined in cast iron and nonalloy steel without separation. The iron is reduced to the ferrous state by ascorbic acid; Fe(II) does not interfere with the determination of aluminum. If the steel contains titanium and vanadium, GOST-11658-65 provides for preliminary removal of iron by extraction with ether, and for the separation of titanium and vanadium by precipitation as cupferronates, i.e., Short's [1162] procedure is adopted.

In the analysis of high-alloy steels, the Aluminon determination of aluminum becomes more complex. The most accurate standard methods include a stage involving separation by electrolysis on a mercury cathode. Thus, according to Bendigo and Bell [555], chromium is first eliminated as chromyl chloride, and then a number of metals are separated by electrolysis on a mercury cathode, and the cupferronates of aluminum and other metals are extracted by chloroform at pH 3.5. Finally, chloroform is expelled from the extract, the extract is strongly acidified, and Ti, Zr, V and certain other metals are separated as cupferronates. Such laborious methods are suitable for umpire analyses or for the determination of the composition of standard samples. For ordinary purposes, simpler methods will also give sufficiently accurate results.

Photometric determination by Chrome Azurol S. In the method of Hosoya et al. [820], the bulk of the iron is eliminated by extraction with methyl isobutyl ketone. Residual iron and certain other metals are masked by thioglycolic acid.

Iron or steel (0.5 g) is dissolved in 20 ml of HCl (1 : 1) after adding 1 ml of nitric acid. The resulting solution is heated and evaporated to dryness. To the residue are added 20 ml of 1 : 1 HCl, and the solution is heated to dissolve the salts. Silica is filtered together with other insoluble substances. The residue is ignited, treated with HF, and fused with potassium pyrosulfate. The solution of the melt is combined with the filtrate.

The combined filtrates are concentrated to 10 ml, transferred to a separatory funnel containing 7 N HCl, and the iron is extracted with methyl isobutyl ketone (two 20-ml portions). The acid layer is poured into a beaker and evaporated to dryness, and the residue is then boiled with 5 ml of aqua regia for 2–3 min. The solution is again evaporated to dryness, the residue is dissolved in 5 ml of 1 : 1 HCl, and the solution is cooled and diluted to 100 ml in a volumetric flask.

To determine aluminum, 10–20 ml of the solution are placed in a 100-ml volumetric flask, 1 ml of a 10% solution of thioglycolic acid and 2 ml of a 0.1% solution of Chrome Azurol S are added, the pH is adjusted to 5.6–5.8 by 0.2 M solutions of acetic acid and sodium acetate (pH-meter), and the solution is made up to the mark with water. The optical density is determined after 20 min at 550 mμ against a control solution prepared in the same manner and containing small amounts of aluminum-free iron. The content of aluminum is read off a calibration curve.

Since thioglycolic acid reduces the color intensity of the aluminum complex, equal amounts of this acid must also be introduced into standard solutions.

Stepin et al. [378] proposed that aluminum be determined in transformer steel without prior separation of the interfering elements, which are masked by thioglycolic acid. It would appear that very small amounts of aluminum cannot be determined by this method; this is because iron will interfere, owing to the formation of the violet-colored complex between Fe(III) and thioglycolic acid.

Astanina and Ponomarev [18] also determine aluminum without preliminary separation, after reducing iron by ascorbic acid. The presence of large amounts of ferrous iron somewhat attenuates the color intensity of the aluminum complex, and therefore iron in an amount corresponding to its content in the sample should also be introduced into the standard solutions.

According to Brockmann and Keller [592], aluminum in steels can be determined after the iron has been reduced by ascorbic acid without separation only if the results need not be very accurate. According to these workers, ascorbic acid not only attenuates the color intensity of the aluminum complex, but also reduces the stability of the coloration with time.

Aluminum in ferroboron can be determined with Chrome Azurol S after the interfering elements have been separated as chloride complexes on AV-17 anion exchanger [378].

Photometric hydroxyquinoline method. In the determination of aluminum in steel according to Kassner and Ozier [864], titanium, vanadium and certain other impurities are bound as peroxide complexes. Iron which has been reduced by sulfite, and many other impurity elements, are complexed by cyanide. Kakita and Jokohama [856] improved the method of Kassner and Ozier, and proposed a rapid method for the determination of aluminum in iron and steel. According to these workers, one determination takes 20 min.

The sample (0.5–1 g) is dissolved by heating with 6–10 ml of 1 : 1 nitric acid and 10 ml of 60% $HClO_4$. The heating is prolonged for a few minutes after the solubilization is complete. The solution is made up to 500 ml in a volumetric flask. A 5–25 ml aliquot is taken for the determination, and 1 ml H_2O_2, 2 ml of a 10% solution of tartaric acid, 10 ml of a saturated solution of sodium sulfite and 5 ml of a 20% solution of KCN are added.

The solution is heated to 70°C to reduce Fe(III), and cooled. It is then transferred to a separatory funnel, and 2 ml of a 50% solution of NH_4NO_3 and 3 ml of a 3% solution of hydroxyquinoline are added. The pH is adjusted to 8.6–9.2 with the aid of HCl or NH_4OH. Fifteen ml of benzene are added and the mixture is vigorously shaken for one minute. The layers are allowed to separate, the aqueous layer is discarded, and the optical density of the benzene layer is measured in a photocolorimeter against a blank solution. The content of aluminum is found from a calibration curve, determined under the same conditions.

For aluminum contents in the range of 0.15–1.7%, the relative error of the method is 1–4% [856].

In the determination of aluminum by hydroxyquinoline, the interfering elements are removed by electrolysis on a mercury cathode [1264], cupferronate extraction [1289], or amyl acetate extraction from conc. HCl (iron).

Other photometric methods. Different variants of the determination of aluminum by Eriochrome Cyanine R have been described. The preliminary separation methods are the same as those described above. The interfering elements can be separated by NaOH in the presence of large amounts of zinc and boric acid to minimize the occlusion of aluminum [809], by electrolysis on a mercury cathode [568, 878, 880, 909, 1245, 1258], by extraction with ether [831, 1258], or by removal as cupferronates [586, 831]. Neither Eriochrome Cyanine R nor other reagents proposed for the determination of aluminum in steels, such as Alizarin S [656, 1024], Stilbazo [4, 5, 1259], etc., have been widely applied.

Determination of aluminum in copper and copper alloys

Gravimetric methods. Aluminum in copper alloys can be determined by the gravimetric hydroxyquinoline method, while masking the interfering elements by KCN, tartaric acid, and Complexone III [682].

One g or 0.2 g of the alloy is dissolved in 15 ml of 4 N HNO_3 and 15 ml of 4 N HCl. Thirty ml of a 50% solution of tartaric acid are added, the solution is neutralized with 7 M ammonia to pH 7–8 (in the presence of large amounts of zinc or cadmium ammonia is added to the disappearance of the precipitate). When cool, sodium sulfite is added (1 g of sulfite for each 0.1 g of sample), the solution is diluted to 100–125 ml with water,

5 g KCN and 10 ml of Complexone III solution (10 g Complexone III are dissolved in 100 ml conc. ammonia) are added. The mixture is boiled for 3 min and a 0.5% solution of hydroxyquinoline is added drop by drop. The required amount of hydroxyquinoline is calculated from the formula

$$x = 0.02 \, V + a/3,$$

where V is the overall volume of the solution in ml, and a is the estimated amount of aluminum, in mg.

The solution is heated, held for 30 min close to its boiling point, and filtered at 70°C. The precipitate is washed with 100 ml of hot 0.1 M ammonia, dried at 135°C, and weighed.

The elements Cu, Zn, Sn, Fe, Pb, Ni, Co, and Sb do not interfere.

Satisfactory results can be obtained by the gravimetric benzoate method for the determination of aluminum, involving preliminary removal of the bulk of the copper by thioglycolic acid, but the method is more complicated than the hydroxyquinoline method described above [521].

Titrimetric methods. Aluminum can be rapidly determined by the titrimetric hydroxyquinoline method [986].

The alloy (0.1, 0.2 or 0.5 g of alloy at aluminum contents of 10, 5 and 1%, respectively) is dissolved in dilute nitric acid. Nitrogen oxides are expelled by boiling, and the solution is diluted to about 100 ml with hot water. Five ml of a cold-saturated solution of Na_2SO_3 and 25 ml (35 ml for brasses) of a masking mixture (100 g NaCN, 50 g Complexone III, and 300 ml of conc. ammonia in 1 liter) are added. Aluminum is precipitated by adding 10 ml of a hydroxyquinoline solution (40 g of reagent are dissolved in 120 ml of glacial CH_3COOH, the solution is diluted to 1 liter, neutralized, and weakly acidified with acetic acid) at the boiling point, and the mixture is stirred until the precipitate has coagulated. The solution with the precipitate is left to stand for about one minute and is then filtered through a large-pored filter. The precipitate is washed five times with hot water. The filtrate is discarded, the precipitate on the filter is dissolved in 20 ml of hot HCl (sp. gr. 1.19), and the filter is washed with water. The filtrate together with the washings are collected in a 300-ml conical flask and cooled, and a 0.1 N $KBrO_3$-KBr solution is added until one drop of methyl orange is decolorized. Ten ml of a 5% solution of KI are added, and the separated iodine is titrated by 0.1 N thiosulfate.

Aluminum in copper alloys can be determined by complexometric methods with Xylenol Orange [260], salicylic or sulfosalicylic acids [354, 976] as indicators.

Photometric methods. The photometric determination by Eriochrome Cyanine R is recommended for the determination of aluminum in copper alloys [250].

Brass (0.1 g) is dissolved in 2 ml HNO_3 (sp. gr. 1.40). The solution is transferred to a 100-ml volumetric flask, 8 ml of 1 : 1 HCl are added, and the solution is made up to the mark. Five ml of this solution are transferred to a 100-ml volumetric flask, 40 ml of water, 0.1 ml of a 4% solution of ascorbic acid and 0.25 ml of a 15% solution of sodium thiosulfate are added, and the solution is shaken; 2.5 ml of a 0.075% solution of Eriochrome Cyanine R and 20 ml of a 40% solution of CH_3COONa are added, and the solution is made up to the mark with water.

Two minutes later the optical density is determined in a photocolorimeter against water, using a green filter; a blank correction is introduced.

The solution of Eriochrome Cyanine R is prepared as follows: 0.15 g of the reagent is dissolved in 50 g of water, 5 g NaCl, 5 g $NaNO_3$ and 0.4 ml of HNO_3 (sp. gr. 1.40) are added, and the solution is diluted to 200 ml with water.

The determination takes 10−12 min; the relative error is 3%.

Dozinel [687] determines aluminum in copper alloys by Eriochrome Cyanine R after masking the interfering elements by thioglycolic acid. Up to 10% Sn and Pb, up to 30% Mn, and up to 1% P, Sb and As do not interfere. The determination is conducted at pH 5.1−5.2; if more than 40 mg Al is determined, the pH should be 4.0. In the presence of Ni and Fe the results are corrected (1% Ni and 1% Fe are equivalent to 0.005 and 0.008% Al, respectively).

Aluminum in copper alloys can be determined by Aluminon, after the interfering elements have been masked by thioglycolic acid [939].

Manganese bronze (0.2 g) is dissolved in 5 ml HCl (sp. gr. 1.19) and 5 ml of 30% H_2O_2. After the sample has dissolved, the solution is boiled for two minutes to expel the excess H_2O_2. The solution is diluted to 200 ml in a volumetric flask. Ten ml of the resulting solution are transferred to a 100-ml volumetric flask, 4 ml of 1 : 24 thioglycolic acid solution and 15 ml of aluminum buffer solution (for preparation, see p. 97) are added, the mixture is stirred, and the flask is placed on a vigorously boiling water bath (a 400-ml beaker) for 5 min. The solution is taken off the bath and placed in cold water after one minute.

When the solution has cooled to room temperature, it is made up to the mark with water and the optical density is measured at 525 mμ against a blank solution; the content of aluminum is read off a calibration curve. The amounts of aluminum taken to establish the calibration curve are 0, 0.01, 0.02, 0.03, 0.04, 0.06, 0.08 and 0.10 mg, one ml of $CuCl_2$ solution (5 mg Cu/ml) is added, the solution is diluted to 10 ml with water, thioglycolic acid is added, and the determination is continued as above.

When aluminum is determined in copper alloys containing tin and zinc ("gunmetal"), many interfering elements are previously removed by electrolysis on a mercury cathode [1086].

Determination of aluminum in zinc and zinc alloys

Titrimetric methods. Aluminum in zinc alloys can be determined by complexometric methods, with the complexes of copper with PAN and sulfosalicylic acid as indicators. However, aluminum must first be separated from interfering elements. It is simpler to determine aluminum in zinc and zinc alloys by photometric methods, with Aluminon or Eriochrome Cyanine R.

One g of zinc is dissolved by gently heating with 15 ml of 1 : 1 HCl. The insoluble residue, consisting mainly of lead, is filtered and washed with cold water. The filtrate and the washings are diluted to 200 ml in a volumetric flask. An aliquot of this solution (5 ml or 10 ml for aluminum contents of 0.08–0.24% and 0.025–0.08%, respectively) is transferred to a 200-ml volumetric flask and diluted with water to 100–150 ml, and 5 ml of a 0.1% solution of Eriochrome Cyanine R are added. After 5–7 min, 15 ml of buffer solution (pH 6) are added (for a 5-ml aliquot; if a 10-ml aliquot has been taken, 20 ml of buffer solution are added). The solution is made up to the mark with water, and after 5 min the optical density is determined at 530 mμ.

Zinc does not interfere if the aluminum content exceeds 0.002%; Fe(II), Cd, Pb and Sn also do not interfere if present in amounts in which they usually occur in zinc [831].

When analyzing zinc alloys, thioglycolic acid must be introduced to mask the interfering elements. The methods employed for the determination of aluminum in copper alloys with Aluminon and Eriochrome Cyanine R can also be used for the determination of aluminum in zinc alloys.

Determination of aluminum in tin and tin alloys

Aluminum in tin is determined photometrically with Aluminon [938], Chrome Azurol [488], Eriochrome Cyanine R [227], and hydroxyquinoline [654]. Tin must be preliminarily removed in all cases by vaporization as bromide or chloride.

One g of tin is placed in a quartz glass beaker, half submerged in cold water. Then 2 ml HCl (sp.gr. 1.19) and 2.5 ml HBr are added, and 1.5–2 ml of bromine are cautiously introduced under a watch glass drop by drop. When the vigorous reaction has abated, the watch glass is taken off and rinsed, and the beaker is placed on a Teflon-covered table under an IR lamp to finally decompose the sample and separate the tin. The solution is evaporated at 85–90°C until the residue is just moist, 1 ml of 1 : 1 HCl is added and the solution is evaporated to dryness. Another 0.5 ml of 1 : 1 HCl is added and the solution is again evaporated to dryness.

The dry residue is dissolved in acetate buffer solution at pH 5.4 and the resulting solution is made up to 10 ml. To this solution are added 0.2 ml of a freshly prepared 5% solution of ascorbic acid and 0.3 ml of a 0.3% solution of Chrome Azurol S. The optical density is measured after 5 min in a photocolorimeter with a green filter, in 2-cm cells. The content of aluminum is found from a calibration curve [488].

Lel'chuk et al. [227] used a photometric method to determine aluminum in high-purity tin with Eriochrome Cyanine R. The sample is decomposed and the tin separated as described above.

After three evaporations with 1 : 1 HCl, the dry residue is dissolved in 6 drops of 1 : 4 HCl and the solution is transferred to a 25-ml volumetric flask with the aid of the buffer solution. Freshly prepared ascorbic acid solution (5%, 0.5 ml) is added, the solution is stirred, and 0.5 ml of a 0.075% solution of Eriochrome Cyanine R, containing 5 g NaCl, 5 g $NaNO_3$, and 0.4 ml HNO_3 (sp. gr. 1.4) in 200 ml, is added. The solution is made up to the mark with buffer solution at pH 8.3 (460 ml of $2M$ acetic acid are mixed with 540 ml of $2M$ ammonium hydroxide). After 7 min, the optical density is determined against water, using a photocolorimeter with a green filter. A correction is introduced for the blank solution.

The sensitivity of the method is 10^{-5}–$5 \cdot 10^{-6}$; the relative error is 10–15%. A determination takes 4–5 hours.

According to the authors, HCl, HBr and Br_2 must be purified from traces of aluminum by triple distillation in a quartz apparatus.

Chrome Azurol S and Eriochrome Cyanine R can clearly be employed to determine aluminum in tin alloys, after preliminary separation of interfering elements.

Luke [938] determined aluminum in tin and tin alloys by a photometric method with Aluminon. Tin, antimony and arsenic are eliminated by evaporation after the addition of HBr and Br_2. If the sample contains more than 5% Pb, the Pb is separated by precipitation as $PbSO_4$. The solution is then electrolyzed on a mercury cathode to remove the interfering elements. A number of metals that remain in the electrolyte are separated from aluminum by cupferronate extraction with chloroform. The pH of the aqueous phase is adjusted to 5, and aluminum is separated by extracting its hydroxyquinolate with chloroform. The chloroform and the bulk of hydroxyquinoline are expelled, the organic residues are decomposed by $HClO_4$, and the residue is dissolved in water. Aluminum is determined with Aluminon using the photometric technique.

Determination of aluminum in nickel and nickel alloys

Aluminum in nickel alloys can be determined by a complexometric method [253], based on the titration of excess Complexone III in the presence of Methylthymol Blue as indicator. Fluoride ions are introduced to enhance the selectivity.

The alloy (0.1 g) is dissolved by heating with HCl (sp. gr. 1.19), and the solution is diluted to 250 ml in a volumetric flask. To determine aluminum, an aliquot containing 4–10 μg Al is taken, 15–30 ml of 0.025 M Complexone III are added, and the solution is boiled for 2–3 min. The solution is neutralized with ammonia to Bromothymol Blue until the green solution turns greenish-blue. Five ml of acetate buffer solution are added (pH 7.0), and the solution is left to stand in a warm place for 10 min. When cool, Methylthymol Blue indicator (mixed with KNO_3 in the ratio of 1 : 100) is added, and the solution is titrated by a solution of lead or zinc until the color of the solution changes from green to violet. Twenty ml of a 4% solution of NaF are added, the solution is boiled for 2–3 min and cooled. Some more indicator is added and the solution is made alkaline with ammonia until it turns green. The liberated Complexone III is titrated by a solution of a zinc or lead salt. One ml of the titrant is equivalent to 0.67426 mg of aluminum.

To prepare the buffer solution, 50 ml of 98% CH_3COOH are diluted to 250 ml and 90 ml of ammonium hydroxide (sp. gr. 0.9) are added. When cool, the solution is diluted to 500 ml with water, a solution of Bromothymol Blue is added and, if required, ammonia or acid until the solution turns greenish-blue (pH 7.0–7.2).

Up to 10% Fe(III), Cu(II) and Co do not interfere; more than 1% Ti interferes. If 4–10 mg Al are determined, the relative error is less than 1%; if 1–3 mg Al are present, the error is 1%.

Aluminum in nickel alloys can be determined by Aluminon [697].

The sample should contain 3–20 μg. The sample is dissolved in the smallest possible volume of 1 : 1 HNO_3, the solution is treated with HCl (sp. gr. 1.19), and evaporated to dryness. The residue is dissolved in doubly distilled water and transferred to a separatory funnel. The pH is adjusted to 5.2 with acetate buffer solution (470 ml NH_4OH (sp. gr. 0.9) and 430 ml glacial acetic acid are diluted to 1 liter with water). Two ml of a 5.7% solution of sodium diethyldithiocarbamate are added and the solution is extracted with 15–20 ml of ethylene chloride, chloroform, or trichloroethylene. The extraction is repeated with 2 ml of diethyldithiocarbamate each time, and 0.5 ml for the final extraction) until the organic phase is perfectly colorless. The aqueous phase is decanted into a beaker, and then heated on a sand bath until the odor of the organic solvent has disappeared. The solution is cooled and transferred to a 50-ml volumetric flask. Then 2 ml of acetate buffer solution and 2 ml of Aluminon solution (0.1 g of Aluminon and 0.5 g of gelatin are dissolved in 50 ml of doubly distilled water), 10 ml of acetate buffer solution and 20 ml of a 2% solution of benzoic acid in methanol are added, and the solution is diluted to 100 ml with doubly distilled water. The flask is placed on a boiling water bath for 20 min. When cool, the solution is made up to the mark with doubly distilled water and the optical density is determined. The content of aluminum is found from a calibration curve. The standard deviation of the method is ±6%.

The following elements do not interfere or are quantitatively separated: Ni, Co, Cu, Mn, Mo, V, Pb, Zn, Cd, Sn and Si. The metals W, Ti and Cr must be separated once again before aluminum is determined.

Kuznetsov and Golubtsova [215] determined aluminum in chromium-nickel alloys with Arsenazo, after the interfering elements had been removed by electrolysis on a mercury cathode and precipitation as cupferronates.

Determination of aluminum in titanium and titanium alloys

Aluminum in metallic titanium can be photometrically determined with Aluminon [545, 649, 655] and Methylthymol Blue [431].

In all these methods titanium and certain other metals present in titanium as impurities are preliminarily separated by filtering as cupferronates, as proposed by Corbett [655].

Metallic titanium (0.5 g) is dissolved by heating in 30 ml of 5 N HCl. The volume of the resulting solution is made up to 20 ml with HCl (sp. gr. 1.19), and then the solution is diluted to 100 ml with water. Two grams of ammonium chloride are added and the solution is cooled to 10°C. The solution is transferred to a 500-ml separatory funnel, and 80 ml of a 9% solution of Cupferron are added. The solution is vigorously shaken for 1 min to coagulate the precipitate. After 5 min, 50 ml of chloroform are added and the solution is shaken for one minute. When the layers have separated, the chloroform layer is decanted, and titanium is again extracted from the aqueous layer by adding 5 ml of Cupferron solution and 25 ml of chloroform. The extraction is repeated until both layers are colorless.

The aqueous layer is filtered through paper, evaporated almost to dryness, and diluted to 20 ml with water. The solution is cooled to 10°C and transferred to a separatory funnel, then 2 ml of Cupferron solution are added, the solution is shaken for one minute and left to stand for 5 min. Ten ml of chloroform are added and the mixture is shaken for one minute. After phase separation, the chloroform layer is decanted into a platinum dish. The extraction is repeated twice, with 10 ml of chloroform each time. The chloroform is expelled by evaporation and the organic matter is ignited at 500°C.

Five ml of HCl (sp. gr. 1.19) are added to the residue in the platinum dish, and the mixture is heated until it has completely dissolved. The solution is transferred to a 50-ml volumetric flask and the dish is rinsed with 30 ml of water. Ten ml of a 20% solution of ammonium acetate and 5 ml of a 0.2% solution of Aluminon are added, and the pH of the solution is adjusted to 4.9–5.0 by adding dilute HCl or dilute ammonium hydroxide (the pH is measured with a pH-meter). The solution is made up to 50 ml and the optical density is measured in a photocolorimeter in 1-cm cells, using a green filter.

The relative error of the method is 10%.*

---

* In our view, it would be expedient to modify the preparation of the colored solution in Corbett's method. It is best to use a hot aluminum solution (p. 97) to accelerate color formation. This procedure is also recommended in two studies on the determination of aluminum in titanium by Aluminon [545, 649].

In this method many interfering metals are separated from aluminum. Nevertheless, Banarjee [545] recommended the addition of thioglycolic acid before Aluminon to mask trace elements inevitably introduced into the solution with the reagents.

In the method of Banarjee, interfering elements are separated, and to the solution thus prepared for the determination of aluminum 1 ml of a 1% solution of thioglycolic acid is added. After 5 min, one drop of a 0.1% solution of Metacresol Purple is added and then 1 : 1 ammonia until the red color has disappeared. The volume of the solution is made up to 25 ml, 15 ml of "composite" Aluminon solution (p. 97) are added, and the determination is continued as above (p. 238).

The determination of aluminum in the presence of Methylthymol Blue is discussed in [431]. In this method also, many interfering elements are removed by means of Cupferron. Complexone III is used to improve the selectivity of the method.

Aluminum in titanium alloys can be determined by a complexometric method, with Xylenol Orange as indicator [1083, 1173].

The alloy is dissolved and an aliquot containing 2–40 mg Ti and 2–70 mg Al is taken. The solution is diluted to 100 ml with water, 20–50 ml of a 20% solution of triethanolamine are added, then $2 M$ NaOH until the mixture is decolorized. The mixture is boiled for one minute and the precipitate is filtered and washed, first with a hot 1% solution of triethanolamine and then twice with water. After the separation of $Ti(OH)_4$, the filtrate is acidified to pH 4 by 1 : 4 HNO_3, excess Complexone III is added and urotropin to pH 5–5.5 The solution is titrated by a solution of $Pb(NO_3)_2$ with Xylenol Orange as indicator, and addition of NaF to improve the selectivity [1083].

Determination of aluminum in magnesium and magnesium alloys

Aluminum in magnesium is best determined by the photometric technique with Aluminon. This method can also be applied to the analysis of magnesium alloys.

The alloy (0.1 g sample; if the sample consists of metallic magnesium, a larger amount must be taken) is dissolved in 5 ml of 1 : 1 HCl, 1–2 drops of 30% H_2O_2 are added, and the mixture is heated until the dissolution is complete. The solution is boiled to a residual volume of 2 ml, cooled, and diluted to 500 ml in a volumetric flask. An aliquot containing 0.02–0.06 mg Al (the total volume of the solution can be used if the sample consists of metallic aluminum) is placed in a 100-ml volumetric flask and is diluted to 10 ml with water. Fifteen ml of an Aluminon buffer solution (p. 97) is added, and the determination is continued as described above for copper alloys [939], see p. 234.

Methods for the determination of aluminum by a 0.2% aqueous, freshly prepared Aluminon solution have also been published; however, the accuracy and reproducibility of such methods are inferior to those obtained when the Aluminon solution, as described on p. 97, is employed.

A differential spectrophotometric determination with Chrome Azurol S can be recommended as a rapid and accurate method [417].

The alloy (0.1 g) is dissolved in 2 ml of 1 : 1 HCl. The solution is diluted to 500 ml in a volumetric flask and 5 ml of solution are withdrawn into a 100-ml volumetric flask. Two ml of a 1% solution of ascorbic acid, 5 ml of 0.1 N HCl, and water to a total volume of 50 ml, are added. Then 5 ml of a 0.12% solution of Chrome Azurol S and 5 ml of 2 N CH$_3$COONa are introduced, and the solution is made up to the mark with water.

After 10 min the optical density is determined at 545 mμ in a 1-cm cell against a solution containing 0.075 mg of aluminum. The content of aluminum is found from a calibration curve. This curve is plotted by taking 0.080, 0.085, 0.090, 0.095 and 0.100 mg of aluminum, the colored solutions are prepared, and the optical densities are determined with reference to a solution containing 0.075 mg of aluminum.

The method is suitable for alloys containing 7.5—10% of aluminum. If the proportion of aluminum in the alloy is smaller, the sample size must be correspondingly increased.

The relative error of the method is about 1%.

Aluminum in magnesium alloys may also be determined by the photometric method, using Eriochrome Cyanine R [66, 1156].

Aluminum in magnesium alloys can be most accurately determined by gravimetric methods, the hydroxyquinoline and the benzoate. In the determination by the hydroxyquinoline method, a correction must be introduced for the iron and zinc contents, while in the benzoate method the only correction to be introduced is for the iron content.

Determination by hydroxyquinoline. The alloy (0.1 g containing 7—10% Al; if the aluminum content is smaller, a larger sample size must be taken) is dissolved in 2 ml of 1 : 1 HCl and the solution is diluted to 100 ml with water. The cold solution is neutralized with ammonia to first turbidity, which is dissolved by adding 2 M HCl drop by drop, and then 5—10 drops in excess. Ten ml of a 2.5% solution of hydroxyquinoline in 5% acetic acid are added, followed by 12 ml of a 10% solution of sodium acetate introduced slowly with stirring. The determination of aluminum is then continued by the gravimetric (p. 28) or volumetric (bromatometric) method.

Determination by ammonium benzoate. The alloy (0.5—1 g, depending on the aluminum content) is dissolved in 20 ml of 1 : 1 HCl. When the alloy has dissolved, 8—10 drops of HNO$_3$ (sp. gr. 1.4) are added and the solution is boiled to remove nitrogen oxides. If the solution is turbid, it is filtered, and the filtrate is collected in a 300-ml beaker. The filter is washed 6—8 times with hot 2% HCl. The cold solution, 100—120 ml in volume, is neutralized with ammonia (1 : 1) to a faint, permanent turbidity, and then 1 ml of glacial CH_3COOH and 25 ml of a 25% solution of NH_4Cl are added. Thirty ml of a 10% solution of ammonium benzoate are introduced slowly, with constant stirring. The solution with the precipitate is heated to boiling, and boiled for not more than five minutes.

The hot solution is filtered through a "white ribbon" filter, the walls of the beaker are thoroughly rinsed, and the precipitate on the filter is washed 10 times with the wash liquor (to 100 ml of a 10% solution of ammonium benzoate are added 80 ml of water, 20 ml of glacial CH_3COOH, and the solution is diluted to 1 liter). The filter with the precipitate is dried and ignited in a platinum or a porcelain crucible at 110–1,200°C for one hour.

The content of aluminum is calculated from the formula

$$\% \, Al = \left[\frac{a \cdot 100}{b} - (\% \, Fe \cdot 1.43) \right] \cdot 0.5291,$$

where a is the weight of the precipitate, g; 1.43 is the conversion factor from Fe to Fe_2O_3; 0.5291 is the conversion factor from Al_2O_3 to Al; and b is the weight of the alloy, g.

The relative error is 2–3% [66, 467].

Complexometric methods are convenient, since the determination of aluminum in the prepared solutions takes only 10–15 min. Of these, direct titration in the presence of PAN indicator and copper complexonate [416], and back-titration of excess Complexone III by a solution of lead nitrate in the presence of Xylenol Orange [66] can be recommended. In the former method a correction must be introduced for the content of zinc, in the latter for the contents of zinc and manganese. Potentiometric titration by sodium hydroxide solution is recommended as a rapid method [340, 450a]. Almost all the methods enumerated above have been described in [66].

Determination of aluminum in uranium, thorium, plutonium, and their alloys

Aluminum in uranium is determined by the photometric hydroxyquinoline method [534, 775, 1072].

Two grams of uranium are dissolved in 4 ml of 12 N HCl, the solution is heated, HNO_3 (sp. gr. 1.42) is added drop by drop to oxidize the uranium, and the solution is made up to the mark in a volumetric flask. To an aliquot of the solution, containing 4–150 μg Al, 25 ml of a 30% solution of ammonium carbonate are added. The pH is adjusted to 8.8–9.2 by ammonia (sp. gr. 0.9) or HCl (sp. gr. 1.19). Five ml of an 8% solution of thioglycolic acid and 5 ml of 0.2 N KCN are added to bind uranium, iron and other elements.

Aluminum is extracted with 10 ml of a 3% chloroform solution of hydroxyquinoline for 90 sec, and the extraction is repeated twice. The extracts are collected in a 50-ml volumetric flask containing about 1 g of sodium sulfate, the solution is made up to the mark with chloroform, and the optical density is measured at 392 mμ against water [1072].

The method proposed by Granger [775] is identical, except that the interfering elements are masked by thioglycolic acid, Complexone III, and a carbonate-cyanide mixture.

To the sample solution are added 5 drops of 92% thioglycolic acid, a sufficient amount of a 4% solution of Complexone III, and a carbonate-cyanide mixture (100 g Na_2CO_3, 40 g $NaHCO_3$, and 30 g KCN (or 23 g NaCN) dissolved in 960 ml of warm water). If the amount of uranium is 2.5–5 g, 25 ml of Complexone III solution and 100 ml of carbonate-cyanide mixture are required. In this way, all the elements are masked except Ti, Zr and V, which are preliminarily separated by extracting their N-benzoyl-N-phenylhydroxylamine complexes by chloroform [775].

Aluminum in uranium alloys can be determined by complexometric methods, e.g., by back-titration by a solution of zinc in the presence of dithizone [833, 1090, 1091]. Uranium and iron are preliminarily separated from aluminum on Deacidite-FF anion exchanger from 9 M HCl solutions.

Aluminum in plutonium alloys is determined by a photometric hydroxy-quinoline method after the separation of plutonium and certain other elements on an anion exchanger as chloride complexes [705, 850], extraction of cupferronates [648] and hydroxyquinaldates [850].

The alloy (0.5 g or less) is dissolved in 5 ml HCl (sp. gr. 1.19); the dissolution is terminated by heating on a sand bath. An aliquot of the solution, containing 100–500 μg Al, is evaporated to a residual volume of not more than 0.5 ml. To the cold solution are added 3 ml HNO_3 (sp. gr. 1.4), and the evaporation is continued to a residual volume of 0.2 ml. Five ml HCl (sp. gr. 1.19) are added and the solution is again heated to the evolution of gases. The solution is cooled to room temperature and transferred to the anion exchanger column.

The column (6 mm in diameter) is filled with Dowex 1 × 8 anion exchanger to a height of 12.5 cm, and is washed with a solution 12 M in HCl and 0.12 M in HNO_3 (three times with 5 ml each time). By applying weak suction, the solution is passed through the column into a 25-ml volumetric flask, at the rate of 40±5 drops per minute. The eluate and the washings (dilute HCl) are evaporated almost to dryness. The residue is diluted with water to about 10 ml and heated to 60–70°C. The aluminum is precipitated by adding 1 ml of a 2% solution of hydroxyquinoline, 1 ml of 1 M KCN, and 3 ml of 2 M ammonium acetate, with stirring. The precipitate is held at 60–70°C for 15 min. It is then cooled to room temperature and filtered with suction. It is washed with small portions of water and dissolved in 15 ml HCl (sp. gr. 1.19). The solution is diluted to

100 ml with water in a volumetric flask, and the optical density is measured at 360 mμ against water.

The relative standard deviation of the method is less than 2%.

In the method of Jones and Phillips [850] plutonium is preliminarily separated on Deacidite-FF anion exchanger in the nitrate form. The interfering elements which remain in the eluate are removed by extraction as hydroxyquinaldates from a solution in [aqueous] ammonia, containing tartaric acid and hydrogen peroxide. Aluminum hydroxyquinaldate is then extracted by chloroform and the optical density of the extract is determined at 390 mμ.

If the content of aluminum in the sample is 0.1 – 1%, the relative error is less than 6% [705].

Aluminum in thorium can be determined by the method of Middleton [967].

First variant – thorium is masked with 4-sulfobenzenearsonic acid. One gram of thorium is dissolved by heating in 50 ml of 1 : 1 HNO_3 and 1 : 2 H_2SiF_6. The excess HNO_3 is evaporated and the solution is diluted to 100 ml in a volumetric flask. An aliquot of the solution, containing 0.25 g Th, is transferred to a beaker containing 25 ml of dilute acetate solution (200 g CH_3COONH_4 and 100 ml of glacial CH_3COOH in 1 liter), 20 ml of a 0.1% solution of o-phenanthroline, 1 ml of a 10% solution of hydroxylamine hydrochloride, and 10 ml of 0.24 M 4-sulfobenzenearsonic acid. The solution is diluted to 100 ml with water and the pH is adjusted to 4.8.

After 15–20 min, the solution is transferred to a separatory funnel, and shaken for 3 min with 20 ml of a 1% solution of hydroxyquinoline in chloroform. After phase separation, the extract is placed in a 25-ml volumetric flask containing 1–2 g of anhydrous Na_2SO_4. The mixture is vigorously shaken and allowed to stand for 15 min to dry the chloroform. The optical density is determined at 390 mμ against a blank solution, which has been run through the analysis in the same way as the sample solution.

Second variant – thorium is masked with acetate solution. Two or three grams of thorium are solubilized as in the first variant. To an aliquot containing about 1 g Th are added 1 ml of hydroxylamine hydrochloride solution and 10 ml of o-phenanthroline solution. After a few minutes, when the formation of the o-phenanthroline complex with Fe(II) is allowed to go to completion, 100 ml of concentrated acetate solution are added.* The pH is adjusted to 4.8 in a separatory funnel, and the solution is extracted with 20 ml of hydroxyquinoline solution in chloroform. The extract is transferred to a separatory funnel containing 25 ml of dilute acetate solution (for composition of this solution, see first variant), and 20 ml of o-phenanthroline solution, 1 ml of hydroxylamine hydrochloride and 50 ml of water are added. The solution is again shaken, the chloroform layer

---

* A solution of 300 g CH_3COONH_4 and 300 ml of glacial CH_3COOH in 1 liter, previously purified by three extractions with 100-ml portions of a 1% solution of hydroxyquinoline in chloroform; the hydroxyquinoline is removed by washing with 100-ml portions of chloroform.

is drawn off and dried, and its optical density is measured. If Zn, Cu, Co and Ni are present, their effect is eliminated by shaking the extract with 100 ml of a solution containing 40 g of ammonium nitrate, 20 g KCN and 10 ml of ammonium hydroxide (sp. gr. 0.9) per liter.

Determination of aluminum in other metals and alloys

Aluminum in lead and antimony is determined by the method of Luke [938], using Aluminon, while aluminum in metallic silver is determined by Stilbazo [150]. Aluminum in calcium can be determined by the photometric method, using hydroxyquinoline [1188].

CHAPTER V

Determination of Impurities in High-Purity Aluminum

Impurities in high-purity aluminum are mostly determined by photometric methods. Polarographic, spectroscopic, and radioactivation methods are also employed. Alkali metals are determined by flame photometry.

Table 16 shows the most important photometric methods for the determination of impurities in high-purity aluminum.

Cumulative amalgam polarography [75, 80, 81, 129, 936] has been proposed for the determination of certain impurities. The impurities to be determined are concentrated by deposition on a stationary mercury electrode, when the resulting amalgam is anodically polarized. The depths of the peaks are a measure of concentration of impurity elements. The following data may illustrate the sensitivity of the method:

Element	Sensitivity,%	Element	Sensitivity,%	Element	Sensitivity,%
Cu	10^{-5}–10^{-6}	Ga	10^{-5}	Pb	10^{-6}
Zn	$6 \cdot 10^{-5}$	In	$1 \cdot 10^{-6}$	Sb	$2 \cdot 10^{-5}$
Cd	$2 \cdot 10^{-6}$	Sn	$3 \cdot 10^{-5}$	Bi	$2 \cdot 10^{-6}$

The relative error is 10–20%.

Lithium can be determined by flame photometry with a sensitivity of $10^{-4}\%$ [1062]. Isopropanol and acetone are introduced into the sample solutions. The determination is carried out in an oxygen-hydrogen flame,

245

246 DETERMINATION OF IMPURITIES IN HIGH-PURITY ALUMINUM

Table 16. Photometric methods for the determination of impurities in high-purity aluminum

Element	Reagent	Sensitivity, %	Relative error, %	References
Cu	Diethyldithiocarbamate*	$2 \cdot 10^{-4}$		[96]
	Dithizone*	$2 \cdot 10^{-4}$		[96]
	Neocuproin* (2,9-dimethyl-1,10-phenanthroline*	$5 \cdot 10^{-4}$	5	[740, 197]
	Nickel dithiophosphate*	$2 \cdot 10^{-5}$	20	[70, 129]
	Oxalyl hydrazide*	$3 \cdot 10^{-4}$	20	[66]
Be	Acetylacetone*	10^{-4}		[386a]
Mg	Titanium Yellow	10^{-4}	15–20	[129]
Ca	Glyoxal-*bis*-(2-hydroxyanil)	$10^{-3}-10^{-4}$	15	[129]
Zn	Dithizone*	$5 \cdot 10^{-5}$	20	[129, 456]
	PAN*		5	[1068]
B	Acetylquinalizarin	10^{-4}	10	[300]
	1,1-Dianthrimide	10^{-4}		[870]
Ga	Rhodamine B		20	[1044]
REE	Arsenazo III			[333]
Si	Reduced silicomolybdic complex	$2 \cdot 10^{-4}$	20	[129]
	Silicomolybdic acid*	$2 \cdot 10^{-4}$	5–15	[96, 768a]
Ti	Diantipyrylmethane	$5 \cdot 10^{-5}$	20	[129, 130]
	Salicylhydroxamic acid*	$2 \cdot 10^{-4}$	10–20	[13]
Zr	Arsenazo III	10^{-4}	15–20	[129]
Sn	Phenylfluorone	$2 \cdot 10^{-5}$	20	[129]
Pb	Dithizone*	10^{-4}	10	[66]
P	Reduced silicomolybdic complex	10^{-4}	25	[129]
V	*p*-Phenetidine**	$2 \cdot 10^{-4}$	10	[911]
	Phosphotungstic acid	$5 \cdot 10^{-5}$	25	[129]
As	Molybdenum Blue	$5 \cdot 10^{-5}$	10–20	[129]
Sb	Methyl Violet*	$2 \cdot 10^{-5}$	10–15	[129]
Cr	Diphenylcarbazide	10^{-5}	20	[129]
U	Arsenazo III*	$5 \cdot 10^{-4}$		[221]
Mn	Potassium periodate	10^{-5}	20	[129]
Fe	*o*-Phenanthroline	10^{-4}	20	[129]
	o-Phenanthroline	$5 \cdot 10^{-4}$		[839]
	Thiocyanate and antipyrine*	$2 \cdot 10^{-4}$	5	[1210]
Ni	α-Furyldioxime*	$5 \cdot 10^{-5}$	20	[66]

* Extraction-photometric method.
** Catalytic oxidation by potassium chlorate.

using the 670.7 mμ line; the reproducibility of the method is 0.0001 –
0.0002% for impurity contents of 0.0003–0.0004%. Flame photometry
can also be employed in the determination of sodium [1136]. To enhance
the sensitivity of the method, the bulk of aluminum is preliminarily separated as $AlCl_3 \cdot 6H_2O$, or as the triethyl bromide; if a one-gram sample is
employed, the sensitivity of the method is $1.2 \cdot 10^{-4}\%$.

In the spectroscopic method, some of the impurities can be determined
directly, without preliminary enrichment, but preliminary concentration is
still required in most cases. Magnesium, copper and silicon can be determined without preliminary enrichment; the sensitivities are $5 \cdot 10^{-5}\%$,
$2 \cdot 10^{-5}\%$ and $2 \cdot 10^{-4}\%$, respectively [129]. Degtyareva et al. [103] determined 34 impurity elements in high-purity aluminum without preliminary
enrichment.

The analysis is carried out by fractional distillation from carbon electrodes in a DC
arc. The spectra are excited once at 7 A for 20 sec, and again at 12 A for 30 sec. In this
way volatile elements can be separated from low-volatility elements, so that the determination of the latter becomes more sensitive, but the sensitivity of determination of the
former remains unimpaired. The spectra are simultaneously recorded in two spectrographs, ISP-22 and ISP-51 with an F-270 camera. Both spectrographs are focused on
the same light source. Gold serves as the internal standard.

To concentrate the impurity elements, the bulk of the aluminum is separated as $AlCl_3 \cdot 6H_2O$ by the method of Gooch and Havens [766], in which
the solution is treated with hydrogen chloride [76, 129, 162, 610]. The
concentrate is evaporated to a small volume and mixed with carbon powder.
The mixture is dried and injected into the discharge zone [76, 129] or
introduced into the discharge zone by the method of Karabash et al. [162].
In the latter method, the residual aluminum is converted to the nitrate, the
solution is evaporated, and the residue ignited to Al_2O_3. Forty mg of the
concentrate (i.e., Al_2O_3 together with the impurities) are then introduced
into the crater of a carbon electrode, wetted with two drops of a 20% alcoholic solution of bakelite, and dried on a hotplate. The bakelite prevents
the sample from being ejected. The following sensitivites can be attained
by this method:

Element	Sensitivity, %	Element	Sensitivity, %	Element	Sensitivity, %
Be, Cd	$6 \cdot 10^{-6}$	Ba	$4 \cdot 10^{-4}$	Mn, Cu	$5 \cdot 10^{-6}$
Mg	10^{-5}	V, Co, Ni, Sb	$4 \cdot 10^{-5}$	Zn	$2 \cdot 10^{-4}$
Ca, Ti	$6 \cdot 10^{-5}$	Mo, Pb, Sn	$2 \cdot 10^{-5}$		

According to [76], the sensitivity of the determination of silicon is $2 \cdot 10^{-4}\%$. The relative error is 20%.

Impurities can be concentrated by evaporation in vacuo; the sample solution is previously evaporated and the aluminum ignited to Al_2O_3 [129]. The sensitivity of determination of Fe and Si is $10^{-4}\%$, of Cu, Mg, Ti and Pb it is $10^{-5}\%$, and of Mn, Sn and Ag, $10^{-6}\%$. Impurity elements can also be concentrated by distillation in high vacuum, without dissolving the sample [438].

Evaporation at $1,100-1,150°C$ with deposition on a cooled copper electrode is suitable for determining up to $4 \cdot 10^{-7}\%$ of zinc in a 5-gram sample; the relative error is 15% [1009].

The impurities in aluminum can also be concentrated before the spectroscopic determination as follows.

Thioacetamide precipitation in alkaline solution, with thallous ions as scavengers [1067]; in this way Ca, Cd, Co, Cr, Cu, Fe, Mg, Mn, Ni, Pb, Ti and Zn can be concentrated. Beryllium is introduced as an internal standard.

Precipitation by ammonium pyrrolidinedithiocarbamate and thionalide in acetic acid medium (pH 4) with thallous ions as scavengers [1067]; Cd, Co, Cu, Fe, Ga, Mo, Ni, Sb, Sn, V and Zn are precipitated. Beryllium is again taken as the internal standard.

Aluminum is distilled as an organometallic compound after treatment with C_2H_5Br [1010]. The residue is dissolved in HNO_3, carbon powder and 30 mg In are added, and the mixture is dried and ignited. The method can be used to determine up to $2 \cdot 10^{-6}-5 \cdot 10^{-5}\%$ Fe, Cu, Cr, Ag, Pb, Mn, Co and Ni.

Concentration of impurities by extracting their pyrrolidinedithiocarbamates and dithizonates by chloroform [879]. The sensitivity of the method is $10^{-5}-10^{-3}\%$; the relative error is 5−15%.

Impurities of Cu, Cd, Ag, Zn, Pb and Au are concentrated on a Bi_2S_3 collector. The sensitivity of the determination is $3 \cdot 10^{-7}\%$ Ag, $2 \cdot 10^{-6}\%$ Cu, Pb, $8 \cdot 10^{-7}\%$ Cd, $6 \cdot 10^{-5}\%$ Zn, $8 \cdot 10^{-6}\%$ Au. The relative error is 20−40% [129].

More than 40 publications are available on the determination of impurities in high-purity aluminum by the radioactivation method. The sample and the etalons are irradiated by a neutron flux of $10^{12}-10^{13}$ neutrons/$cm^2 \cdot sec$ in a nuclear reactor. The activities of the radioisotopes thus formed are measured by means of a scintillation γ-spectrometer. The duration of the irradiation varies between several hours and several weeks, depending on the type of impurities to be determined. It is a very common practice to preliminarily separate the impurity elements into groups by different techniques: precipitation on carriers, extraction, ion-exchange chromatography. The impurities can be determined by γ-spectrometry even without chemical separation. The determination of any given element can be made selective by a suitable choice of the duration of irradiation and duration of cooling [595]. Prussin et al. proposed a technique for activation analysis which does not involve the decomposition of the sample; he used Ge(Li) radiation detectors with a high resolving power [1093].

The sensitivity of the determination of isolated elements by the activation method is as follows: $10^{-3}\%$ Ti, $8 \cdot 10^{-4}\%$ O_2, $10^{-4}\%$ Fe, Si, $5 \cdot 10^{-5}\%$ C, Zr, V, $10^{-6}-10^{-8}\%$ Cu, Ag, Au, Zn, Cd, Sc, Ga, Sn, Th, U, P, As, Sb, Mn, Co. The relative error is $10-50\%$.

Iron, cobalt and titanium can be determined by irradiation in a cyclotron with 17 meV protons. They can also be determined by irradiation with cyclotron deuterons, with α-particles (for oxygen and carbon), with He-3 nuclei (determination of oxygen and zirconium), and with hard betatron γ-rays (determination of oxygen).

Impurity elements in high-purity aluminum can be determined by the mass-spectrographic technique [717, 965] with a sensitivity of 10^{-6} at.% and a relative error of $15-20\%$, and by the X-ray fluorescence method, by which vanadium can be determined with a sensitivity of $2 \cdot 10^{-4}\%$ [911]. Hydrogen is determined by a vacuum-melting method. Carbon can be determined by igniting the sample, and then titrating potentiometrically by a solution of $Ba(OH)_2$ [714]; the sensitivity of the method is $10^{-4}\%$.

BIBLIOGRAPHY

1. ABARBARCHUK, I.L., K.P. KOSTITSYNA, and E.M. SKOBETS. – *Pochvovedenie,* No. 2 : 114. 1962.
2. AVAYA, MUËSI, MOMODZIMA. – *Bunseki Kagaku,* **6** : 503. 1957; *RZhKhim.,* No. 8 : 24887. 1958.
3. AGAFONOVA, V.I. and I.P. RYAZONOV. – *Izvestiya Vuzov SSSR. Khimiya i Khimicheskaya Tekhnologiya,* **10** : 1200. 1967.
4. AGRINSKAYA, N.A. – *Zavodskaya Laboratoriya,* **23** : 279. 1957.
5. AGRINSKAYA, N.A. and V.I. PETRASHEN'. – *Trudy Novocherkasskogo Politekhnicheskogo Instituta,* **31** : 45, 63. 1955.
6. AGRINSKAYA, N.A. and V.I. PETRASHEN'. – *Ibid.,* **72** : 13. 1959.
7. ADAMOVICH, A.P., O.V. MORGUL'-MESHKOVA, and B.V. YUTSIS. – *Zhurnal Analiticheskoi Khimii,* **17** : 678. 1962.
8. AIZENBERG, N.N. – *Trudy Vsesoyuznogo Instituta Nauchno-Issledovatel'skikh i Proektnykh Rabot Ogneupornoi Promyshlennosti,* No. 37 : 115. 1965.
9. AKSEL'RUD, N.V. and Ya.A. FIALKOV. – *Ukr. Khim. Zh.,* **16** : 283. 1950.
10. ALEKPEROV, R.A., S.S. GEIBATOVA, and Z.A. AKHUNDOVA. – *Azerb. Khim. Zh.,* No. 6 : 123. 1964.
11. ALIMARIN, I.P., F.A. ABDEL'RAZIK, and A.I. KAMENEV. – *Zavodskaya Laboratoriya,* **34** : 160. 1968.
12. ALIMARIN, I.P., T.A. BOL'SHOVA, N.I. ERSHOVA, and A.D. KISELEVA. – *Vestnik MGU, Khimiya,* No. 4 : 55. 1967.
13. ALIMARIN, I.P., N.P. BORZENKOVA, and N.A. ZAKARINA. – *Zavodskaya Laboratoriya,* **27** : 958. 1961.
14. ALIMARIN, I.P. and I.M. GIBALO. – *Zh. Analit. Khim.,* **11** : 389. 1956.
14a. ALYKOV, N.M., B.I. KAZAKOV, and A.I. CHERKESOV. In: *"Fiziko-Khimicheskie issledovaniya prirodnykh sorbentov i ryada analiticheskikh sistem,"* p. 87. Saratov, Izd. Saratovskogo Universiteta. 1967.
15. ARIKOVA, I. – *Bunseki Kagaku,* **9** : 806. 1960; *RZhKhim.,* **12D** : 79. 1961.
16. ARTEM'EVA, V.Ya. – *Zavodskaya Laboratoriya,* **30** : 1331. 1964.
17. ARTEM'EVA, V.Ya. – *Ibid.,* **33** : 426. 1967.

18. ASTANINA, A.S. and A.I. PONOMAREV. – In: *"Problemy bol'shoi metallurgii i fizicheskoi khimii novykh splavov,"* p. 300. Moskva, "Nauka." 1965.

19. AKHMEDLI, M.K. and E.A. BASHIROV. – *Uchenye Zapiski Azerbaidzhanskogo Universiteta,* **25**, No. 7. 1955.

20. BABAEVA, A.E. and O.N. EVSTAF'EVA. – *Zh. Analit. Khim.,* **13** : 304. 1958.

21. BABACHEV, G.N. – *Zh. Analit. Khim.,* **13** : 716. 1958.

22. BABACHEV, G.N. and A. NIKOLAEVA. – *Stroitel'nye Materialy i Silikatna Promyshlennost',* **7** : 8. 1966.

23. BABACHEV, G.N., M. PETROVA, and L. RAEVA. – *Nauchnoe Soobshchenie Gosudarstvennogo Vsesoyuznogo Nauchno-Issledovatel'skogo Instituta Tsementnoi Promyshlennosti,* No. 17 (48) : 60. 1964.

24. BABENYSHEV, V.M. and O.M. KUZNETSOVA. – *Zh. Analit. Khim.,* **15** : 568. 1960.

25. BABKO, A.K. – *Zh. Prikl. Khim.,* **12** : 1560. 1939.

26. BABKO, A.K. – *Naukovi Zapysky. Kyyivs'kyi Derzhavnyi Universytet im. Shevchenko,* **2** : 59. 1936.

27. BABKO, A.K. – *Ibid.,* **3** : 49. 1937.

28. BABKO, A.K., A.I. VOLKOVA, and T.E. GET'MAN. – *Zh. Analit. Khim.,* **22** : 1004. 1967.

29. BABKO, A.K. and G.I. GRIDCHINA. – *Zh. Prikl. Khim.,* **36** : 1722. 1963.

30. BABKO, A.K. and A.I. DUBOVENKO. – *Zh. Neorgan. Khim.,* **2** : 1924. 1957.

31. BABKO, A.K. and A.I. ZOLINA. – *Byulleten' Vsesoyuznogo Nauchno-Issledovatel'skogo Instituta Mineral'nogo Syr'ya,* No. 11 (187) : 4. 1958.

32. BABKO, A.K. and T.T. MIKHAL'CHIKHIN. – *Ukr. Khim. Zh.,* **22** : 676. 1956.

33. BABKO, A.K. and T.K. NAZARCHUK. – *Ukr. Khim. Zh.,* **20** : 678. 1954.

34. BABKO, A.K. and A.T. PILIPENKO. *Kolorimetricheskii analiz (Colorimetric Analysis).* Moskva–Leningrad, Goskhimizdat. 1951.

35. BABKO, A.K., A.T. PILIPENKO, I.V. PYATNITSKII, and O.P. RYABUSHKO. *Fiziko-khimicheskie metody analiza (Physicochemical Methods of Analysis).* Moskva, "Vysshaya shkola." 1968.

36. BABKO, A.K. and T.N. RYCHKOVA. – *Zh. Obshch. Khim.,* **18** : 1617. 1948.

37. BABKO, A.K., Z.N. CHALAYA, and M.A. YAKUMOVA. Patent of the USSR, No. 154433. 1963; *Byulleten' Izobretenii,* No. 9. 1963.

38. BADEEVA, T.I., L.A. MOLOT, N.S. FRUMINA, and K.G. PETRIKOVA. – *Uchenye Zapiski Saratovskogo Universiteta,* **75** : 100. 1962.

39. BARBASH, T.L., N.P. PETROVA, and E.V. SILAEVA. – *Trudy VNIISO,* **2** : 108. 1964.

40. BARBASH, T.L. and T.F. KHANOVA. – *Trudy VNIISO,* **1** : 56. 1964.

41. BAUSOVA, N.V. – *Trudy Instituta Metallurgii Ural'skogo Filiala AN SSSR,* No. 3 : 125. 1959.

42. BASHKIRTSEVA, A.A. Candidate Thesis. Ural'skii Politekhnicheskii Institut. Sverdlovsk. 1956.

43. BASHKIRTSEVA, A.A. – *Trudy Ural'skogo Politekhnicheskogo Instituta,* No. 130 : 89. 1963.

44. BASHKIRTSEVA, A.A. and L.D. PRUDNIKOVA. – *Zavodskaya Laboratoriya*, **26** : 1107. 1960.

45. BASHKIRTSEVA, A.A. and E.M. YAKIMETS. – *Zavodskaya Laboratoriya*, **25** : 1166. 1959.

46. BASHKIRTSEVA, A.A. and E.M. YAKIMETS. – *Trudy Ural'skogo Politekhnicheskogo Instituta*, No. 58 : 76. 1957.

47. BEZUGLYI, V.D. – *Zavodskaya Laboratoriya*, **25** : 277. 1959.

48. BELETSKII, M.S., N.K. DRUZHININA, and V.G. YANKOVSKAYA. – In: *"Titan i ego splavy,"* No. 8 : 247. Moskva, Izd. AN SSSR. 1962.

48a. BEL'KEVICH, Ya.P. *Rukovodstvo po spektral'nomu analizu metallov (Textbook of Spectroscopic Analysis of Metals).* Leningrad, Sudpromgiz. 1955.

49. BELYAVSKAYA, T.A. and M.K. CHMUTOVA. – *Nauchnye Doklady Vysshei Shkoly. Khimiya i Khimicheskaya Tekhnologiya*, No. 2 : 305. 1958.

50. BELYAEVA, N.I. – *Pochvovedenie*, No. 2 : 106. 1966.

51. BERG, R. *Das o-Oxychinolin "Oxin."* Stuttgart, F. Enke. 1935. [Russian translation from German. 1937].

52. BERKOVICH, M.T., A.M. SIRINA, and N.L. LAGUNOVA. – *Trudy Ural'skogo Nauchno-Issledovatel'skogo Khimicheskogo Instituta*, No. 11 : 26. 1964.

53. BERLINOV, Kh. – *Stroitel'nye Materialy i Silikatna Promyshlennost'*, **6**, No. 2 : 19. 1965.

54. BIL'TYUKOVA, E.P. and A.M. ESECHKO. *Steklo (Glass).* – *Byulleten' Gosudarstvennogo Nauchno-Issledovatel'skogo Instituta Stekla*, No. 4(113) : 27. 1961.

54a. BOGDANOVA, V.T. and K.I. TAGANOV. – In: *"Sovremennye metody analiza v metallurgii."* Moskva, Metallurgizdat. 1955.

55. BOGINA, L.A. and I.P. MARTYUKHINA. – *Kauchuk i Rezina*, No. 10 : 34. 1961.

56. BOZHEVOL'NOV, E.A. – *Trudy Vsesoyuznogo Nauchno-Issledovatel'skogo Instituta Khimicheskikh Reaktivov*, No. 22 : 70. 1958.

57. BOZHEVOL'NOV, E.A. and G.V. SEREBRYAKOVA. – In: *"Metody analiza khimicheskikh reaktivov i preparatov,"* No. 11 : 15, 19. Moskva, Izdanie IREA. 1965.

58. BOZHEVOL'NOV, E.A. and V.M. YANISHEVSKAYA. – *Zh. Vses. Khim. Ob-stva im. D.I. Mendeleeva*, **5** : 356. 1960.

59. BONDAREVA, E.G., M.A. SHARDAKOVA, and G.G. KOBYAK. – *Uchenye Zapiski Permskogo Universiteta*, No. 141 : 279. 1966.

60. BOROVIK, S.A., A.V. BABAEVA, N.I. USHAKOVA, and R.I. RUDYI. – *Zh. Analit. Khim.*, **13** : 580. 1958.

61. BRITTON, H.T.S. *Hydrogen Ions.* London, Chapman and Hall. 1929.

62. BRODSKAYA, V.M. – In: *"Metody khimicheskogo analiza mineral'nogo syr'ya,"* No. 8 : 5. Moskva, "Nedra." 1965.

63. BRUILE, E.S. – *Zh. Prikl. Khim.*, **39** : 1192. 1966.

64. BRUILE, E.S. and N.S. DOMBROVSKAYA. – *Zh. Prikl. Khim.*, **36** : 2305. 1963.

65. BUDANOVA, L.M. and R.S. VOLODARSKAYA. – *Zavodskaya Laboratoriya*, **19** : 157. 1953.

66. BUDANOVA, L.M., R.S. VOLODARSKAYA, and N.A. KANAEV. *Analiz alyumi-nievykh i magnievykh splavov (Analysis of Aluminum and Magnesium Alloys).* Moskva, "Metallurgiya." 1966.
67. BUDESHINSKII, B. – *Zh. Analit. Khim.,* **18** : 1071. 1963.
68. BUZLANOVA, M.M. and N.A. KUROCHKINA. – *Zavodskaya Laboratoriya,* **31** : 947. 1965.
69. BUSEV, A.I. – *Zh. Analit. Khim.,* **8** : 299. 1953.
70. BUSEV, A.I. and N.P. BORZENKOVA. – *Zavodskaya Laboratoriya,* **27** : 13. 1961.
71. BUSEV, A.I., A.G. PETRENKO, and I.A. BYKHOVSKAYA. – *Zavodskaya Laboratoriya,* **27** : 659. 1961.
71a. BUSEV, A.I., N.V. SUKHORUKOVA, L.K. TERZEMAN, and N.L. BABENKO. – *Vestnik Moskovskogo Universiteta. Khimiya,* No. 2 : 119. 1968.
71b. BUYANOV, N.V., T.A. POPOVA, M.V. BABAEV, and V.F. NAZARENKO. – *Zavodskaya Laboratoriya,* **18** : 299. 1952.
72. VAKAMATSU, S. – *Bunseki Kagaku,* **9** : 238. 1960; *RZhKhim.,* No. 20 : 80742. 1960.
73. VAKAMATSU, S. – *Tetsu to khagane,* **44** : 1067. 1958; *RZhKhim.,* No. 9 : 3107. 1959.
74. VAKAMATSU, S. – *Tetsu to khagane,* **47** : 298. 1961; *RZhKhim.,* **6D** : 155. 1962.
75. VASIL'EVA, L.N. and E.N. VINOGRADOVA. – *Zh. Analit. Khim.,* **18** : 454. 1963.
76. VASIL'EVA, L.S. and T.F. GROMOSHINSKAYA. – *Zh. Prikl. Spektrosk.,* **1** : 324. 1964.
77. VASIL'EV, K.A. – *Zavodskaya Laboratoriya,* **6** : 432. 1937.
78. VERBITSKAYA, V.A., V.V. STEPIN, I.A. ONORINA, and L.S. STUDENSKAYA. – *Trudy VNIISO,* **2** : 52. 1965
79. VERNYI, E.A. and V.N. EGOROV. – *Zh. Analit. Khim.,* **15** : 24. 1960.
79a. VESELOVSKAYA, I.M. – *Zavodskaya Laboratoriya,* **15** : 940. 1949.
80. VINOGRADOVA, E.N. and L.N. VASIL'EVA. – *Zh. Analit. Khim.,* **17** : 579. 1962.
81. VINOGRADOVA, E.N., L.N. VASIL'EVA, and K. IOBST. – *Zavodskaya Labora-toriya,* **27** : 525. 1961.
82. VITAL'SKAYA, N.M. and N.F. PANTAEVA. – *Kauchuk i Rezina,* No. 6 : 53. 1962.
83. VLADIMIROVA, V.V. and O.A. SONGINA. – *Informatsiya Giredmeta,* No. 10 : 84. 1959.
84. VLASOV, N.A. and E.A. MORGEN. – *Izvestiya Fiziko-Khimicheskogo Nauchno-Issledovatel'skogo Instituta pri Irkutskom Universitete,* **6** : 200. 1964.
85. VOLKOV, I.I. and E.A. OSTROUMOV. – *Zh. Analit. Khim.,* **19** : 1223. 1964; *Trudy Instituta Okeanologii AN SSSR,* **79** : 81. 1965.
85a. VOLKOV, I.N. and E.A. OSTROUMOV. – *Zh. Analit. Khim.,* **23** : 863. 1968.
86. WÊNG YÜAN-K'AI. – *Yaohsüeh Hsüehpao,* **7**, No. 3 : 99. 1959; *RZhKhim.,* No. 3 : 8828. 1960.

87. GAIBAKYAN, G.S. – *Izvestiya AN ArmSSR, Biologicheskie i Sel'skokhozyai-stvennye Nauki*, **9** : 67. 1956; *RZhKhim.*, No. 5 : 15767. 1957.
88. GAPONENKOV, G.K. and Z.I. PROTSENKO. – *Zh. Neorgan. Khim.* **9** : 841. 1964.
89. HILLEBRAND, W.F., G.E. LUNDELL, H.A. BRIGHT, and J.I. HOFFMAN. *Practical Textbook of Inorganic Analysis.* New York, Wiley. 1929.
90. GOLOVATYI, R.N. – *Ukr. Khim. Zh.*, **24** : 653. 1958.
91. GOLOVATYI, R.N. and M.N. KOTOVSKAYA. – *Ukr. Khim. Zh.*, **25** : 791. 1959.
92. GOLOVINA, A.P., I.P. ALIMARIN, D.I. KUZNETSOV, and A.D. FILYUGINA.– *Zh. Analit. Khim.*, **21** : 163. 1966.
93. GOLUBTSOVA, R.B. – *Zh. Analit. Khim.*, **12** : 420. 1957.
94. GORDEEVA, M.N. and V.D. PROSVIRYAKOV. – *Uchenye Zapiski LGU*, No. 297 : 5. 1960.
95. GORELOVA, A.A. and L.Ya. POLYAK. – *Zavodskaya Laboratoriya*, **25** ; 285. 1959.
96. GORODENTSEVA, T.B., G.S. DOLGORUKOVA, K.F. VOROZHBITSKAYA, V.A. VERBITSKAYA, L.S. STUDENSKAYA, and V.S. SHVAREV. – *Trudy VNIISO*, **2** : 34. 1965.
97. AMO, GOSHIMA. *Giru daigaku gakugei gakuou kenkyu hakoku*, **1** : 496. 1957; *RZhKhim.*, No. 23 : 77162. 1958.
98. TAKAYAMA, GOTO. *Nihon kindzoku gakhashi*, **20** : 212. 1956; *RZhKhim.*, No. 6 : 19644. 1957.
99. GOKHSHTEIN, Ya.P. – *Zavodskaya Laboratoriya*, **5** : 28, 158. 1936.
99a. GRANBERG, I.A., K.A. SUKHENKO, K.A. RAZVYAZKINA, R.G. LIBERMAN, and R.G. SEREBRYANAYA. – *Zavodskaya Laboratoriya*, **17** : 1093. 1951.
100. GRISHCHENKO, M.P. and A.B. SHAEVICH. – *Izvestiya AN SSSR, Seriya Fizicheskaya*, **19** : 203. 1955.
101. DAVYDOV, A.L. and V.S. DEVEKKI. – *Zavodskaya Laboratoriya*, **10** : 134. 1941.
102. DANDZUKA, T. and K. UENO. – *Bunseki kagaku*, **8** : 126. 1959; *RZhKhim.*, No. 11 : 42359. 1960.
103. DEGMYAREVA, O.F., L.G. SINITSYNA, and A.E. PROSKURYAKOVA. – *Zh. Analit. Khim.*, **18** : 510. 1963.
104. DENISOVA, N.E. and E.V. TSVETKOVA. – *Zavodskaya Laboratoriya*, **27** : 656. 1961.
105. DOLABERIDZE, L.R., Yu.V. POLITOVA, L.T. GVELESIANI, and A.G. DZHA-LIASHVILI. – *Zavodskaya Laboratoriya*, **30** : 1439. 1964.
106. DOLABERIDZE, L.R., Yu.V. POLITOVA, L.T. GVELESIANI, and A.G. DZHA-LIASHVILI. – *Trudy Kavkazskogo Instituta Mineral'nogo Syr'ya*, No. 5 (7) : 81. 1963.
107. DOROSH, V.M. – *Sbornik Nauchnykh Trudov Irkutskogo Nauchno-Issledovatel'-skogo Instituta Redkikh Metallov*, No. 8 : 38. 1959.

107a. DUBINSKII, I.G. and V.F. ERKO. – *Zavodskaya Laboratoriya*, **25** : 1483. 1959.
108. DUBOVA, O.A., K.P. BEGEL'FED, P.A. SAZONOVA, and K.M. FUNTIKOVA. – *Steklo i Keramika*, No. 9 : 49. 1960.
109. DUBROVO, S.K. – *Zh. Prikl. Khim.*, **25** : 1151. 1952.
110. DYMOV, A.M. and V.V. KORENEVA. – *Izvestiya Vuzov SSSR. Chernaya Metallurgiya*, No. 3 : 192. 1961.
111. EVZLINA, B.B. – *Soobshcheniya o Nauchno-Issledovatel'skikh Rabotakh i Novoi Tekhnike Nauchno-Issledovatel'skogo Instituta po Udobreniyam i Insektofungisidam*, No. 8 : 87. 1958.
112. EGOROVA, L.G. – *Trudy Instituta Metallurgii i Obogashcheniya. AN KazSSR*, **12** : 151. 1965.
113. ERKO, V.F. and N.I. BUGAEVA. – *Fizicheskii Sbornik. L'vovskii Gosudarstvennyi Universitet im. I. Franko*, 4(9) : 490. 1958.
113a. ERMAKOVA, M.D. and Z.Ya. YAKOVLEVA. – *Zavodskaya Laboratoriya*, **23** : 592. 1957.
114. ERMOLAEVA, E.V. and L.A. KOROBKA. – *Byulleten' Nauchno-Tekhnicheskoi Informatsii Vsesoyuznogo Nauchno-Issledovatel'skogo Instituta Ogneuporov*, **2** : 84. 1957.
115. ZHAROVSKII, F.G. – *Naukovi Zapysky. Kyyivs'kyi Derzhavnyi Universitet im. Shevchenko*, **16**, No. 15 : 147. 1957.
116. ZHDANOV, A.K. and N.V. KOZHEVNIKOV. – *Zavodskaya Laboratoriya*, **18** : 529. 1952.
117. JÊN CHING-HUI, T'AO TSUNG-HSIANG. – Huahsüeh shihchien, **13** : 210. 1958; *RZhKhim.*, No. 15 : 53124. 1959.
118. ZHIVOPISTSEV, V.P., I.S. KALMYKOVA, and L.P. PYATOSIN. – *Uchenye Zapiski Permskogo Universiteta*, **25**, No. 2 : 108. 1963.
119. ZHIVOPISTSEV, V.P. and A.A. MININ. – *Zavodskaya Laboratoriya*, **26** : 1346. 1960.
120. ZHUKOVSKAYA, S.S. – *Zavodskaya Laboratoriya*, **3** : 102. 1934.
121. ZHUKOVSKAYA, S.S. and S.T. BALYUK. – *Zavodskaya Laboratoriya*, **3** : 485. 1934.
122. ZABIYAKO, V.I. and I.B. BULYCHEVA. – *Trudy Ural'skogo Nauchno-Issledovatel'skogo Khimicheskogo Instituta*, No. 11 : 82. 1964.
123. ZABIYAKO, V.I. and M.B. SALTYKOVA. – *Zavodskaya Laboratoriya*, **29** : 652. 1963.
124. ZABIYAKO, V.I. and M.B. SALTYKOVA. – *Trudy Ural'skogo Nauchno-Issledovatel'skogo Khimicheskogo Instituta*, No. 11 : 5. 1964.
125. ZAIDEL', A.N. *Osnovy spektral'nogo analiza (Fundamentals of Spectroscopic Analysis)*. Moskva, "Nauka." 1965.
126. ZAIDEL', A.N., V.K. PROKOF'EV, S.M. RAISKII, and E.Ya. SHREIDER. *Tablitsy spektral'nykh linii (Tables of Spectral Lines)*. Moskva, Fizmatgiz. 1962.
127. ZAN'KO, A.M. – *Dopovidi Akademiyi Nauk Ukrayins'koyi RSR*, **6** : 27. 1940.
128. ZAN'KO, A.M. and G.A. BUTENKO. – *Zavodskaya Laboratoriya*, **5** : 415. 1936.

129. ZINCHENKO, V.A. and T.A. POTAPOVA. – In: *"Metody analiza veshchestv vysokoi chistoty."* Moskva, "Nauka." 1965.
130. ZINCHENKO, V.A. and S.I. RUDINA. – *Zavodskaya Laboratoriya*, 27 : 956. 1961.
131. ZOLOTAVIN, V.L. and Yu.I. SANNIKOV. – *Trudy Ural'skogo Politekhnicheskogo Instituta*, No. 81 : 228. 1959.
132. ZOLOTUKHIN, V.K. – *Zh. Analit. Khim.*, 12 : 271. 1957.
133. IVAMOTO, T. – *Bunseki kagaku*, 10 : 190. 1961; *RZhKhim.*, 1D : 41. 1962.
134. IVANOVA, Z.I., P.N. KOVALENKO, M.G. ERISTOVA, and L.S. IPATOVA. – In: *"Fiziko-khimicheskie metody analiza i kontrolya proizvodstva,"* p. 164. Rostovskii Universitet. Rostov-na-Donu. 1961.
135. IVANOVA, I. – *Godishnik na Nauchnoizsledovatelskiya Institut po Metallurgiya i Obogat.*, 2 : 167. 1961.
136. IVANOV, B.T. and S.M. BEZYAIKO. – *Zavodskaya Laboratoriya*, 15 : 511. 1949.
137. IVASAKI, I. and T. OMORI. – *Kogë ësui*, No. 54 : 30. 1963; *RZhKhim.*, 11G : 90. 1964.
138. IZHAK, I.G. – *Zavodskaya Laboratoriya*, 30 : 1449. 1964.
139. IUDA, M. – *Bunseki kagaku*, 9 : 209. 1960; *RZhKhim.*, No. 21 : 84427. 1960.
140. INOUE, T. – *Kogë ësui*, No. 34 : 34. 1961; *RZhKhim.*, 15D : 50. 1962.
141. ISIBASI, M., T. SIGEMATSU, and I. NISIKAVA. – *Bunseki kagaku*, 6 : 568. 1957; *RZhKhim.*, No. 9 : 28419. 1958.
142. ISIBASI, M., T. SIGEMATSU, and I. NISIKAVA. – *Nihon kagaku dzassi*, 81 : 259. 1960; *RZhKhim.*, No. 21 : 84424. 1960.
143. ISIKI, I. – *Bunseki kagaku*, 14 : 1120. 1965; *RZhKhim.*, 13G : 103. 1966.
144. ITIKUNI, M. – *Bunseki kagaku*, 13 : 1040. 1964; *RZhKhim.*, 13G : 113. 1965.
145. ITO, S., K. KHAYASI, and Ya. UDZUMASA. – *Bunseki kagaku*, 12 : 257. 1963; *RZhKhim.*, 21G : 41. 1963.
146. IONEYAMA, I. and T. KAMADA. – *Bunseki kagaku*, 10 : 187. 1961; *RZhKhim.*, 21D : 51. 1961.
147. IORDANOVA, N. and Z. PROINOVA. – *Stroitel'nye Materialy i Silikatna Promyshlennost'*, 8, No. 21. 1967.
148. IOSIMURA, T., Kh. NOGUTI, and Kh. KHARA. – *Kindzoku khëmen gidzyutsu*, 18 : 333. 1967; *RZhKhim.*, 10G : 59. 1967.
149. KABANNIK, G.T. and T.N. NAZARCHUK. – *Zavodskaya Laboratoriya*, 28 : 546. 1962.
150. KABANOVA, O.L. and M.A. DEDYUSHENKOVA. – *Zh. Analit. Khim.*, 18 : 780. 1963.
151. KABANOV, B.N. and L.Ya. POLYAK. – *Zh. Analit. Khim.*, 13 : 538. 1958.
152. KAVAGUTI, Kh., T. NAKADZIMA, and M. TAKAKHASI. – *Denki sikensë ikho*, 22, No. 3 : 175. 1958; *RZhKhim.*, No. 1 : 1008. 1959.
153. KAZAKOV, B.I., V.V. EMZHIN, and A.I. CHERKESOV. – *Zavodskaya Laboratoriya*, 33 : 697. 1967.

154. KAZAKOV, B.I. and Yu.V. PUSHINOV. – In: *"Novye issledovaniya po analiti-cheskomu primeneniyu organicheskikh reagentov,"* p. 15. Saratov. 1967.

155. KAZANTSEV, E.I., K.V. SOKOLOVA, and L.K. SOSNOVSKIKH. – *Trudy Ural'skogo Politekhnicheskogo Instituta*, No. 14887. 1966.

156. KAKITA, Ya., E. SUDO, and M. NAMIKI. – *Nihon kindzoku gakkashi*, **24** : 363. 1960; *RZhKhim.*, **11 D** : 174. 1961.

157. KAKITA, Ya., M. KHOSOYA, and M. AMANO. – *Nihon kindzoku gakkashi*, **21** : 551. 1957; *RZhKhim.*, No. 11 : 35897. 1958.

158. KALINA, V.I. and N.F. LEVE. – *Byulleten' Nauchno-Tekhnicheskoi Informatsii Ukrainskogo Nauchno-Issledovatel'skogo Instituta Metallov*, No. 3 : 92. 1957.

159. KAMENSKII, Ya. M. – *Zavodskaya Laboratoriya*, **24** : 755. 1958.

160. KAMENSKII, Ya.M. and K.I. NIKIFOROVA. – *Zavodskaya Laboratoriya*, **28** : 413. 1962.

161. KAPUSTINA, A.I., I.M. KONSTANTINOVA, and Z.I. MIKHEEVA. – *Trudy Vostochno-Sibirskogo Filiala AN SSSR*, No. 43 : 90. 1962.

162. KARABASH, A.G., Sh.I. PEIZULAEV, R.L. SLYUSAREVA, and V.M. MESH-KOVA. – *Zh. Analit. Khim.*, **14** : 598. 1959.

163. KAROLEV, A.N. and M.K. KOICHEV. – *Zavodskaya Laboratoriya*, **25** : 546. 1959.

164. KASHKOVSKAYA, E.A. and I.S. MUSTAFIN. – *Zavodskaya Laboratoriya*, **24** : 1189. 1958.

165. KINOSITA, S. – *Bunseki kagaku*, **14** : 1154. 1965; *RZhKhim.*, **16 G** : 127. 1966.

166. KLASSOVA, N.S. – *Zh. Analit. Khim.*, **20** : 1747. 1950.

167. KLEINER, K.E. – *Zh. Obshch. Khim.*, **20** : 1747. 1950.

168. KLIMOV, V.V., O.S. DIDKOVSKAYA, and V.I. KOZACHENKO. – *Zavodskaya Laboratoriya*, **28** : 952. 1962.

169. KOVALENKO, P.N. – *Uchenye Zapiski Rostovskogo-na-Donu Universiteta*, **40** : 45. 1958.

170. KOVALENKO, P.N. – *Uchenye Zapiski Rostovskogo-na-Donu Universiteta*, **60** : 87. 1959.

171. KOGAN, D.I. – *Nauchnye Trudy Irkutskogo Nauchno-Issledovatel'skogo Instituta Redkikh Metallov*, No. 11 : 48. 1963.

172. KODZIMA. – *Bunseki kagaku*, **6** : 369. 1957; *RZhKhim.*, No. 4 : 11031. 1958.

173. KOZLOVA, A.A. and M.A. PORTNOV. – *Zavodskaya Laboratoriya*, **9** : 287. 1940.

174. KOKORIN, A.I. and V.M. ROPOT. – *Uchenye Zapiski Kishinevskogo Universiteta*, **56** : 105. 1960.

175. KOLBOVSKII, Yu.Ya. and M.K. KRIZHANOVSKAYA. – *Fizicheskii Sbornik. L'vovskii Gosudarstvennyi Universitet im. I. Franko*, 4(9) : 402. 1958.

176. KOLECHKOVA, A.F. and N.M. OSTANINA. – *Trudy Vsesoyuznogo Instituta Nauchno-Issledovatel'skikh i Proektnykh Rabot Ogneupornoi Promyshlennosti*, No. 37 : 55. 1965.

177. KOLOBOVA, K.K. – *Ibid.*, No. 37 : 74. 1965.
178. KOLTHOFF, I.M. and E.B. SANDELL. *Textbook of Quantitative Inorganic Analysis.* New York. Macmillan and Co. 1943.
178a. KOMAROVSKII, A.G. *Analiz zharoprochnykh splavov spektral'nym metodom (Spectroscopic Analysis of Refractory Materials).* Moskva, Mashgiz. 1952.
179. KOMISSARENKO, V.S. – *Zavodskaya Laboratoriya,* **22** : 453. 1956.
180. KONKIN, V.L. – *Byulleten' Nauchno-Tekhnicheskoi Informatsii Ukrainskogo Nauchno-Issledovatel'skogo Instituta Metallov,* No. 2 : 76. 1957.
181. KONKIN, V.D. Candidate Thesis. – Ukrainskii Nauchno-Issledovatel'skii Institut Metallov. 1953.
182. KONKIN, V.D. and V.I. ZHIKHAREVA. – *Zavodskaya Laboratoriya,* **27** : 143. 1961.
183. KONSTANTINOVA, B.P. and O.V. OSHURKOVA. – *Izvestiya AN SSSR, Seriya Khimicheskaya,* No. 8 : 1648. 1967.
184. KOPYLOVA, V.P. and T.N. NAZARCHUK. – *Zh. Analit. Khim.,* **20** : 892. 1965.
185. KORENMAN, I.M. – In: *"Peredovye metody khimicheskoi tekhnologii i kontrolya proizvodstva,"* p. 339. Izd. Rostovskogo Universiteta. 1964.
186. KORENMAN, I.M. and T.D. VERBITSKAYA. – *Trudy po Khimii i Khimicheskoi Tekhnologii,* No. 2(10) : 248. Gorki. 1964.
187. KORENMAN, I.M. and I.A. GRISHIN. – *Trudy po Khimii i Khimicheskoi Tekhnologii,* No. 2 : 383. Gorki. 1958.
188. KORENMAN, I.M. and V.S. EFIMYCHEV. – *Trudy po Khimii i Khimicheskoi Tekhnologii,* **4**, No. 2 : 323. Gorki. 1961.
189. KORENMAN, I.M. and G.V. NAZAROVA. – *Trudy po Khimii i Khimicheskoi Tekhnologii,* No. 2(10) : 255. Gorki. 1964.
190. KORENMAN, I.M. and A.M. NOVIKOVA. – *Trudy po Khimii i Khimicheskoi Tekhnologii,* No. 2(10) : 259. Gorki. 1964.
191. KORENMAN, I.M. and U. FAIZIEV. – *Trudy po Khimii i Khimicheskoi Tekhnologii,* No. 2(13) : 164. Gorki. 1965.
192. KORENMAN, I.M. and U. FAIZIEV. – *Ibid.,* No. 2(16) : 301. 1966.
193. KORENMAN, I.M., F.S. FRUM, and E.K. KARZHENKINA. – *Uchenye Zapiski Gor'kovskogo Universiteta,* No. 4 : 131. 1953.
194. KORENMAN, I.M. and S.F. CHELYSHEVA. – *Zh. Analit. Khim.,* **18** : 145. 1963.
195. KORENMAN, I.M. and S.F. CHELYSHEVA. – *Trudy po Khimii i Khimicheskoi Tekhnologii,* No. 3(14) : 103. Gorki. 1965.
196. KORENMAN, I.M., F.R. SHEYANOVA, and N.I. OSIPOVA. – *Trudy po Khimii i Khimicheskoi Tekhnologii,* No. 2 : 395. Gorki. 1962.
197. KOROLEV, V.V. and E.E. VAINSHTEIN. – *Zh. Analit. Khim.,* **13** : 627. 1958.
198. KOSTERNAYA, A.F. and N.A. ZAVOROVSKAYA. – *Sbornik Nauchnykh Rabot Institutov Okhrany Truda VTsSPS,* No. 1 : 82. 1962.
199. KOSTITSYNA, K.P. and E.K. SKOBETS. – *Zavodskaya Laboratoriya,* **29** : 1059. 1963.
200. KOSTROMIN, A.I. and M.Kh. AKHMADIEV. – *Zavodskaya Laboratoriya,* **29** : 402. 1963.

201. KOSTROMIN, A.I. and A.I. KRUGLOV. − *Uchenye Zapiski Kazanskogo Universiteta*, **124**, No. 3 : 173. 1965.
202. KOCHARYAN, A. − *Sbornik Studencheskikh Nauchnykh Trudov Erevanskogo Universiteta*, No. 8 : 197. 1958; *RZhKhim.*, No. 23 : 77233. 1958.
203. KRASIL'NIKOVA, L.N. and L.I. MAKSAI. − *Sbornik Nauchnykh Trudov VNIITsvetmeta*, No. 1 : 3. 1956.
204. KRASIL'NIKOVA, L.N. and L.I. MAKSAI. − *Ibid.*, No. 3 : 87. 1958.
205. KRASYUKOVA, N.G. − In: *"Materialy po geologii i poleznym iskopaemym Buryatskoi ASSR,"* No. 10 : 238. 1966.
206. *Kratkaya khimicheskaya entsiklopediya (Concise Chemical Encyclopedia).* Vol. 1. Moskva, Izd. "Sovetskaya entsiklopediya." 1961.
207. KRAUS, K.A. and F. NELSON. − *Proceedings of the International Conference on the Peaceful Uses of Atomic Energy*, Geneva. 1955. Vol. 7, p. 837.
207a. KROTOVA, I.K. and M.M. KOSHELEVA. − In: *Sbornik trudov nauchnogo instituta po udobreniyam i insektofungisidam*, p. 3. Moskva, Goskhimizdat. 1962.
208. KRUGLOVA, M.N., T.B. GORODENTSEVA, G.S. DOLGORUKOVA, K.F. VOROZHBITSKAYA, L.S. STUDENSKAYA, T.L. BARBASH, and E.V. SILAEVA. − *Trudy VNIISO*, **2** : 75. 1965.
209. KRUPSKII, N.K., A.M. ALEKSANDROVA, and A.I. KHIZHNYAK. − *Pochvovedenie*, No. 10 : 93. 1961.
210. KRYUKOVA, T.A., S.I. SINYAKOVA, and T.V. AREF'EVA. *Polyarograficheskii analiz (Polarographic Analysis)*, p. 254. Moskva, Goskhimizdat. 1959.
211. KUDELYA, E.S. − *Avtomaticheskaya Svarka*, No. 6 : 28. 1959.
212. KUDELYA, E.S. − *Zavodskaya Laboratoriya*, **26** : 1128. 1960.
212a. KUDELYA, E.S. *Spektral'nyi analiz metallov i splavov (Spectroscopic Analysis of Metals and Alloys)*. Kiev, Gostekhizdat UkrSSR. 1961.
213. KUZNETSOV, V.I. − *Zh. Obshch. Khim.*, **17** : 175. 1947.
214. KUZNETSOV, V.I. and R.B. GOLUBTSOVA. − *Zavodskaya Laboratoriya*, **21** : 1422. 1955.
215. KUZNETSOV, V.I. and R.B. GOLUBTSOVA. − *Ibid.*, **22** : 161. 1956.
216. KUZNETSOV, V.I. and D.A. DRAPKINA. − *Trudy Vsesoyuznogo Nauchno-Issledovatel'skogo Instituta Khimicheskikh Reaktivov*, No. 21 : 18. 1956.
217. KUZNETSOV, V.I., G.G. KARANOVICH, and D.A. DRAPKINA. − *Zavodskaya Laboratoriya*, **16** : 787. 1950.
218. KUL'BERG, L.M. and L.A. MOLOT. − *Ukr. Khim. Zh.*, **21** : 256. 1955.
219. KUL'BERG, L.M. and I.S. MUSTAFIN. − DAN SSSR, **77** : 285. 1951.
220. KUNENKOVA, E.N. and E.A. OSTROUMOV. − *Zh. Analit. Khim.*, **19** : 955. 1964.
221. KURAKHA, T., M. SAKAKUBARA, S. SIBUYA, and M. OGURA. − *Bunseki kagaku*, **15** : 569. 1966; *RZhKhim.*, **16** : 130. 1967.
222. LAZAREV, A.I. − In: *"Sovremennye metody analiza v metallurgii,"* p. 182. Moskva, Metallurgizdat. 1955.

223. LEBEDEV, O.P. and I.I. VLASOVA. – *Sbornik Nauchnykh Trudov Nauchno-Issledovatel'skogo Gornorudnogo Instituta UkrSSR*, **7** : 293. 1963.

223a. LEBEDEVA, M.I., B.I. ISAEVA, and V.N. PROKHOROV. – *Trudy Tambovskogo Instituta Khimicheskogo Mashinostroeniya*, No. 2 : 56. 1968.

224. LEBED', N.B. and R.P. PANMALER. – *Zavodskaya Laboratoriya*, **31**:163. 1965.

225. LEVITIN, R.Z. and V.I. SMIRNOVA. – *Fizicheskii Sbornik. L'vovskii Gosudarstvennyi Universitet im. I. Franko*, No. 4 (9) : 497. 1958.

226. LEL'CHUK, Yu.L., V.B. SOKOLOVICH, and O.V. DRELINA. – *Izvestiya Tomskogo Politekhnicheskogo Instituta*, **128** : 101. 1964.

227. LEL'CHUK, Yu.L., V.B. SOKOLOVICH, and E.A. KURYSHEVA. – *Ibid.*, **128** : 106. 1964.

228. LENSKAYA, K.K. and E.I. VERZHBOVSKAYA. – *Sbornik Trudov TsNIIChERMETa*, **24** : 185. 1962.

228a. LIVSHITS, E.V. and L.N. MOSOVA. – *Zavodskaya Laboratoriya*, **25** : 952. 1962.

229. LI K'AI-YÜAN. – *Huahsüeh t'ungpao*, No. 2 : 43. 1961; *RZhKhim.*, **6 D** : 84. 1962.

230. LISENKO, N.F. Candidate Thesis. – Saratovskii Gosudarstvennyi Universitet. 1967.

231. LISENKO, N.F. and I.S. MUSTAFIN. – *Zh. Analit. Khim.*, **22** : 25. 1967.

232. LISENKO, N.F., I.S. MUSTAFIN, and L.A. MOLOT. – *Izvestiya vuzov SSSR. Khimiya i Khimicheskaya Tekhnologiya*, **5** : 712. 1962.

233. LIKHOED, L.S. and N.I. NOSENKO. – *Fizicheskii Sbornik. L'vovskii Gosudarstvennyi Universitet im. I. Franko*, No. 4(9) : 497. 1958.

234. LOBANOV, E.M. and G.G. MINGALIEV. – In: *"Aktivatsionnyi analiz elementnogo sostava geologicheskikh ob"ektov,"* pp. 77, 93. Tashkent, "Fan." 1967.

235. LOBANOV, E.M., A.I. CHANYSHEV, A.G. DUTOV, M.G. ASHIROV, and A. KHUDAIBERGANOV. – *Trudy 1-go Vsesoyuznogo koordinatsionnogo soveshchaniya po aktivatsii analizu*. 1962. p. 94. Tashkent, "Nauka." 1964.

236. LOBANOV, E.M., A.I. CHANYSHEV, T.I. CHANYSHEVA, Yu.N. TALANIN, L.V. NAMIKHIN, and V.A. KIREEV. – In: *Yadernaya fizika i ee primenenie,"* Part 1, p. 38. Tashkent, "Fan." 1966.

237. LUKIN, A.M., G.S. PETROVA, N.B. ETINGEN, L.A. MOLOT, and A.S. ARKHANGEL'SKAYA. Patent of the USSR, 213405. 1967; *Byulleten' Izobretenii*, No. 10. 1968.

238. LUR'E, Yu. Yu. and N.A. FILIPPOVA. – *Zavodskaya Laboratoriya*, **13** : 539. 1947.

239. LUR'E, Yu.Yu. and N.A. FILIPPOVA. – *Ibid.*, **14** : 159. 1948.

240. L'VOV, B.V. *Atomno-absorbtsionnyi spektral'nyi analiz (Atomic Absorption Spectroscopic Analysis)*. Moskva, "Nauka." 1966.

241. LIU KUO-CHÜN. – *K'ohsüeh t'ungpao*, No. 20 : 637. 1958; *RZhKhim.*, No. 15, 53123. 1959.

242. LIU CHIN, CHIN P̂ENG-YING. – *Chienchu ts'ailiao ch'unye*, No. 16 : 31. 1959; *RZhKhim.*, No. 18 : 73059. 1960.

243. LIAO CH'EN-CHIEN. — *Wuhan tahsüeh tz'üan k'ohsüeh hsüehpao,* No.1 : 24. 1959; *RZhKhim.,* No. 3 : 8853. 1960.

244. LYASHENKO, A.G. — *Obogashchenie Rud,* No. 5 (23) : 23. 1959.

245. MAKSIMOVA, O.S. and V.A. KUCHINSKII. — *Uchenye Zapiski Politekhnichesko-go Instituta,* 6 : 272. 1962.

246. MALEVANNYI, V.A. and V.A. SHUMINA. — *Lakokrasochnye Materialy i ikh Primenenie,* No. 4 : 75. 1967.

247. MALENSKAYA, V.P. — *Trudy po Khimii i Khimicheskoi Tekhnologii,* No. 2 : 321. Gorki. 1960.

248. MALININA, R.D. and E.V. PTUSHKINA. — *Sbornik Trudov TsNIIChERMETa,* No. 19 : 61. 1960.

249. MAL'TSEV, V.F. — *Byulleten' Nauchno-Tekhnicheskoi Informatsii Vsesoyuznogo Nauchno-Issledovatel'skogo Trubnogo Instituta,* No. 3 : 108. 1957.

250. MAL'TSEV, V.F., L.P. LUK'YANENKO, and D.M. KUKUI. — *Zavodskaya Laboratoriya,* 27 : 807. 1961.

251. MAMEDOV, N.A. Author's Summary of Candidate Thesis. — Azerbaidzhanskii Industrial'nyi Institut, Baku. 1953.

251a. MAEKAVA, S. and I. IONEYAMA. — *Nihon kindzoku gakkashi,* 25 : 651. 1961; *RZhKhim.,* 20 D : 46. 1962.

251b. MAEKAVA, S. and K. KATO. — *Bunseki kagaku,* 17 : 70. 1968; *RZhKhim.,* 21G : 106. 1968.

252. MARGOLIS, L.D. — *Zavodskaya Laboratoriya,* 28 : 290. 1962.

253. MEDVEDEVA, A.P. and M.I. MAZEL'. — In: *"Spektral'nye i khimicheskie metody analiza materialov,"* p. 166. Moskva, Metallurgizdat. 1964.

253a. MELAMED, Sh. G. and A.M. SALTYKOVA. — *Zavodskaya Laboratoriya,* 23 : 573. 1957.

254. *Metody analiza veshchestv vysokoi chistoty (Analysis of High-Purity Materials).* Moskva, "Nauka." 1965.

255. MINEVA, G. — *Khimiya i Industriya (Sofia),* 32 : 117. 1960.

256. MITUZAS, I. — *Nauchnye Trudy Vuzov Litvy. Khimiya i Khimicheskaya Tekhnologiya,* No. 1 : 38. 1961; *RZhKhim.,* 20G : 58. 1963.

257. MIURA. — *Nihon kindzoku gakkashi,* 19 : 310. 1955; *RZhKhim.,* No. 10 : 29336. 1956.

258. MIKHAILENKO, M.I. — In: *"Issledovaniya v oblasti farmatsii,"* p. 105. Odessa. 1959.

259. MIKHAILOV, M. and B. BENEV. — *Minnoe Delo i Metallurgiya,* 16 : 27. 1961.

260. MIYADZIMA, Ts. — *Bunseki kagaku,* 13 : 362. 1964; *RZhKhim.,* 2G : 43. 1965.

261. MIYADZIMA, Ts. — *Bunseki kagaku,* 13 : 1042. 1964; *RZhKhim.,* 14G : 124. 1964.

262. MOLOT, L.A. Candidate Thesis. — Saratovskii Gosudarstvennyi Universitet. 1952.

263. MOLOT, L.A., A.S. ARKHANGEL'SKAYA, M.I. TRUSOVA, and I.S. MUSTA-FIN. — *Zavodskaya Laboratoriya,* 34 : 408. 1968.

264. MOLOT, L.A. and L.M. KUL'BERG. – *Zh. Analit. Khim.*, **11** : 198. 1956.
265. MOLOT, L.A. and L.M. KUL'BERG. – *Uchenye Zapiski Saratovskogo Universiteta*, **42** : 79. 1955.
265a. MOLOT, L.A., I.S. MUSTAFIN, and A.P. ANDRONOVA. – In: *"Primenenie organicheskikh reagentov v analize,"* p. 61. Saratovskii Universitet, Saratov. 1967, 1968.
266. MOLOT, L.A., I.S. MUSTAFIN, and R.F. ZAGREBINA. – *Izvestiya Vuzov SSSR. Khimiya i Khimicheskaya Tekhnologiya*, **9** : 873. 1966.
267. MOLOT, L.A., I.S. MUSTAFIN, A.N. IVANOVA, K.G. PETRIKOVA, V.S. KOVA-LEVA, and L.A. AGRANOVSKAYA. – In: *"Peredovye metody khimicheskoi tekhnologii i kontrolya proizvodstva,"* p. 329. Rostov-na-Donu Universitet. 1964.
268. MOLOT, L.A., I.S. MUSTAFIN, and N.K. NEMKOVA. – *Izvestiya Vuzov SSSR. Khimiya i Khimicheskaya Tekhnologiya*, **10** : 1060. 1967.
269. MOLOT, L.A., I.S. MUSTAFIN, and N.S. FRUMINA. – *Trudy Komissii po Analiticheskoi Khimii*, **11** : 231. 1960.
270. MORACHEVSKII, Yu.V. and Z.S. BASHUN. – *Izvestiya AN SSSR, Otdelenie Khimicheskikh Nauk*, **10** : 1185. 1956.
271. MORACHEVSKII, Yu.V. and M.N. GORDEEVA. – *Zavodskaya Laboratoriya*, **23** : 1066. 1957.
272. MORACHEVSKII, Yu.V. and V.N. ZAITSEV. – *Trudy Komissii po Analiticheskoi Khimii*, **15** : 260. 1965.
273. MORACHEVSKII, Yu.V. and V.N. ZAITSEV. – *Uchenye Zapiski LGU, Seriya Khimicheskikh Nauk*, **18** : 272, 134. 1959.
274. MORACHEVSKII, Yu.V. and A.I. NOVIKOV. – *Trudy Komissii po Analiticheskoi Khimii*, **19(12)** : 121. 1958.
275. MORACHEVSKII, Yu.V. and A.I. NOVIKOV. – *Uchenye Zapiski LGU, Seriya Khimicheskikh Nauk*, **18**, No. 272 : 112. 1959.
276. MORII, F. – *Nihon kagaku dzassi*, **82** : 1507. 1961; *RZhKhim.*, **20D** : 31. 1962.
277. MORKOVCHINA, K.S. and A.M. PINDUS. – *Zavodskaya Laboratoriya*, **28** : 805. 1962.
278. MOROZOVA, O.V., Z.E. MEL'CHAKOVA, and V.V. STEPIN. *Trudy VNIISO*, **2** : 60. 1965.
279. MOROSHKINA, T.M. and T.F. MALININA. – *Zh. Analit. Khim.*, **16** : 245. 1961.
280. MORRISON, G.H. and H. FREISER. *Solvent Extraction in Analytical Chemistry.* New York, Wiley. 1957.
281. MOMODZIMA, K. and Kh. KHASITANI. – *Bunseki kagaku*, **6** : 642. 1957; *RZhKhim.*, No. 9 : 28394. 1958.
282. MOMODZIMA, K., Kh. KHASITANI, and K. KATSUYAMA. – *Bunseki kagaku*, **9** : 517. 1960; *RZhKhim.*, **3D** : 142. 1961.
283. MUSAKIN, A.P. – *Zavodskaya Laboratoriya*, **9** : 507. 1940.
284. MUSTAFIN, I.S. and E.A. KASHKOVSKAYA. – *Zh. Analit. Khim.*, **13** : 215. 1958.
284a. MUSTAFIN, I.S. and N.F. LISENKO. – *Zh. Analit. Khim.*, **17** : 1052. 1962.

285. MUSTAFIN, I.S. and L.O. MATVEEV. – *Zavodskaya Laboratoriya,* 24 : 259. 1958.
285a. MUSTAFIN, I.S., L.A. MOLOT, and A.S. ARKHANGEL'SKAYA. – *Zh. Analit. Khim.,* 22 : 1808. 1967.
285b. MUSTAFIN, I.S., L.A. MOLOT, and G.M. BORISOVA. – In: *"Primenenie organicheskikh reagentov v analize,"* p. 71. Saratovskii Universitet, Saratov. 1967, 1968.
286. MUKHINA, Z.S. – *Zavodskaya Laboratoriya,* 17 : 289. 1951.
287. MUKHINA, Z.S., E.I. NIKITINA, L.M. BUDANOVA, R.S. VOLODARSKAYA, L.Ya. POLYAK, and A.A. TIKHONOVA. *Metody analiza metallov i splavov (Analysis of Metals and Alloys).* Moskva, Oborongiz. 1959.
288. NAVYAZHSKAYA, E.A. and V.S. SPORYKHINA. – *Lakokrasochnye Materialy i ikh Primenenie,* No. 4 : 52. 1962.
289. NAZARENKO, V.A. and E.A. BIRYUK. – In: *"Sovremennye metody analiza v metallurgii,"* p. 188. Moskva, Metallurgizdat. 1955.
290. NAZARENKO, V.A. and E.A. BIRYUK. – *Sbornik Trudov Giredmeta,* 2 : 77. 1959.
291. NAZARENKO, V.A., E.A. BIRYUK, and E.N. POLUEKTOVA. – *Radiokhimiya,* 5 : 497. 1963.
292. NAZARENKO, V.A., E.A. BIRYUK, and R.V. RAVITSKAYA. – *Ukr. Khim. Zh.,* 34 : 408. 1968.
293. NAZARENKO, V.A., TKHU LYAM NGOK, and R.M. DRANITSKAYA. – *Zh. Analit. Khim.,* 22 : 1175. 1967.
294. NAZARENKO, V.A., M.B. SHUSTOVA, R.Ts. RAVITSKAYA, and M.P. NIKO-NOVA. – *Zavodskaya Laboratoriya,* 28 : 537. 1962.
295. NAZARCHUK, T.N. and L.N. MEKHANOSHINA. – *Zh. Analit. Khim.,* 20 : 260. 1965.
295a. NAZARCHUK, T.N. and L.N. MEKHANOSHINA. – *Zh. Analit. Khim.,* 20 : 260. 1965.
296. NAKAGAVA, G. – *Nihon kagaku dzassi,* 81 : 747. 1960; *RZhKhim.,* 5D : 41. 1961.
297. NAKA, K., G. MATSUMAE, and Ya. TANAKA. – *Bunseki kagaku,* 12 : 928. 1963; *RZhKhim.,* 13G : 103. 1964.
298. NAKA, K., G. MATSUMAE, and Ya. TANAKA. – *Nagoya kogë gidzyutsu sikensë kokoku,* 13 : 300. 1964; *RZhKhim.,* 7G : G114. 1965.
299. NAMIKI, Kh. and Kh. KHODANO. – *Kogë ësui,* 18 : 24. 1960; *RZhKhim.,* 13 D : 66. 1961.
300. NEMODRUK, A.A., P.N. PALEI, and HO HUAN-I. – *Zavodskaya Laboratoriya,* 28 : 406. 1962.
301. NEKHAMKINA, M.A. Candidate Thesis. – *IONKh AN SSSR,* Moskva. 1947.
302. NECHAEVA, E.A. and E.S. LAPIDUS. – *Zavodskaya Laboratoriya,* 25 : 544. 1959.

303. NEGISI, R. – *Bunseki kagaku*, **11** : 202. 1962; *RZhKhim.*, **6G** : 144. 1963.

303a. NIKITINA, O.I. – *Trudy Ukrainskogo Nauchno-Issledovatel'skogo Instituta Metallov*, No. 6 : 283. 1960.

304. NIKITINA, O.I., L.L. GUDYRINA, and L.P. KOLOMIETS. – *Zh. Prikl. Spektrosk.*, **1** : 319. 1964.

305. NIKOLAEV, G.I. – *Zh. Analit. Khim.*, **19** : 63. 1964.

306. NIKOLAEV, G.I. – *Ibid.*, **20** : 445. 1965.

307. NIKOLAEV, G.I. and V.B. ALESKOVSKII. – *Zh. Analit. Khim.*, **18** : 816. 1963.

307a. NISIDA, Kh. – *Bunseki kagaku*, **12** : 56. 1963; *RZhKhim.*, **5G** : 92. 1964.

307b NISIKAVA, Ya., K. KHIRAKI, K. MORISIGE, N. TSUTIYAMA, and I. SIGEMATSU. – *Bunseki kagaku*, **17** : 1092. 1968; *RZhKhim.*, **5G** : 170. 1969.

307c. NISIKAVA, Ya., K. KHIRAKI, and Ts. SIGEMATSU. – *Bunseki kagaku*, **16** : 692. 1964; *RZhKhim.*, **4G** : 82. 1969.

308. NOVIKOV, A.I. – *Zh. Neorgan. Khim.*, **4** : 2161. 1959.

309. NOVIKOV, A.I. – *Izvestiya Vuzov SSSR. Khimiya i Khimicheskaya Tekhnologiya*, **3** : 239. 1960.

310. NOSAKA, Kh., K. NAGASIMA, and M. FUDZISIRO. – *Titaniumu*, **7** : 101. 1959; *RZhKhim.*, No. 11 : 42450. 1960.

310a. NEDLER, V.V. *Spektral'nyi analiz stalei i splavov (Spectroscopic Analysis of Steels and Alloys)*. Moskva, ONTI. 1961.

311. OVSENYAN, E.N. and V.M. TARAYAN. – *Nauchnye Trudy Erevanskogo Universiteta*, **53** : 85. 1956.

312. OVSENYAN, E.N. and M.G. EKIMYAN. – *Izvestiya AN ArmSSR. Seriya Fiziko-matematicheskikh Nauk (Fiziko-matematicheskie, Estestvennye i Tekhnicheskie Nauki)*, **8**, No. 5 : 41. 1955.

313. ODA, N. and M. IDOKHARA. – *Bunseki kagaku*, **10** : 246. 1961; *RZhKhim.*, **1D** : 128. 1962.

314. ODA, N., M. IDOKHARA, and K. TSUKASIMA. – *Nihon kindzoku gakkashi*, **25** : 781. 1961; *RZhKhim.*, **22 D** : 48. 1962.

315. OKA, MUTO, VADA. – *Seisan kenkyu*, **9**, No. 7 : 303. 1957; *RZhKhim.*, No. 15 : 49976. 1958.

316. OKUDA, I. and I. SATO. – *Fudzi seitetsu chikho*, **9** : 207. 1960; *RZhKhim.*, **10D** : 31. 1961.

317. ONUKI, S., K. VATANUKI, and Yu. IOSINO. – *Bunseki kagaku*, **15** : 924. 1966; *RZhKhim.*, **9G** : 157. 1967.

318. OSTROUMOV, E.A. *Novye metody khimicheskogo analiza s primeneniem organicheskikh osnovanii (New Methods of Chemical Analysis with Application of Organic Bases)*. Moskva, Gosgeolizdat. 1952.

319. OSTROUMOV, E.A. *Novye metody khimicheskogo analiza s primeneniem piridina i geksametilentetramina (New Methods of Chemical Analysis with Pyridine and Hexamethylenetetramine)*. Moskva, Izd. geologicheskoi literatury. 1940.

320. OSTROUMOV, E.A. *Primenenie organicheskikh osnovanii v analiticheskoi khimii (Use of Organic Bases in Analytical Chemistry)*. Moskva, Izd. AN SSSR. 1959.

321. OSTROUMOV, E.A. and I.I. VOLKOV. – *Trudy Instituta Okeanologii AN SSSR*, **47** : 150. 1961.

322. OSTROUMOV, E.A. and I.I. VOLKOV. – *Zh. Analit. Khim.*, **15** : 719. 1960; **17** : 461. 1962; **18** : 52, 1452. 1963; **19** : 216, 1073. 1964.

323. OTO, KHAMAZUTI, MATSUMOTO, IOSINAKA, and NAKAO. – *Nihon kindzoku gakkashi*, **20** : 315. 1956; *RZhKhim.*, No. 17 : 57784. 1957.

324. PAVLINOVA, O.V. and L.A. AL'BOTA. – *Uchenye Zapiski Chernovitskogo Universiteta*, **21** : 81. 1956.

325. PAVLOV, N.N. and A.R. KUZNETSOV. – *Zavodskaya Laboratoriya*, **29** : 1059. 1963.

326. PAVLOV, N.N. and A.R. KUZNETSOV. – *Nauchnye Trudy Moskovskogo Tekhnologicheskogo Instituta Legkoi Promyshlennosti*, No. 28 : 61. 1963.

327. PAVLOVSKAYA, M.P., I.M. REIBEL', L.N. AIZENBERG, and R.S. AIZENBERG. – *Trudy Kishinevskogo Sel'skokhozyaistvennogo Instituta*, **26** : 149. 1962.

328. PASOVSKAYA, G.B. – *Zh. Analit. Khim.*, **12** : 760. 1957.

329. PASOVSKAYA, G.B. – *Ibid.*, **20** : 392. 1965.

330. PASOVSKAYA, G.B. – *Izvestiya Vuzov SSSR. Khimiya i Khimicheskaya Tekhnologiya*, **5** : 43. 1962.

331. PASOVSKAYA, G.B. – *Ibid.*, **8** : 345. 1965.

331a. PAKHOLKOV, V.S. and V.V. RYLOV. – *Trudy Ural'skogo Politekhnicheskogo Instituta*, No. 148 : 68. 1966.

331b. PASHEVKIN, B.P. – *Zavodskaya Laboratoriya*, **25** : 1468. 1959.

332. PEN'KOVA, E.F. – *Zavodskaya Laboratoriya*, **18** : 548. 1952.

332a. PEN'KOVA, E.F. and P.Ya. YAKOVLEV. – *Zavodskaya Laboratoriya*, **18** : 655. 1952.

332b. PETRENKO, A.G. – In: *"Khimicheskii analiz tsvetnykh i redkikh metallov,"* p. 3. SO AN SSSR. Novosibirsk. 1964.

332c. PETROVA, G.S., A.M. LUKIN, N.B. ETINGEN, L.A. MOLOT, and A.S. ARKHANGEL'SKAYA. – *Zh. Analit. Khim.*, **24** : 1332. 1969.

333. PINAEVA, S.N. and L.M. BUDANOVA. – *Zavodskaya Laboratoriya*, **32** : 1209. 1966.

333a. PLASTININ, V.V. – *Zavodskaya Laboratoriya*, **25** : 577. 1959.

334. PLOTNIKOV, V.I. – *Zh. Neorgan. Khim.*, **4** : 2775. 1959.

335. PLOTNIKOV, V.I. – *Sbornik Trudov Vsesoyuznogo Nauchno-Issledovatel'skogo Gorno-Metallurgicheskogo Instituta Tsvetnykh Metallov*, No. 5 : 65. 1959.

336. POVOLOTSKAYA, G.L. – *Metallurgicheskaya i Khimicheskaya Promyshlennost' Kazakhstana, Nauchno-tekhnicheskii Sbornik*, No. 6(16) : 66. 1961.

337. POVOLOTSKAYA, G.L., D.A. VAL'DMAN, and N.M. KHANINA. – *Ibid.*, No. 5(15) : 71. 1961.

338. PODCHAINOVA, V.N. – *Trudy Ural'skogo Politekhnicheskogo Instituta*, No. 57 : 38. 1956.

339. PODCHAINOVA, V.N., N.G. ANTONOVA, and M.L. ROS'. – In: *"Metody analiza chernykh i tsvetnykh metallov,"* p. 91. Sverdlovsk–Moskva, Metallurgizdat. 1953.

340. POLYAK, L.Ya. – *Zavodskaya Laboratoriya*, **12** : 268. 1946.
341. POLYAKOVA, R.S. and Z.V. SOLOV'EVA. – *Sbornik Trudov TsNIIChERMETa* No. 37 : 118. 1964.
342. POLUEKTOV, N.S. *Metody analiza po fotometrii plameni (Flame Photometric Analysis)*. Moskva, Izd. "Khimiya." 1967.
343. PONOMAREV, A.I. *Metody khimicheskogo analiza silikatnykh i karbonatnykh porod (Chemical Analysis of Carbonate and Silicate Rocks)*, p. 93. Moskva, Izd. AN SSSR. 1961.
344. PONOMAREV, A.I. and A.Ya. SHESKOL'SKAYA. – *Zh. Analit. Khim.*, **11** : 102. 1956.
345. POPOVA, O.I. and I.N. GODOVANNAYA. – *Zh. Analit. Khim.*, **20** : 355. 1965.
346. PROKOP'EVA, A.I. – *Fizicheskii Sbornik. L'vovskii Gosudarstvennyi Universitet im. I. Franko*, No. 4(9) : 446. 1958.
346a. PROKOP'EVA, A.N. and K.I. TAGANOV. – *Zavodskaya Laboratoriya*, **15** : 299. 1949.
347. PRIBIL, R. *"Komplexony v chemické analyse."* Nakl. *Československé akad. ved*, Praha. 1953.
348. PYATNITSKII, I.V. – *Naukovi Zapysky. Kyyivs'kyi Derzhavnyi Universitet im. Shevchenko*, **14** : 131. 1955.
349. RAD'KO, V.A. – *Trudy Ural'skogo Politekhnicheskogo Instituta*, No. 163 : 98. 1967.
350. RAD'KO, V.A. and E.M. YAKIMETS. – *Zavodskaya Laboratoriya*, **27** : 1464. 1961.
351. RAD'KO, V.A. and E.M. YAKIMETS. – *Zavodskaya Laboratoriya*, **27** : 1465. 1961.
352. RAINES, M.M. and Yu.A. LARIONOV. – *Zavodskaya Laboratoriya*, **14** : 1000. 1948.
353. RODIONOVA, S.K. and A.N. SHTEINBERG. – In: *"Problemy bol'shoi metallurgii i fizicheskoi khimii novykh splavov,"* p. 277. Moskva, "Nauka." 1965.
353a. ROZENBERG, I.V. and M.I. LOMBERG. *Kolichestvennyi spektral'nyi analiz bronz (Quantitative Spectroscopic Analysis of Bronzes)*. Moskva, Oborongiz. 1953.
354. ROZENBERG, M.I. – *Zavodskaya Laboratoriya*, **24** : 1060, 1166. 1958.
355. RYABCHIKOV, D.I. and V.E. BUKHTIAROV. – *Zh. Analit. Khim.*, **7** : 377. 1952.
356. RYABCHIKOV, D.I. and V.F. OSIPOVA. – *Zh. Analit. Khim.*, **11** : 278. 1956.
357. RYABCHIKOV, D.I. and M.M. SENYAVIN. – *Zh. Analit. Khim.*, **8** : 195. 1953.
358. RYAZANOV, I.P. and A.I. CHERKESOV. – *Uchenye Zapiski Saratovskogo Universiteta*, **25** : 3. 1951.
359. SAKSIN, V.F. and A.A. SOBOLEVA. – *Uchenye Zapiski Yaroslavskogo Tekhnologicheskogo Instituta*, **9** : 113. 1966.
359a. SALIKHOV, V.D. and M.Z. YAMPOL'SKII. – *Uchenye Zapiski Kurskogo Pedagogicheskogo Instituta*, **23** : 236. 1966.
360. SANDELL, E.B. *Colorimetric Determination of Traces of Metals*. New York, Interscience. 1959.

361. SEABORG, G.I., I. PERLMAN, and J. HOLLANDER. *Tables of Isotopes.* – *Rev. Mod. Phys.,* **20** : 585–667. 1948.

361a. SIDZË, I. and I. TAKEUTI. – *Bunseki kagaku,* **17** : 61. 1968; *RZhKhim.,* **3G** : 65. 1969.

361b. SIDZË, I. and I. TAKEUTI. – *Bunseki kagaku,* **17** : 323. 1968; *RZhKhim.,* **1G** : 71. 1969.

362. SILAEVA, E.V. and L.V. KAMAEVA. – *Trudy VNIISO,* **1** : 70. 1964.

363. SILAEVA, E.V. and L.M. KRYUCHKOVA. – *Trudy VNIISO,* **2** : 88. 1965.

364. SKOBETS', E.M., I.L. ABARBARCHUK, and K.P. KOSTITSYNA. – *Dopovidi Ukrayins'koyi Akademiyi Sil's'kohospodars'kikh Nauk,* No. 3 : 9. 1960.

365. SKOBETS', E.M., I.L. ABARBARCHUK, and K.P. KOSTITSYNA. – *Naukovi Pratsi, Ukrayins'ka Akademiya Sil's'kohospodars'kikh Nauk,* **11** : 125. 1960.

366. SLAVINSKII, M.P. *Fiziko-khimicheskie svoistva elementov (Physicochemical Properties of Elements).* Moskva, Metallurgizdat. 1952.

367. SOBOLEVA, T.A. – In: *"Metody analiza chernykh i tsvetnykh metallov,"* p. 31. Sverdlovsk–Moskva, Metallurgizdat. 1953.

368. SOLODOVNIKOV, P.P. – *Zh. Analit. Khim.,* **16** : 237. 1961.

369. SOLOMIN, G.A. and N.G. FESENKO. – *Gidrokhimicheskie Materialy,* **33** : 128. 1961.

370. SOLOMIN, G.A. and N.G. FESENKO. – In: *"Sovremennye metody analiza prirodnykh vod,"* p. 57. Moskva, Izd. AN SSSR. 1962.

371. SOKOLOVICH, V.B., Yu.L. LEL'CHUK, and B.N. BESPROZVANNYKH. – *Izvestiya Tomskogo Politekhnicheskogo Instituta,* **128** : 112. 1964.

372. SOCHEVANOVA, M.M. – *Zavodskaya Laboratoriya,* **29** : 143. 1963.

373. SOCHEVANOVA, M.M. and V.G. SOCHEVANOV. – *Zavodskaya Laboratoriya,* **26** : 543. 1960.

374. *Spravochnik khimika (Handbook for Chemists).* Vol. 1. Leningrad–Moskva, Goskhimizdat. 1963.

375. *Spravochnik khimika (Handbook for Chemists).* Vol. 3. Moskva–Leningrad, "Khimiya." 1965.

376. *Spravochnik khimika (Handbook for Chemists).* Vol. 2. Moskva–Leningrad, "Khimiya." 1965.

377. *Spravochnik khimika (Handbook for Chemists).* Vol. 4. Moskva–Leningrad, "Khimiya." 1965.

377a. *Stali (legirovannye i vysokolegirovannye) (Steels (Alloy and High-Alloy Steels)).* – *Metody Khimicheskogo Analiza,* GOST 12357-66. Moskva. 1967.

378. STEPIN, V.V., V.I. KURBATOVA, M.N. KRUGLOVA, L.M. KRYUCHKOVA, and V.A. VERBITSKAYA. – *Trudy VNIISO,* **1** : 62. 1964.

379. STOLYAROV, K.P. and N.N. GRIGOR'EV. – *Zh. Analit. Khim.,* **17** : 565. 1962.

380. STREKALOVA, O.S. Candidate Thesis. – Gosudarstvennyi Universitet, Kazan. 1956.

381. STREL'NIKOVA, N.P. – *Zavodskaya Laboratoriya,* **23** : 1308. 1957.

382. STREL'NIKOVA, N.P. and V.N. PAVLOVA. – *Zavodskaya Laboratoriya,* **26** : 425. 1960.

383. STRUKOVA, M.P. and I.V. KIRILLOVA. – *Zh. Analit. Khim.*, **21** : 1236. 1966.
384. SUVOROVSKAYA, N.A., M.M. VOSKRESENSKAYA, and T.A. MEL'NIKOVA. – *Nauchnye Soobshcheniya Instituta Gornogo Dela*, **10** : 148. 1961.
385. SUVOROVSKAYA, N.A. and G.I. LOPATINA. – *Ibid.*, **6** : 30. 1960.
386. SUDZUKI, T. – *Bunseki kagaku*, **13** : 524. 1964; *RZhKhim.*, **9G** : 97. 1965.
386a. SUDO, Z. and Kh. OGAVA. – *Bunseki kagaku*, **13** : 406. 1964; *RZhKhim.*, **3G** : 130. 1965.
387. SUKHENKO, K.A. *Spektral'nyi analiz stalei i splavov (Spectroscopic Analysis of Steels and Alloys)*. Moskva, Oborongiz. 1954.
387a. SUKHENKO, K.A., F.I. FILATOV, P.P. GALONOV, K.A. MOISEEVA, and L.D. METELINA. – In: *"Fotoelektricheskie metody spektral'nogo analiza,"* p. 44. Moskva, Oborongiz. 1961.
388. TAKAO, L. and S. MIËSI. – *Tetsu to khagane*, **47** : 145. 1961; *RZhKhim.*, **18D** : 55. 1955.
389. TAKIURA, TAKUNO, TANGE. – *Yakugaku dzassi*, **75** : 724. 1955; *RZhKhim.*, No. 11 : 32716. 1956.
390. TAKEUTI, Ts. and M. SUDZUKI. – *Bunseki kagaku*, **10** : 58. 1961; *RZhKhim.*, **23D** : 51. 1961.
391. TALINOV, Sh. and Z.G. SOFEIKOVA. – *Zavodskaya Laboratoriya*, **13** : 816. 1947.
392. TALINOV, Sh. and I. TEODOROVICH. – *Zavodskaya Laboratoriya*, **15** : 1031. 1949.
393. TANAKA. – *Bunseki kagaku*, **6** : 13. 1957; *RZhKhim.*, No. 20 : 66353. 1957.
394. TANAKA, MASAYA. – *Nihon kagaku dzassi*, **84** : 582. 1963; *RZhKhim.*, **4G** : 65. 1964; **85** : 119. 1964; *RZhKhim.*, **22**, No. 7 : 49. 1964.
395. TANAKA, K., I. NAKAGAVA, and S. KHONDA. – *Bunseki kagaku*, **10** : 1148. 1961; *RZhKhim.*, **15D** : 49. 1962.
396. TANAKA, K. and K. YAMAĚSU. – *Bunseki kagaku*, **13** : 540. 1964; *RZhKhim.*, **10G** : 64. 1965.
397. TANANAEV, I.V. and S.T. ABILOV. – *Zh. Prikl. Khim.*, **15** : 61. 1942.
398. TANANAEV, I.V. and A.D. VINOGRADOVA. – *Zh. Analit. Khim.*, **14** : 487. 1959.
399. TANANAEV, I.V. and A.D. VINOGRADOVA. – *Zh. Neorgan. Khim.*, **2** : 2455. 1957.
400. TANANAEV, I.V. and M.I. LEVINA. – *Zavodskaya Laboratoriya*, **11** : 804. 1945.
401. TANANAEV, I.V. and Yu.L. LEL'CHUK. – *Zh. Analit. Khim.*, **2** : 93. 1943.
402. TANANAEV, I.V. and Sh. TALINOV. – *Zavodskaya Laboratoriya*, **8** : 23. 1939.
403. TANANAEV, I.V. and P.Ya. YAKOVLEV. – *Zavodskaya Laboratoriya*, **14** : 1155. 1950.
404. TANANAEV, N.A. – *Zavodskaya Laboratoriya*, **4** : 1348. 1935.
405. TANANAEV, N.A. and V.N. TIKHONOV. – *Nauchnye Doklady Vysshei Shkoly. Metallurgiya*, No. 1 : 259. 1959.
406. TARASOV, N.Ya. and P.V. MOSHKOVSKII. – *Trudy Nauchno-Tekhnicheskogo Obshchestva Chernoi Metallurgii, Ukrainskoe Respublikanskoe Pravlenie*, **4** : 98. 1956.

407. TARAYAN, V.M. – *Zavodskaya Laboratoriya*, **8** : 273. 1929.
408. TARAYAN, V.M. and E.N. OVSEPYAN. – *Nauchnye Trudy Erevanskogo Universiteta*, **53** : 75. 1956.
409. TARTAKOVSKII, V.Ya. – *Zavodskaya Laboratoriya*, **4** : 1023. 1935.
410. TASKAEVA, T.P. and G.K. PARYGINA. – In: *"Doklady mezhvuzovskoi nauchnoi konferentsii po spektroskopii i spektral'nomu analizu,"* p. 69. Tomskii Gosudarstvennyi Universitet. 1960.
411. TATSYO, I. – *Nihon kagaku dzassi*, **84** : 502. 1963; *RZhKhim.*, **12G** : 10. 1964.
412. TEMYANKO, S.V. and I.V. KHOZOVA. – In: *"Sovremennye metody analiza v metallurgii,"* p. 28. Moskva, Metallurgizdat. 1955.
413. TERENT'EVA, E.A. – *Zavodskaya Laboratoriya*, **28** : 807. 1962.
414. TINOVSKAYA, E.S. – *Zh. Analit. Khim.*, **5** : 345. 1950.
415. TINOVSKAYA, E.S. – In: *"Khimiya, tekhnologiya i primenenie proizvodnykh piridina i khinolina,"* p. 253. Riga, Izd. AN LatvSSR. 1960.
415a. TIMKOVA, N.F. – *Nauchnye Doklady Vysshei Shkoly. Biologicheskie Nauki*, No. 9 : 126. 1968.
416. TIKHONOV, V.N. – *Zh. Analit. Khim.*, **17** : 422. 1962.
417. TIKHONOV, V.N. – *Zh. Analit. Khim.*, **19** : 1204. 1964.
418. TIKHONOV, V.N. – *Ibid.*, **20** : 941. 1965.
419. TIKHONOV, V.N. – *Ibid.*, **20** : 1219. 1965.
420. TIKHONOV, V.N. – *Ibid.*, **21** : 275. 1966.
421. TIKHONOV, V.N. – *Ibid.*, **21** : 829. 1966.
422. TIKHONOV, V.N. – *Ibid.*, **21** : 1172. 1966.
423. TIKHONOV, V.N. – *Ibid.*, **22** : 658. 1967.
424. TIKHONOV, V.N. Candidate Thesis. – Ural'skii Politekhnicheskii Institut. Sverdlovsk. 1961.
425. TIKHONOV, V.N. – *Nauchnye Trudy Tul'skogo Pedagogicheskogo Instituta, Khimiya*, No. 1 : 3. 1967.
426. TIKHONOV, V.N. – *Ibid.*, No. 1 : 9. 1967.
427. TIKHONOV, V.N. – *Ibid.*, No. 1 : 19. 1967.
428. TIKHONOV, V.N. – *Trudy Vsesoyuznogo Nauchno-Issledovatel'skogo Alyuminiego-Magnievogo Instituta*, No. 47 : 145. 1961.
429. TIKHONOV, V.N. – *Ibid.*, No. 51 : 78. 1963.
430. TIKHONOV, V.N. and M.Ya. GRANKINA. – *Zavodskaya Laboratoriya*, **29** : 653. 1963.
431. TIKHONOV, V.N. and M.Ya. GRANKINA. – *Ibid.*, **32** : 278. 1966.
432. TIKHONOV, V.N. and A.P. NIKITINA. – *Ibid.*, **28** : 662. 1962.
433. TONOSAKI, K. and M. OTOMO. – *Nihon kagaku dzassi*, **80** : 41. 1959; *RZhKhim.*, No. 3 : 8813. 1960.
434. TRAMM, R.S. and K.S. PEVZNER. – *Zavodskaya Laboratoriya*, **31** : 163. 1965.
435. TUMANOV, A.A. and V.S. EFIMYCHEV. – *Zh. Analit. Khim.*, **22** : 700. 1967.
436. TURCHINSKII, M.L. and L.Z. NAIMAN. – *Zavodskaya Laboratoriya*, **30** : 673. 1964.

437. UVAROVA, K.A., M.S. PIKH, and K.P. SIKORA. – *Trudy IREA*, No. 29 : 139. 1966.
438. UDZUMASA, Ya., K. KHAYASI, and Yu. NURISI. – *Bunseki kagaku*, 14 : 902. 1965; *RZhKhim.*, 22G : 73. 1966.
439. WU MU-AI, HSÜ SHU-MIN. – *Wuhan tahsüeh tz'üan k'ohsüeh hsüehpao*, No. 1 : 38. 1959; *RZhKhim.*, No. 18 : 73058. 1960.
440. WUNO. – *Nihon sio gakkaisi*, 11, No. 3 : 139. 1957; *RZhKhim.*, No. 18 : 43075. 1958.
441. USATENKO, Yu.I. and G.E. BEKLESHOVA. – *Zavodskaya Laboratoriya*, 19 : 147. 1953.
442. USATENKO, Yu.I. and G.E. BEKLESHOVA. – *Ibid.*, 20 : 266. 1954.
443. USATENKO, Yu.I., G.E. BEKLESHOVA, E.I. GRENBERG, M.Ya. GENIS, and E.E. KARPUSHA. – *Zavodskaya Laboratoriya*, 21 : 26. 1955.
444. USATENKO, Yu.I., E.I. GRENBERG, and V.I. KOPELIOVICH. – *Zavodskaya Laboratoriya*, 18 : 1063. 1952.
445. USATENKO, Yu.I. and L.I. GUREEVA. – *Zavodskaya Laboratoriya*, 22 : 781. 1956.
446. FAINBERG, S.Yu. and A.A. BLYAKHMAN. – *Sbornik Nauchnykh Trudov Gintsvetmeta*, No. 12 : 119. 1956.
447. FEDOROV, A.A. and F.A. OZERSKAYA. – *Sbornik TsNIIChERMETa*, No. 31 : 195. 1963.
448. FEDOROV, A.A. and G.P. SOKOLOVA. – In: *"Novye metody analiza na metallurgicheskikh i metalloobrabatyvayushchikh zavodakh,"* p. 17. Dnepropetrovsk, Izd. "Metallurgiya." 1964.
449. FEDOROV, A.A. and G.P. SOKOLOVA. – *Sbornik Trudov TsNIIChERMETa*, No. 24 : 128. 1962.
450. FEDOROV, A.A. and G.P. SOKOLOVA. – *Ibid.*, No. 31 : 162. 1963.
450a. FEDOROVA, M.V., G.I. VAKHRINA, L.N. VAROVA, and A.I. SHCHUPLE-TSOVA. – In: *"Puti tekhnicheskogo progressa v magnievoi promyshlennosti,"* p. 98. Perm, Permskoe knizhnoe izdatel'stvo. 1959.
451. FIOLETOVA, A.F. – *Zh. Analit. Khim.*, 14 : 739. 1959; 17 : 302, 520. 1962.
452. FIALKOV, Ya. A., V.V. GRIGOR'EVA, N.K. DAVIDENKO, and N.G. PERISHKINA. – *Farmatsevtychnyi Zhurnal*, No. 1 : 10. 1959.
453. FOGEL'SON, E.I. and N.V. KALMYKOVA. – *Zavodskaya Laboratoriya*, 11 : 359. 1945.
454. FRATKINA, G.P. – *Zavodskaya Laboratoriya*, 24 : 1373. 1958.
455. FUKAMUTI, Kh., R. IDENO, M. YANAGIDA, and G. INOTSUME. – *Bunseki kagaku*, 10 : 1929. 1961; *RZhKhim.*, 11D : 79. 1962.
456. KHARKOVER, M.Z. and K.F. VOROZHBITSKAYA. – *Trudy VNIISO*, 3 : 75. 1967.
457. KHASITANI, Kh. – *Nihon gensirëku gakaisi*, 4 : 287. 1962; *RZhKhim.*, 7G : 126. 1963.

457a. KHASITANI, Kh., K. KATSUYAMA, T. SAGAVA, and K. MOTODZIMA. – *Bunseki kagaku*, 16 : 596. 1967; *RZhKhim.*, 17G : 139. 1968.

458. KHASITANI, Kh. and K. MOTODZIMA. – *Bunseki kagaku*, 7 : 478. 1958; *RZhKhim.*, No.16 : 56918. 1959.

459. KHASITANI, Kh. and K. YAMAMOTO. – *Nihon kagaku dzassi*, 80 : 727. 1959; *RZhKhim.*, No.9 : 34556. 1960.

460. KHLOPIN, N.Ya. – *Zavodskaya Laboratoriya*, 14 : 156. 1948

461. KHONDZĖ, T. – *Ibaraki daigaku kogakubu kenkyu syukho*, 7 : 41. 1959; *RZhKhim.*, 16D : 68. 1961.

462. KHONMA, N., Ts. SIMODZAKI. – *Kenkyu dzitsuëki khokoku*, 12 : 379. 1963; *RZhKhim.*, 11G : 140. 1964.

463. HUANG HUI-SHIN. – *Huahsüeh shihchien*, 13, No.4 : 168. 1958; *RZhKhim.*, No.6 : 19098. 1959.

464. HUANG CHENG-CH'ÜAN, HUANG KANG-CH'ÜAN. – *Wuhan tahsüeh tz'üan k'ohsüeh hsüehpao*, No.2 : 48. 1960; *RZhKhim.*, No.24 : 96158. 1960.

464a. KHUDYAKOVA, T.A. – *Trudy Gor'kovskogo Politekhnicheskogo Instituta*, 13, No.5 : 37. 1957.

465. TSANIT, V.L. and G. KISHA. – *Glasn. hem. Društ. Beogr.*, 28, Nos. 3–4 : 143. 1963.

466. TSAP, M.L. Patent USSR, No. 155647. – *Byulleten' Izobretenii*, No.13. 1963.

467. *Tsvetnye metally i splavy. Metody ispytanii (Nonferrous Metals and Alloys. Testing Methods).* Moskva, Standartgiz. 1959.

468. TSEKHANOVICH, V.N. – *Trudy Voronezhskogo Universiteta*, 32 : 149. 1953.

469. CHIEN TS'UI-LIN. – *Huahsüeh hsüehpao*, 23 : 324. 1957; *RZhKhim.*, No.12 : 39357. 1958.

470. TSIMBLER, M.E., V.I. DERENOVSKII, and N.S. PROSYANIK. – *Trudy Kievskogo Instituta Inzhenerno-Vodnogo Khozyaistva*, No.8 : 159. 1959.

471. TSINBERG, S.L. – *Zavodskaya Laboratoriya*, 2 : 13. 1933.

472. TSINBERG, S.L. – *Ibid.*, 4 : 1161. 1935.

473. TSYVINA, B.S. and O.V. KON'KOVA. – *Zavodskaya Laboratoriya*, 25 : 403. 1959.

474. CHERKESOV, A.I. – *DAN SSSR*, 118 : 309. 1958.

475. CHERKESOV, A.I., N.M. ALYKOV, and B.I. KAZAKOV. – In: *"Fiziko-khimicheskie issledovaniya prirodnykh sorbentov i ryada analiticheskikh sistem,"* No.2 : 102. Saratov. 1967.

476. CHERKESOV, A.I., B.I. KAZAKOV, and P.A. SHILIN. – *Ibid.*, p.18.

477. CHERKESOV, A.I., B.I. KAZAKOV, and V.I. SHCHEPKO. – *Zavodskaya Laboratoriya*, 34 : 786. 1968.

478. CHERNIKHOV, Yu.A. and B.M. DOBKINA. – *Zavodskaya Laboratoriya*, 25 : 131. 1959.

479. CHERNIKHOV, Yu.A., B.M. DOBKINA, and L.M. KHERSONSKAYA. – *Zavodskaya Laboratoriya*, 21 : 638. 1955.

480. CHANG T'I-HÊNG, TSAO FÊNG-YÜN. – *Huahsüeh shihchien*, **14** : 181. 1959; *RZhKhim.*, No. 4 : 13072. 1960.
481. CHIRKOV, S.K. – *Zavodskaya Laboratoriya*, **14** : 783. 1948.
482. CHULKOV, Ya.I. and L.A. SOLOV'EVA. – *Vestnik Tekhnicheskoi i Ekonomicheskoi Informatsii Nauchno-Issledovatel'skogo Instituta Tekhniko-Ekonomicheskikh Issledovanii Gosudarstvennogo Komiteta Soveta Ministrov SSSR po Khimii*, No. 10 : 32. 1960.
483. CHUNOSOV, V.I. – *Tsement*, No. 5 : 22. 1955.
483a. SHAEVICH, A.B. and S.B. SHUBINA. *Promyshlennye metody spektral'nogo analiza (Industrial Methods of Spectroscopic Analysis).* Moskva, "Metallurgiya." 1965.
484. SHAFRAN, I.G., M.Z. PARTASHNIKOVA, and T.I. RACHKOVA. Patent USSR, No. 131135. 1960. – *Byulleten' Izobretenii*, No. 16. 1960.
485. SHAKHTAKHTINSKII, G.B. and I.A. MAMEDOV. – *Azerb. Khim. Zh.*, No.1 : 45. 1959.
486. SHAKHTAKHTINSKII, G.B. and I.A. MAMEDOV. – *Trudy Azerbaidzhanskogo Industrial'nogo Instituta*, No. 14 : 59. 1956.
487. SHAKHTAKHTINSKII, G.B. and I.A. MAMEDOV. – *Ibid.*, No. 19 : 273. 1957.
488. SHVAIGER, M.I. and E.I. RUDENKO. – *Zavodskaya Laboratoriya*, **26** : 936. 1960.
489. SHVANGIRADZE, R.R. and T.A. MOZGOVAYA. – *Zh. Analit. Khim.*, **12** : 708. 1957.
489a. SHVARTS, L.M. and N.S. NILOVA. – *Izvestiya AN SSSR, Seriya Fizicheskaya*, **19** : 96. 1955.
490. SHVARTSENBAKH, G. – In: *"Kompleksometriya."* Moskva, Goskhimizdat. 1958.
491. SHEMYAKIN, F.M. – *Zavodskaya Laboratoriya*, **11** : 986. 1934.
492. SHEMYAKIN, F.M. and S.I. BARSKAYA. – *Zavodskaya Laboratoriya*, **16** : 278. 1950.
493. SHEMYAKIN, F.M. and N.I. BELYAKOV. – *Zavodskaya Laboratoriya*, **20** : 552. 1954.
494. SHEYANOVA, F.R. and V.P. MALENSKAYA. – *Zavodskaya Laboratoriya*, **23** : 907. 1957.
495. SHEYANOVA, F.R. and V.P. MALENSKAYA. – *Trudy Komissii po Analiticheskoi Khimii*, **11** : 243. 1960.
496. SHEYANOVA, F.R. and V.P. MALENSKAYA. – *Trudy po Khimii i Khimicheskoi Tekhnologii*, No. 3 : 552. Gorki. 1958.
497. SHEYANOVA, F.R. and V.P. MALENSKAYA. – *Ibid.*, No. 3 : 560. Gorki. 1959.
498. SHIKHOVA, M. and M. KATSAROVA. – *Stroitel'nye Materialy i Silikatna Promyshlennost'*, **4**, No. 3 : 24. 1963.
499. SHISHKINA, N.I. Candidate Thesis. – Ural'skii Nauchno-Issledovatel'skii Institut Chernykh Metallov. 1955.

500. SHNAIDERMAN, S.Ya. — *Izvestiya Kievskogo Politekhnicheskogo Instituta,* **29** : 122. 1960.

501. SHPILEV, F.S. and V.I. OGOLOVA. — *Trudy Dagestanskogo Sel'skokhozyai-stvennogo Instituta,* **11** : 233. 1959.

502. ŠUŠIDJ, S.K. — *Zbornik Radova. Poljoprivrednog Fakulteta, Universitet u Beogradu,* **4** : 93. 1956; *RZhKhim.,* No. 2 : 4283. 1959.

503. SHCHERBAKOV, V.G. and Z.K. STEGENDO. — *Sbornik Trudov Vsesoyuznogo Nauchno-Issledovatel'skogo Instituta Tverdykh Splavov,* No. 5 : 281. 1964.

504. SHCHIGOL', M.B. and N.B. BURCHINSKAYA. — *Zh. Analit. Khim.,* **11** : 106. 1956.

505. EGEL', L.E. *Rudy chernykh, tsvetnykh i redkikh metallov i ikh promyshlennoe znachenie (Ores of Ferrous, Nonferrous and Rare Metals and Their Industrial Significance),* p. 80. Moskva, Gosgortekhizdat. 1962.

506. ENDO, I. — *Bunseki kagaku,* **7** : 611. 1958; *RZhKhim.,* No. 16 : 56830. 1959.

506a. ENDO, I., Kh. OKHATA, and Yu. NAKAKHARA. — *Bunseki kagaku,* **16** : 364. 1961; *RZhKhim.,* **19G** : 139. 1968.

507. ENDO, I. and N. TAKAGI. — *Bunseki kagaku,* **8** : 491. 1959; *RZhKhim.,* No. 17 : 69226. 1960.

508. ENDO, I. and N. TAKAGI. — *Ibid.,* **8** : 829. 1959; *RZhKhim.,* No. 17 : 69227. 1960.

509. ENDO, I. and N. TAKAGI. — *Ibid.,* **9** : 998. 1960; *RZhKhim.,* **15D** : 118. 1961.

510. YUASA, G. — *Bunseki kagaku,* **11** : 1269. 1962; *RZhKhim.,* **22G** : 86. 1963.

510a. YUDELEVICH, I.G. and V.G. KOVALEVA. — *Zavodskaya Laboratoriya,* **24** : 754. 1958.

511. CH'IN-YŬN, Yu. — *Huahsüeh shihchien,* **13**, No. 5 : 226. 1958; *RZhKhim.,* No. 15 : 53185. 1959.

512. YAKOVLEVA, E.F. — *Sbornik Trudov TsNIIChERMETa,* No. 24 : 58. 1962.

513. YAKOVLEV, P.Ya. Candidate Thesis. *Elektrostal', zavod "Elektrostal'."* 1951.

514. YAKOVLEV, P.Ya., R.L. MALININA, and L.K. RAGINSKAYA. — *Sbornik Trudov TsNIIChERMETa,* No. 49 : 67. 1966.

514a. YAKOVLEV, P.Ya., A.A. FEDOROV, and N.V. BUYANOV. *Analiz materialov metallurgicheskogo proizvodstva (Analysis of Materials of the Metallurgical Industry),* p. 290. Moskva, Metallurgizdat. 1961.

514b. YANAGISAVA, M. and Ts. TAKEUTI. — *Kogë kagaku dzassi,* **70** : 2254. 1967; *RZhKhim.,* **22G** : 76. 1968.

515. YATSIMIRSKII, K.B. and V.P. VASIL'EV. *Konstanty nestoikosti kompleksnykh soedinenii (Instability Constants of Complex Compounds).* Moskva, Izd. AN SSSR. 1959.

516. ABDEL RAHEEM, A.A., A.S. MOUSTAFA, and ABDEL-AZIZ Amin. — *Z. analyt. Chem.,* **175** : 19. 1960.

517. ABRAHAMCZIK, E. — *Mikrochemie,* **33** : 209. 1948.

517a. ABSOLON, K. — *Rudy,* **12** : 184. 1964.

518. AGUILA, J. — *Talanta,* **14** : 1195. 1967.

519. AIKENS, D.A. and F. BAHBAH. – *Analyt. Chem.*, **39** : 646. 1967.
520. ALFONSI, B. – *Anal. chim. Acta*, **20** : 277. 1959.
521. ALFONSI, B. and M. BUSSI. – *Anal. chim. Acta*, **22** : 383. 1960.
522. ALTEN, F., B. WANDROWSKI, and E. HILLE. – *Angew. Chem.*, **48** : 273. 1935.
523. ALTEN, F., H. WEILAND, and E. KNIPPENBERG. – *Z. analyt. Chem.*, **96** : 91. 1934.
524. ALTEN, F., H. WEILAND, and H. LOOFMAN. – *Angew. Chem.*, **46** : 668. 1933.
525. AMOS, M.D. and P.E. THOMAS. – *Anal. chim. Acta*, **32** : 139. 1965.
526. AMOS, M.D. and J.B. WILLIS. *High Temperature Premixed Flames in Atomic Absorption Spectroscopy.* – V Australian Spectroscopy Conference. Perth. 1965.
527. ANGELIS, G. and B.M. PETRONIO. – *Rass. chim.*, **13** : 3. 1961.
528. ANIL, K. and K. SEN ASIT. – *Talanta*, **13** : 1313. 1966.
529. ANTON, A. – *Analyt. Chem.*, **32** : 725. 1960.
530. APOSTOLACHE, S. – *Revtă Chim.*, **11** : 664. 1960.
531. ARNOLD, I.O. *Steel Works Analysis*, p. 199. London. 1907.
532. ARTIGAS, J. and M. CINTAT. – *An. R. Soc. esp. Fis. Quim.*, **B59** : 511. 1963.
533. ASENSI, M.G. – *An. R. Soc. esp. Fis. Quim.*, **B58** : 525. 1962.
534. ASHBROOK, A.W. and G.M. RITCEY. – *Can. J. Chem.*, **39** : 1109. 1961.
535. ATACK, F.A. – *J. Soc. Chem. Ind., Lond.*, **34** : 936. 1915.
536. ATKINS, D.H.F. and E.N. JENKINS. – *Atomic Energy Res. Establ.*, No. C/R 2161. 1960.
537. AUSTIN, G.J. – *Analyst, Lond.*, **72** : 443. 1947.
538. BABACĔV, G.N. – *Chim. analyt.*, **48** : 258. 1966.
539. BABACĔV, G.N. – *Tonindustriezeitung*, **90** : 214. 1966.
540. BACON, A. – *Analyst, Lond.*, **77** : 90. 1952.
541. BACON, A. and H.C. DAVIS. – R.A.E. Report No. M 7889A. 1948.
542. BAILEY, T.H. and S.J. LYLE. – *Talanta*, **12** : 563. 1965.
543. BALDI, F. and R. PASSERI. – *Metallurgia ital., Suppl.*, **48**, No. 1 : 8. 1956.
544. BALIS, E.W., L.B. BRONK, H.G. PFEIFFER, W.W. WELBON, E.H. WINSLOW, and P.D. ZEMANY. – *Analyt. Chem.*, **34** : 1731. 1962.
545. BANARJEE, D.K. – *Analyt. Chem.*, **29** : 55. 1957.
546. BANARJEE, D.K., A.M. SUNDARAM, and H.D. SHARMA. – *Anal. chim. Acta*, **10** : 256. 1954.
547. BANKS, C.V. and R.E. EDWARDS. – *Analyt. Chem.*, **27** : 947. 1955.
548. BÁNYAI, E., E.B. GERE, and L. ERDEY. – *Talanta*, **4** : 133. 1960.
549. BARRACINA-GOMES, L. GASCÓ-SÁNCHER, and FERNANDEZ-CELLINI. – *An. R. Soc. esp. Fis. Quim.*, Ser. **B56** : 861. 1960.
550. BARTON, C.J. – *Analyt. Chem.*, **20** : 1068. 1948.
550a. BARTURA, J. and W. BODENHEIMER. – *Israel J. Chem.*, **6** : 61. 1968.
551. BASSET, C. – *Acta chem. scand.*, **6** : 910. 1952.
552. BEHR, A., M.L. BLANCHET, and L. MALAPRADE. – *Chim. analyt.*, **42** : 501. 1960.

553. BELLOMO, A. and E. BRUNO. – *Atti Soc. pelorit. Sci. fis. mat. nat.,* **5** : 327. 1960.
554. BĚLOHLÁVEK, O. – *Hutn. Listy,* **14** : 809. 1959.
555. BENDIGO, B.B. and R.K. BELL. – *J. Res. Natn. Bur. Stand.,* **A64,** No. 3 : 285. 1960.
556. BENEDETTI-PICHLER, A. – *Mikrochemie (Pregl.-Festschr.),* p. 6. 1929.
557. BENNET, H., W.G. HAWLEY, and R.P. EEARLEY. – *Trans. Br. Ceram. Soc.,* **61** : 201. 1962.
558. BENSCH, H., O. HELMBOLDT, M. KÖSTER, K. HÜBNER, and H. PROTZER. – *Z. Erzbergb. Metallhüttwes.,* **20** : 522. 1957.
559. BERG, B. – *Z. analyt. Chem.,* **71** : 369. 1927.
560. BERG, R. – *Z. analyt. Chem.,* **71** : 378. 1927.
561. BERG, R. – *Ibid.,* **76** : 197. 1929.
562. BERMEJO, F. and A. MARGALET. – *Infción Quim. analit. pura apl. Ind.,* **18** : 35. 1964.
563. BHARGOVA, O.P., W. HINES, and GRANTY. – *Analyt. Chem.,* **40** : 413. 1968.
564. BHAT, A.N., R.D. GUPTA, and B.D. JAIN. – *Proc. Indian Acad. Sci.,* **A63** : 356. 1966.
565. BIEBER, B. and J. DREXLEROVÁ. – *Hutn. Listy,* **21** : 50. 1966.
565a. BISHOP, J.R. – *Analyst, Lond.,* **81** : 291. 1956.
566. BISHOP, J.R. and H. LIEBMANN. – *Analyst, Lond.,* **78** : 117. 1953.
567. BISQUE, R.E. – *J. Sedim. Petrol.,* **31** : 113. 1961.
568. BLAIR, D., R. POWER, D.L. GRIFFITHS, and J.H. WOOD. – *Talanta,* **7** : 80. 1960.
569. BLOCH, L. – *Anal. chim. Acta,* **32** : 233. 1960.
570. BLUM, W. – *J. Am. Chem. Soc.,* **38** : 1281, 1290, 1295. 1916.
571. BOASE, D.G. and J.K. FOREMAN. – *Talanta,* **3** : 82. 1960.
572. BOASE, D.G. and J.K. FOREMAN. – *Talanta,* **8** : 187. 1961.
573. BOBTELSKY, M. and I. BAR-GADDA. – *Anal. chim. Acta,* **9** : 446. 1953.
574. BOBTELSKY, M. and I. BAR-GADDA. – *Ibid.,* **9** : 525. 1953.
575. BOBTELSKY, M. and J.M.E. GOLDSCHMIDT. – *Bull. Res. Coun. Israel,* **A7** : 121. 1958.
576. BOBTELSKY, M. and G. WELLWART. – *Anal. chim. Acta,* **10** : 151. 1954.
577. BOCK, R. – *Z. analyt. Chem.,* **133** : 110, 115. 1951.
578. BOCK, R. and F. UMLAND. – *Angew. Chem.,* **67** : 420. 1950.
579. BODE, H. – *Z. analyt. Chem.,* **142** : 414. 1954; **143** : 182. 1954; **144** : 90, 165. 1955.
580. BOK, L.D.C. and V.C.O. SCHULER. – *Jl. S. Afr. Chem. Inst.,* **11** : 1. 1958.
581. BOLLETER, W.T. – *Analyt. Chem.,* **31** : 201. 1959.
582. BOLOMEY, R.A. and L. WISH. – *J. Am. Chem. Soc.,* **72** : 4483. 1950.
583. BORREL, M. and R. PARIS. – *Anal. chim. Acta,* **4** : 267. 1950.
584. BOSHOLM, J. – *J. prakt. Chem.,* **29** : 65. 1965.
585. BOX, F.W. – *Analyst, Lond.,* **71** : 317. 1946.

586. BRAVO, A.C.S. and C.R. DIAZ. – *Pol. Acad. Cienc. fis., mat. y natur.*, **24** : 67. 1961.
587. BREŠNÝ, B. and A. KURCOVÁ. – *Hutn. Listy*, **18** : 204. 1960.
588. BRHÁČEK, L. – *Chemické Listy*, **52** : 1820. 1958.
589. BRHÁČEK, L. – *Colln. Czech. Chem. Commun.*, **24** : 2811. 1959.
590. BRIL, J. – *Mikrochim. Acta*, No. 2 : 212. 1958.
591. BRITTON, H.T.S. – *J. Chem.*, **127** : 2157. 1925.
592. BROCKMANN, H. and H. KELLER. – *Arch. Eisenhüttwes.*, **35** : 367. 1964.
593. BROOKES, H.E. and C.A. JOHNSON. – *J. Pharm. Pharmac.*, **7** : 836. 1955.
594. BROWNELL, G.M., K. BRAMADAT, R.A. KNUTSON, and A.C. TURNOCK. – *Trans. R. Soc. Can., Sec.*, 4, **51** : 19. 1957.
595. BRUM, D. – *Nukleonik*, **3**, No. 7 : 318. 1961; *RZhKhim.*, **14D** : 15. 1962.
596. BUCK, L. – *Chim. analyt.*, **47** : 10. 1965.
597. BUDAN, F. – *Radex Rdsch.*, No. 2 : 129. 1966.
598. BUDEWSKI, O. and L. SIMOVA. – *Doklady. Bŭlgarska Akademiya na Naukite*, **14** : 179. 1961.
599. BURKE, K.E. – *Analyt. Chem.*, **38** : 1608. 1966.
600. BURKE, K.E. and C. DAVIS. – *Analyt. Chem.*, **36** : 172. 1964.
601. BURRIEL, M.F. and T.C. BOLLI. – *An. R. Soc. esp. Fis. Quim.*, **B50** : 957. 1954.
602. BUSCH, W. – *Z. anorg. allg. Chem.*, **161** : 169. 1927.
603. BYKOWSKI, W. – *Chemia analit.*, **6** : 265. 1961.
604. BYKOWSKI, W. – *Przegl. Elektron.*, **83** : 231. 1962.
605. CĂDARIU, I. and T. GOINA. – *Studia Univ. Babes Bolyai., Ser. Chem.*, **7** : 15. 1962.
606. CĂDARIU, I. and T. GOINA. – *Studia Univ. Babes Bolyai, Ser. Chem.*, **8** : 27. 1963.
607. CĂDARIU, I. and L. ONICIU. – *Studia Univ. Babeş Bolyai, Ser. Chem.*, **5** : 39. 1960.
608. CĂDARIU, I. and L. ONICIU. – *Studii Cerc. ştiinţ. Cluj.*, Ser. 1, **5** : 95. 1954.
609. CĂDARIU, I. and L. ONICIU. – *Studii Cerc. ştiinţ. Cluj.*, **12** : 69. 1961.
610. CALDARARU, H. – *Revtă Chim.*, **14** : 399. 1963.
611. CALDWELL, J.R. and H.V. MOYER. – *J. Am. Chem. Soc.*, **57** : 2372. 1935.
612. CALVET, J. – *Compt. rend.*, **195** : 148. 1932.
613. CAMP, L. – *Iron Age*, **65** : 17. 1900.
614. CANIĆ, V. and T. KISS. – *Tehnika, Beogr.*, **19**, No. 1; *Hem. Ind.*, **18** : 5–7. 1964.
615. CAPACHO-DELGADO, L. and D.C. MANNING. – *Analyst, Lond.*, **92** : 553. 1964.
616. CAPITAN-GARCIA, F. and M. ROMAN. – *Revta Univ. ind. Santander*, **9** : 17. 1967.
617. CAPIZZI, F.M. – *Annali Chim.*, **51** : 563. 1961.
618. CARNOT, G. – *Compt. rend.*, **111** : 914. 1890.
619. CARRUTHERS, C. – *Ind. Engng. Chem. Analyt. Edn.*, **15** : 1102. 1943.

620. CARTER, R.H. – *Ind. Engng. Chem.*, **20**:1195. 1928.
621. CATTRALL, R.W. – *Aust. J. Chem.*, **14**:163. 1961.
622. CAVIGNAC, H. – *Compt. rend.*, **158**:948. 1914.
623. CELLEJA, J. and P.J.M. FERNANDEZ. – *Cemento-Horm.*, **25**:651. 1959.
624. CERRAI, E. and G. GHERSINI. – *J. Chromat.*, **16**:258. 1964.
625. CHABLO, A. – *Chemia analit.*, **9**:501. 1964.
626. CHAKRABATRI, C.L., G.R. LYLES, and F.B. DOWLING. – *Anal. chim. Acta*, **29**:489. 1963.
627. CHALMERS, R.A. and M.A. BASIT. – *Analyst, Lond.*, **92**:680. 1964.
628. CHALUPNY, K. and K. BREISCH. – *Z. angew. Chem.*, **35**:233. 1922.
629. CHANCEL, J.E. – *Z. analyt. Chem.*, **3**:391. 1864.
630. CHARLOT, G. – *Anal. chim. Acta*, **1**:223. 1947.
631. Chemical Analysis Sub-Committee of the British Ceramic Research Association. – *Trans. Br. Ceram. Soc.*, **51**:438. 1952.
632. CHENERY, E.M. – *Analyst, Lond.*, **73**:501. 1948.
633. CHENG, K.L. – *Proc. Intern. Symp. on Microchemistry.* Oxford–London– New York–Paris. 1960. p.465.
634. CHENG, K.L. and R.H. BRAY. – *Analyt. Chem.*, **27**:782. 1955.
635. CHENG, K.L. and F.J. WARMUTH. – *Chemist Analyst*, **48**:74. 1959.
636. CHIACCHIERINI, E. and G. ASCENZO. – *Annali Chim.*, **56**:1485. 1966.
637. CHIRNSIDE, R.C. – *Analyst, Lond.*, **59**:278. 1934.
638. CHIRNSIDE, R.C., C.F. PRITCHARD, and H.P. ROOKSBY. – *Analyst, Lond.*, **66**:399. 1941.
639. CHOLAK, J., D.M. HUBBARD, and R.V. STORY. – *Ind. Engng. Chem. Analyt. Edn.*, **15**:57. 1943.
640. CHWASTOWSKA, J. – *Proc. Analyt. Chem. Conf.* Budapest. 1966. p.45.
641. CHWASTOWSKA, J. and J. MINCZEWSKI. – *Chemia analit.*, **9**:791. 1964.
642. CIMERMAN, C., A. ALON, and J. MASHAL. – *Talanta*, **1**:314. 1958.
643. CLASSEN, A. – *Z. anorg. allg. Chem.*, **142**:257. 1925.
644. CLASSEN, A. and L. BASTINGS. – *Analyst, Lond.*, **92**:614. 1967.
645. CLASSEN, A. and L. BASTINGS. – *Ibid.*, **92**:618. 1968.
646. CLASSEN, A., L. BASTINGS, and J. VISSER. – *Anal. chim. Acta*, **10**:373. 1954.
647. CLENNEL, J.E. – *J. Inst. Metals*, **28**:253. 1922.
648. CLEVELAND, J.M. and P.D. NANCE. – Rep. Congr. Atom. Energy Commn. U.S., REP-53. 1955.
649. CODELL, M. and G. NORWITZ. – *Analyt. Chem.*, **25**:1437. 1953.
650. COLIN, R.H. and D.A. GARDNER. – *Analyt. Chem.*, **21**:701. 1949.
651. COLLAT, J.W. and L.B. ROGERS. – *Analyt. Chem.*, **27**:961. 1955.
652. CONGTON, L.A. and J.A. CARTER. – *Chem. News, Lond.*, **128**:98. 1924; *Chem. ZentBl.*, **95**, 1:2458. 1924.
653. COONEY, B.A. and J.H. SAYLOR. – *Anal. chim. Acta*, **21**:276. 1959.
654. COPPINS, W.C. and W. PRICE. – *Metallurgia*, **48**:149. 1953.

655. CORBETT, J.A. – *Analyst, Lond.,* **78** : 20. 1953.
656. CORBETT, J.A. and B.D. GUERIN. – *Analyst, Lond.,* **91** : 490. 1966.
657. COREY, R.H. and H.W. ROGERS. – *J. Am. Chem. Soc.,* **49** : 216. 1927.
658. COURSIER, J. and J. SAULNIER. – *Anal. chim. Acta,* **14** : 62. 1956.
659. CRAFT, C.H. and G.R. MAKEPEACE. – *Ind. Engng. Chem. Analyt. Edn.,* **17** : 206. 1945.
660. CRIŞAN, I.A. and E. BERCEA. – *Revtă Chim.,* **18** : 370. 1967.
661. CRIŞAN, I.A. and A. LAKATOS. – *Revtă Chim.,* **17** : 557. 1966.
662. CROSS, C.F. – *Chem. News, Lond.,* **39** : 161. 1879.
663. CRUSE, K. and G. NETTESHEIM. – *Z. analyt. Chem.,* **152** : 19. 1956.
664. CUERRIN, G., M.V. SHELDON, and C.N. REILLEY. – *Chemist Analyst,* **49**, No. 2 : 36. 1960.
665. CULP, S.L. – *Chemist Analyst,* **56** : 29. 1967.
666. CURTMAN, L.J. and N. DUBIN. – *J. Am. Chem. Soc.,* **34** : 1485. 1912.
667. CYLLUM, M. – *Can. J. Chem.,* **34** : 915. 1956.
668. GZAJKA, Z. – *Przegl. Odlewn.,* **16** : 169. 1966.
669. CZERNIEG, J. and Z. GREGOROWICZ. – *Zesz. nauk. Politech. śląsk.,* No. 108 : 59. 1964.
670. DAGNALL, R.M., T.S. WEST, and P. JOUNG. – *Analyst, Lond.,* **90** : 13. 1965.
671. DAGNALL, R.M., R. SMITH, and T.S. WEST. – *Talanta,* **13** : 609. 1965.
672. DANNEIL, A. – *Tech.-wiss. Abh. Osram Ges.,* **7** : 350. 1958.
673. DARBEY, A. – *Am. Dyestuff Reptr.,* **42** : 453. 1953.
674. DAS, B. and S. ADITYA. – *J. Indian Chem. Soc.,* **36** : 473. 1959.
675. DAVENPORT, W.H. – *Analyt. Chem.,* **21** : 710. 1949.
676. DEBRAS-GUEDON, J. – *Bull. Soc. fr. Céram.,* No. 62 : 7. 1964.
677. DEBRAS-GUEDON, J. and I. VOINOVITCH. – *Compt. rend.,* **249** : 242. 1959.
678. DEHN, H., G. GRITZNER, and V. GUTMAN. – *Mikrochim. Acta,* No. 3 : 422, 426. 1958.
679. DEMPIR, J. – *Stavivo,* **38** : 28. 1960.
680. DEMPIR, J. – *Stavivo,* **41** : 183. 1963.
681. DETERDING, H.C. and R.G. TAYLOR. – *Ind. Engng. Chem. Analyt. Edn.,* **18** : 127. 1946.
682. DETMUR, D.A. and H.C. ALLER. – *Recl. Trav. chim. Pays-Bas Belg.,* **75** : 1429. 1956.
683. DEY, A.K. – *Mikrochim. Acta,* Nos. 2–4 : 414. 1964.
684. DINNIN, J.I. and G.R. KINSER. – *Prof. Pap. U.S. Ged. Surv.,* No. 424-B : 329. 1961.
685. DIPPEL, C.P. – *Silikattechnik,* **17** : 293. 1966.
686. DOWLING, F.B., C.L. CHAKRABATRI, and G.R. LYCES. – *Anal. chim. Acta,* **28** : 392. 1963.
687. DOZINEL, C. – *Chim. analyt.,* **38** : 244. 1956.
688. DREFAHI, G. and A. GEISSLER. – *Z. analyt. Chem.,* **160** : 34. 1958.
689. DÜBEL, W. and W. FLURSCHÜTZ. – *Chem. Technol.,* **12** : 538. 1960.

690. DUBEY, S.N. and R.C. MEHROTRA. – *J. Indian Chem. Soc.*, **42** : 685. 1965.
691. DUBEY, S.N. and R.C. MEHROTRA. – *J. Inorg. Nucl. Chem.*, **26** : 1543. 1964.
692. DUPUIS, T. and C. DUVAL. – *Anal. chim. Acta*, **3** : 191. 1949.
693. DUTT, N.K. and P. BOSE. – *J. Indian Chem. Soc.*, **30** : 431. 1953.
694. DUTT, N.K. and P. BOSE. – *Z. anorg. allg. Chem.*, **295** : 131. 1958.
695. DUVAL, C. *Inorganic Thermogravimetric Analysis*, p.105. Amsterdam. 1953.
696. DVOŘÁK, J. and E. NYKETOVÁ. – *Mikrochim. Acta*, No. 6 : 1082. 1966.
697. ECKERT, G. – *Z. analyt. Chem.*, **153** : 261. 1956.
698. EDWARD, W.T. – *Analyst, Lond.*, **73** : 556. 1948.
699. EEGRIEWE, E. – *Z. analyt. Chem.*, **76** : 438. 1929; **108** : 268. 1937.
700. ELLIOT, C. and J.W. ROBINSON. – *Anal. chim. Acta*, **13** : 209, 235. 1955.
701. ERDEY, L. and F. PAULIK. – *Acta chim. hung.*, **7** : 45. 1955.
702. ERWARDS, W.T. – *Analyst, Lond.*, **73** : 556. 1948.
703. ESHELMAN, H.C., A.J. DEAN, O. MINES, and T.S. RAINS. – *Analyt. Chem.*, **31** : 183. 1959.
704. EVANS, H.B. – *Analyst, Lond.*, **92** : 685. 1967.
705. EVANS, H.B. and H. HASHITANI. – *Analyt. Chem.*, **36** : 2032. 1964.
706. EVELETH, D.F. and V.C. MYERS. – *J. Biol. Chem.*, **113** : 449. 1936.
707. FALCHI, G. and F. TONANI. – *Periodico Miner.*, **31** : 389. 1962.
708. FAUCHERRE, J., F. FROMAGE, and D. NOIZET. – *Compt. rend.*, **C 262** : 1520. 1966.
709. FEIGL, F. and D. GOLDSTEIN. – *Analyt. Chem.*, **29** : 456. 1957.
709a. FERLIN, C. and B. LELIEVRE. – *Bull. Soc. chim. Fr.*, p. 3415. 1968.
710. FERRARI, C. – *Annali Chim. appl.*, **27** : 479. 1937.
711. FIJALKOWSKI, J. – *Chemia analit.*, **4** : 455. 1959.
712. FILIPOV, D. and N. KIRTCHEVA. – *Doklady. Bulgarska Akademiya na Naukite*, **17** : 467. 1964.
713. FISEL, S., I. GABE, and M. PONI. – *Studii Cerc. Chim., Fil. Iasi Chim.*, **13** : 33. 1962.
714. FISCHER, J. and W. SCHMIDT. – *Z. Erzbergb. Metallhüttwes.*, **9** : 25. 1956.
715. FISCHER, W. and W. SEIDEL. – *Z. anorg. allg. Chem.*, **247** : 333. 1941.
716. FISCHER, W. and N. ZUMBUSCH. – *Z. anorg. allg. Chem.*, **252** : 249. 1944.
717. FITZNER, E. – *Aluminium (BRD)*, **40** : 741. 1964.
718. FLASCHKA, H. and H. ABDINE. – *Mikrochim. Acta*, No. 1 : 37. 1955.
719. FLASCHKA, H. and H. ABDINE. – *Z. analyt. Chem.*, **152** : 77. 1956.
720. FLASCHKA, H. and W. FRANSCHITZ. – *Z. analyt. Chem.*, **144** ; 421. 1955.
721. FLASCHKA, H., K. HAAR, and J. BAZEN. – *Mikrochim. Acta*, No. 4 : 345. 1953.
722. FLECK, H.R. and A.M. WARD. – *Analyst, Lond.*, **58** : 388. 1933.
723. FLOERENCE, T.M. – *Analyt. Chem.*, **34** : 496. 1962.
724. FLOERENCE, T.M. – *Analyt. Chem.*, **37** : 704. 1965.
725. FLOERENCE, T.M. and D.B. IZARD. – *Anal. chim. Acta*, **25** : 386. 1961.
726. FLOERENCE, T.M., F.J. MILLER, and H.E. ZITTEL. – *Analyt. Chem.*, **38** : 1065. 1966.

727. FOGLINO, M.L. and G.P. SPAGLIARDI. − *Metallurgia ital.*, **50** : 372. 1958.
728. FOLIN, O. and V. CIOCALTEU. − *J. Biol. Chem.*, **73** : 627. 1927.
729. FORNASERI, M. and B. TURI. − *Metallurgia ital.*, **58** : 245. 1966.
730. FOURNE, L., N. DESCHAMPS, and P. ALBERT. − *Compt. rend.*, **254** : 1640. 1962.
731. FRAME, H.D., H.H. STRAIN, and J. SHERMA. − *Analyt. Chem.*, **34** : 170. 1962.
732. FRERU, J.N. − *Z. analyt. Chem.*, **95** : 119. 1933.
733. FRESENIUS, R. and G. JANDER. − *Handbuch der analytischen Chemie*, Part 3, **111** : 310. Berlin. 1942.
733a. FREUND, H. and F. MINER. − *J. Analyt. Chem., Wash.*, **25** : 564. 1953.
734. FRICKE, R. − *Kolloidzeitschrift*, **49** : 241. 1929.
735. FRICKE, R. and K. MEYRING. − *Z. anorg. allg. Chem.*, **188** : 127. 1930.
736. FRIEDHEIM, C. and P. HASENCLEVER. − *Z. analyt. Chem.*, **44** : 606. 1965.
737. FRIESE, G. − *Z. angew. Geol.*, **6** : 279, 615. 1960.
738. FRITZ, J.S., J.E. ABBINK, and M.A. PAYNE. − *Analyt. Chem.*, **33** : 1381. 1961.
739. FROMAGE, F. − *Ann. Univ. Assoc. rég. étude et rech. scient.*, **2** : 8. 1963−1964.
740. FUELTON, J.W. and J. HESTINGS. − *Analyt. Chem.*, **28** : 174. 1956.
741. FUNK, H. and H. WAITER. − *Z. analyt. Chem.*, **67** : 88. 1924−1925.
742. FURLANI, D.A. − *Gazz. chim. ital.*, **90** : 1380. 1960.
743. GAD, G. and K. NAUMAN. − *Gas-u. WassFach.*, **80** : 58. 1937.
744. GAGE, D.G. − *Analyt. Chem.*, **28** : 1773. 1956.
745. GAMSJAGER, H. and E. SCHWARZ-BERGKAMPF. − *Mikrochim. Acta*, Nos. 1−2 : 194. 1962.
746. GARZON, R.L. − *Infción. Quim. analit.*, **14** : 64. 1960.
747. GASSNER, K. − *Z. analyt. Chem.*, **152** : 417. 1956.
748. GEGUS, E. − *Acta chim. hung.*, **28** : 65. 1961.
749. GELIS, P. − *Chim. analyt.*, **49** : 30. 1967.
750. GENTRY, C.H.R. and L.G. SHERRINGTON. − *Analyst, Lond.*, **71** : 432. 1946.
751. GERMAN, A. and G. GLANTĂ. − *Revtă Chim.*, **14** : 424. 1963.
752. GEYER, R. and R. BORMANN. − *Z. Chemie*, **7** : 30. 1964.
753. GHEOCALESCU, Ş. − *Metallurgia si constr. mas.*, **12** : 247. 1960.
754. GIEBLER, G. − *Z. analyt. Chem.*, **184** : 401. 1961.
755. GILCHRIST, R. − *J. Res. Natn. Bur. Stand.*, **30** : 89. 1943.
756. GINSBURG, L., K. MILLAR, and L. GORDON. − *Analyt. Chem.*, **29** : 46. 1957.
757. GIUFFRÉ, L. and F.M. CAPIZZI. − *Annali Chim.*, **49** : 1834. 1959.
758. GLASNER, A. and S. SARAH. − *Israel J. Chem.*, **3** : 143. 1965.
759. GLEMSER, O., E. RAULF, and K. GIESEN. − *Z. analyt. chem.*, **141** : 86. 1954.
760. GLOOR, K. and R. PENTSCH. − *Revue Matér. Constr. Trav. publ.*, No. 470, 304. 1954.
761. GMELIN, L. *Handbuch der anorganischen Chemie*, System No. 35. *Aluminium*, Part A. Sec. 1. Berlin. 1934−1935; Part B. 1934.
762. GÖKE, G. *Lab. −Praxis*, **11** : 37. 1959.

763. GOLDSTEIN, G., D.L. MANNING, and O. MENIS. – *Talanta,* 2 : 52. 1959.
764. GOLDSTEIN, G., D.L. MANNING, and H.E. ZITTEL. – *Analyt. Chem.,* 35 : 17. 1963.
765. GOINA, T. and I. RISTCA. – *Studia Univ. Babeş Bolyai, Ser. Chem.,* 10 : 65, 69. 1965.
766. GOOCH, F.A. and F.S. HAVENS. – *Am. J. Sci.,* 2 : 416. 1896.
767. GOON, E., J. PETLEY, W.H. McCULLEN, and S.E. WIBERLEY. – *Analyt. Chem.,* 25 : 608. 1953.
768. GORDON, L. and L. GINSBURG. – *Analyt. Chem.,* 29 : 38. 1957.
769. GOTKOWSKA, A. – *Chemia analit.,* 10 : 749. 1965.
770. GOTÔ, H. – *Sci. Rep. Tôhoku Univ.,* 26 : 391. 1937.
771. GOTÔ, H. – *Ibid.,* 26 : 418. 1938.
772. GOTÔ, H., Y. KAKITA, and M. HOSOYA. – *J. Japan Inst. Metals,* 24 : 32. 1960.
773. GOTTSCHALK, G. – *Z. analyt. Chem.,* 172 : 192. 1960.
774. GOVINDARAJU, M.K. – *Publs. Group. Avanc. Meth. Spectrogr.,* pp. 221, 208. July, Sept. 1960.
775. GRANGER, C.O. Developm. and Engng. Group. U.K. Atomic Energy Author. Rept. No. 219 (c). 1960.
776. GRANT, J. – *J. Appl. Chem.,* 14 : 525. 1964.
777. GREEN, H. – *Metallurgia,* 57 : 157. 1958.
778. GREEN, H. – *Ibid.,* 71 : 243. 1965.
778a. GREGOROWICZ, Z. – *Chemia analit.,* 3 : 783. 1959.
779. GRIMALDI, F.S. and H. LEVINE. – *Bull. Geol. Surv.,* No. 992 : 39. 1953.
780. GROOT, C., R.M. PEEKEMA, and V.H. TROUTNER. – *Analyt. Chem.,* 28 : 1571. 1956.
780a. GROSSKREUTZ, W., D. SCHULTZE, and V.T. WILKE. – *Z. analyt. Chem.,* 232 : 278. 1967.
781. GUERRESCHI, L. and R. ROMITA. – *Ricerca scient.,* 27 : 3361. 1957.
782. GUTMANN, V. and G. SCHÖBER. – *Mikrochim. Acta,* No. 5 : 959. 1962.
783. GWYER, A.G.C. and N.D. PULLEN. – *Analyst, Lond.,* 57 : 701. 1932.
784. HAAR, K. and J. BAZEN. – *Anal. chim. Acta,* 10 : 23. 1954.
785. HAAR, K. and UMLAND. – *Z. analyt. Chem.,* 191 : 81. 1962.
786. HABRECHTE, M. – *Hutn. Listy,* 18 : 138. 1963.
787. HADORN, H. – *Mitt. Geb. Lebensmittelunters u. Hyg.,* 48 : 314. 1934.
788. HALASZ, A., A. JÁNOSI, and K. VILLÁNYI. – *Veszpr. vegyip. Egy. Közl.,* 5 : 151. 1961.
788a. HAMMET, L.P. and S.T. SOTTERY. – *J. Am. Chem. Soc.,* 47 : 142. 1925.
789. HARÁNCZYK, C. and E. KOGUT. – *Szkło i Ceram.,* 8 : 318. 1957.
790. HATFIELD, W.D. – *Ind. Engng. Chem.,* 16 : 233. 1924.
791. HAYWOOD, F.W., F. HARRISON, and A.R. WOOD. – *J. Soc. Chem. Ind., Lond.,* 62 : 187. 1943.
792. HAZAN, I., F. FEIK, and J. KORKISCH. – *Z. analyt. Chem.,* 210 : 171. 1965.
793. HECZKO, T. – *Chemikerzeitung,* 58 : 1032. 1934.
794. HEEB, R. – *Z. analyt. Chem.,* 170 : 95. 1959.

795. HEGEDÜS, A.J. – *Kohász. Lap.*, **9** : 335. 1954.
796. HEGEDÜS, A.J. – *Mikrochim. Acta*, Nos. 5–6 : 831. 1963.
796a. HEGEMANN, F. and V. CAIMANN. – *Glas-u. Hochvak.-Tech.*, **29** : 239. 1956.
797. HEGEMANN, F. and W. HERT. – *Ber. dt. keram. Ges.*, **35** : 258. 1958.
798. HEGEMANN, F., W. HERT, and W. SMIDT. – *Glas.-u. Hochvak.-Tech.*, **31** : 81. 1958.
799. HEGEMANN, F. and O. OSTERRIED. – *Ber. dt. keram. Ges.*, **40** : 424. 1963.
799a. HEGEMANN, F. and C. SYBEL. – *Glas.-u. Hochvak.-Tech.*, **28** : 190, 307. 1955.
800. HEGEMANN, F. and G. WILK. – *Ber. dt. keram. Ges.*, **39** : 483. 1962.
801. HÉJJA, A. – *Vegyip. kut. Intéz. Közl.*, **1** : 8. 1953.
802. HENRY, S. and P. HANISET. – *Industrie chim. belge*, **27** : 24. 1962.
803. HERMAN, P., Z. HAINSKI, J. LOUVERIER, and I.A. VOINOVITCH. – *Publs. Group. Avanc. Méth. Spectrogr.*, p. 455. Oct.–Dec. 1961.
804. HESS, W.H. and E.D. CAMPBELL. – *J. Am. Chem. Soc.*, **21** : 776. 1899; *Chem. ZentBl.*, **70**, 11 : 631. 1899.
805. HEYROVSKÝ, J. – *J. Chem. Soc.*, **117** : 20. 1920.
806. HEYROVSKÝ, J. – *Colln. Czech. Chem. Commun.*, **25** : 3120. 1960.
807. HIBBS, L.E. and D.H. WILKINS. – *Talanta*, **2** : 16. 1959.
808. HILL, U.T. – *Analyt. Chem.*, **28** : 1419. 1956.
809. HILL, U.T. – *Ibid.*, **31** : 429. 1959.
810. HILL, U.T. – *Ibid.*, **38** : 654. 1966.
811. HINEK, R.J. and L.J. WRANGELL. – *Analyt. Chem.*, **28** : 1520. 1956.
812. HINRICHSEN, W. – *Z. anorg. allg. Chem.*, **58** : 83. 1908.
813. HOLADAY, D.A. – *J. Am. Chem. Soc.*, **62** : 989. 1940.
814. HOLLER, A.C. and J.P. YEAGER. – *Ind. Engng. Chem. Analyt. Edn.*, **14** : 719. 1942.
815. HOLLINGSHEAD, R.G.W. *Oxine and its Derivatives*, Vol. 1. London. 1954; Vol. 3. 1956.
816. HOLZBECHER, Z. – *Chemické Listy*, **47** : 680. 1953.
817. HOLZBECHER, Z. – *Ibid.*, **52** : 430, 1822. 1958.
817a. HOLZBECHER, Z. and P. PULKRAB. – *Colln. Czech. Chem. Commun.*, **27** : 1142. 1962.
818. HOPPER, G. *Lab. –Praxis*, **8**, No. 11 : 122. 1956.
819. HORTON, A.D. and P.F. THOMASON. – *Analyt. Chem.*, **28** : 1326. 1956.
820. HOSOYA, M., Y. KAKITA, and H. GOTÔ. – *Sci. Rep. Res. Insts. Tôhoku Univ.*, **A13**, No. 4 : 206. 1961.
821. HOTYNSKA, B. – *Rudy Metale Niezel.*, **10** : 54. 1965.
822. HOUDA, M., J. KÖRBL, V. BAŽANT, and R. PŘIBIL. – *Chemické Listy*, **51** : 2259. 1957.
823. HOUDA, M., J. KÖRBL, V. BAŽANT, and R. PŘIBIL. – *Colln. Czech. Chem. Commun.*, **24** : 700. 1959.
824. HOUGHTON, G.U. – *Analyst, Lond.*, **68** : 208. 1943.
825. HOWICK, L.C. and J.L. JONES. – *Talanta*, **8** : 445. 1961.

826. HOWICK, L.C. and J.L. JONES. − *Talanta*, 9 : 1037. 1962.
827. HOWICK, L.C. and W.W. TRIGG. − *Analyt. Chem.*, 33 : 302. 1961.
827a. HOZDIC, C. − *J. Ass. Off. Analyt. Chemists*, 49 : 1187. 1966.
828. HUMMEL, R.A. and E.B. SANDELL. − *Anal. chim. Acta*, 7 : 308. 1952.
829. HUNTER, A.H. and N.T. COLEMAN. − *Soil Sci.*, 90 : 214. 1960.
830. HYMAN, H.M. − *Appl. Spectrosc.*, 16 : 129. 1962.
831. IKENBERRY, L.C. and A. THOMAS. − *Analyt. Chem.*, 23 : 1806. 1951.
832. IMELIK, B., M.V. MATHIEU, M. PRETTRE, and S. TECHNER. − *J. Chem. Phys.*, 51 : 651. 1954.
833. Ind. Group U.K. Atomic Energy Author., IGO-AM/S-159. 1959.
834. IRVING, R.J. − *Talanta*, 12 : 1046. 1965.
835. ISHIBASHI, M., T. FUJINAGA, and M. SATO. − *Bull. Inst. Chem. Res. Kyoto Univ.*, 37 : 267. 1959.
836. ISHII, H. and H. EINAGA. − *Bull. Chem. Soc. Japan*, 39 : 1721. 1966.
837. ISHIMORI, T., K. KIMURA, T. FUJINO, and H. MURAKAMI. − *J. Atom. Energy Soc. Japan*, 4 : 117. 1962.
838. IWAMOTO, R.T. − *Anal. chim. Acta*, 19 : 272. 1958.
839. JACKSON, H. and D.S. PHILLIPS. − *Analyst, Lond.*, 87 : 712. 1962.
840. JACKSON, W.A. − *J. Agric. Fd. Chem.*, 7 : 628. 1959.
841. JANDER, G. and O. RUPERTI. − *Z. anorg. allg. Chem.*, 153 : 233, 253. 1926.
842. JANDER, G. and R. VEBER. − *Z. anorg. allg. Chem.*, 131 : 266. 1923.
843. JARMAN, L. and M. MATTIC. − *Talanta*, 9 : 219. 1962.
844. JAXA-BYKOWSKA, E. and W. JAXA-BYKOWSKI. − *Acta chim. hung.*, 30 : 335. 1962.
845. JEAN, M. − *Anal. chim. Acta*, 10 : 526. 1954.
846. JEWSBURY, A. and G.H. OSBORN. − *Anal. chim. Acta*, 3 : 642. 1949.
847. JILEK, A. and J. LUKAS. − *Colln. Czech. Chem. Commun.*, 2 : 63. 1930; *Chem. ZentBl.*, 101, I : 3466. 1930.
848. JOHANNSEN, W., E. BOBOWSKI, and P. WEHBER. − *Metall.*, 10 : 211. 1956.
849. JOKELLEOVÁ, D. − *Silikáty*, 11 : 37. 1967.
850. JONES, I.G. and G. PHILLIPS. − *Analyt. Meth. A.E.R.E.*, No. 12 : 2879. 1960.
851. JUNG, E. − *Z. Pfl. Ernähr. Düng. Bodenk.*, 26 : 1. 1932.
852. JUNGREIS, E. and A. LERNER. − *Anal. chim. Acta*, 25 : 199. 1961.
853. JURCZYK. J. − *Chemia analit.*, 10 : 441. 1965.
854. JURCZYK, J. − *Z. analyt. Chem.*, 210 : 324. 1965.
855. KADLEC, J. − *Hutn. Listy*, 17 : 59. 1962.
856. KAKITA, J. and G. JOKOHAMA. − *Sci. Rep. Res. Insts. Tôhoku Univ.*, A8 : 332. 1956.
857. KAKITA, J., E. SUDÔ, and M. NAMIKI. − *Ibid.*, A13 : 199. 1961.
858. KAMBARA, T. and H. HASHITANI. − *Analyt. Chem.*, 31 : 567. 1959.
859. KAMEMATO, G. and S. JAMAGISHI. − *Bull. Chem. Soc. Japan*, 36 : 1411. 1963.
860. KAMPF, L. − *Ind. Engng. Chem. Analyt. Edn.*, 13 : 72. 1941.

861. KAR, B.C. – *J. Scient. Ind. Res.,* Sec.B, **13** : 855. 1954.

862. KARCH, Z. – *Szklo i Ceram.,* 7 : 317. 1956.

863. KASSNER, B. and W. AGERMANN. – *Z. Chemie,* 7 : 438. 1967.

864. KASSNER, J.L. and M.A. OZIER. – *Analyt. Chem.,* **23** : 1453. 1951.

865. KASSNER, J.L. and M.A. OZIER. – *J. Am. Ceram. Soc.,* **33** : 250. 1950.

866. KAWAMURA, F. and H. NAMIKI. – *Bull. Fac. Engng. Yokohama Natn. Univ.,*
     8 : 261. 1959.

867. KEATTCH, C.J. – *Talanta,* **11** : 543. 1964.

868. KEKEDY, L. and G. BOLOGH. – *Studia Univ. Babeş Bolyai, Ser. Chem.,* **9** : 101.
     1964.

869. KENYON, O.A. and H.A. BEWICK. – *Analyt. Chem.,* **24** : 1826. 1952.

870. KERIN, D. – *Mikrochim. Acta,* No. 5 : 670. 1964.

871. KINDU, P.C. – *Naturwissenschaften,* **48** : 644. 1961.

872. KINNUNEN, J. and B. MERIKANTO. – *Chemist Analyst,* **42** : 16. 1953.

873. KINNUNEN, J. and B. MERIKANTO. – *Chemist Analyst,* **44** : 50. 1955.

874. KIRBY, J.R., R.M. MILBURN, and J.H. SAYLOR. – *Anal. chim. Acta,* **26** : 458.
     1962.

875. KIRKBRIGHT, G.F., T.S. WEST, and C. WOODWARD. – *Analyt. Chem.,* **37** :
     137. 1965.

876. KISS, T.A. – *Z. analyt. Chem.,* **208** : 334. 1965.

876a. KISS, T.A. and T. DORIC. – *Chemy Ind.,* No. 45 : 1567. 1968.

877. KLEIN, P. and V. SKŘIVÁNEK. – *Chemické Listy,* **56** : 72. 1962.

877a. KLINGER, P. – *Arch. EisenhüttWes.,* **13** : 21. 1939.

878. KLINGER, P., W. KOCK, and G. BLASCHCZYK. – *Angew. Chem.,* **53** : 243.
     1940.

879. KOCH, O.G. – *Mikrochim. Acta,* No. 1 : 92. 1958.

880. KOCH, W. – *Arch. EisenhüttWes.,* **12** : 69. 1938.

881. KOCH, W. – *Stahl u. Eisen,* **58** : 952. 1938.

882. KODZU, T. – *J. Chem. Soc. Japan,* **54** : 682. 1933; *C. A.,* **24** : 5675. 1933.

883. KOHLSCHÜTTER, H. and H. GETROST. – *Z. analyt. Chem.,* **167** : 264. 1959.

884. KOHLSCHÜTTER, H., H. GETROST, G. HOFMANN, and H. STAMM. –
     Z. analyt. Chem., **166** : 262. 1959.

885. KOHLSCHÜTTER, H., S. MIEDTANK, and H. GETROST. – *Z. analyt. Chem.,*
     192 : 381. 1963.

886. KOLLO, C. and N. GEORGIAN. – *Bull. Soc. Rôman.,* **6** : 111. 1924.

887. KOLTHOFF, I.M. – *Chem. Weekblad.,* **24** : 606. 1924.

888. KOLTHOFF, I.M. – *J. Am. Pharm. Ass.,* **17** : 360. 1928.

889. KOLTHOFF, I.M. – *Z. anorg. allg. Chem.,* **112** : 185. 1920.

890. KOLTHOFF, I.M. and J. LINGANE. – *J. Polarography.* New York, Interscience
     Publishers, Inc. 1952.

891. KOLTHOFF, I.M. and E.B. SANDELL. – *J. Am. Chem. Soc.,* **50** : 1900. 1928.

892. KOLTHOFF, I.M., V.A. STENGER, and B. MOSKOVITZ. – *J. Am. Chem. Soc.,*
     56 : 812. 1934.

893. KONOPICKY, K. and W. SCHMIDT. – *Z. analyt. Chem.*, **174** : 262. 1960.
894. KÖRBL, J. and R. PŘIBIL. – *Chemist-Analyst*, **45**, No.4 : 102. 1956.
895. KÖRBL, J. and R. PŘIBIL. – *Chemické Listy*, **51** : 1061. 1957.
896. KÖRBL, J., R. PŘIBIL, and A. EMR. – *Chemické Listy*, **50** : 1440. 1956.
897. KORKISCH, J. and S.S. AHLUWALIA. – *Anal. chim. Acta*, **34** : 308. 1966.
898. KORKISCH, J. and A. FARAG. – *Z. analyt. Chem.*, **166** : 81. 1959.
899. KORKISCH, J. and I. HAZAN. – *Analyt. Chem.*, **36** : 2308. 1964.
900. KOVARIK, M. and M. MOUČKA. – *Z. analyt. Chem.*, **150** : 416. 1956.
901. KRAJINA, A. and J. DOLEŽAL. – *Talanta*, **14** : 1433. 1967.
902. KRASSNOWSKI, O.W. – *Z. analyt. Chem.*, **79** : 175. 1930.
903. KRISHNAMURTY, K. and C. VENKATESWARLU. – *Revue Trav. Chim.*, **71** : 668. 1952.
904. KRISTIANSEN, H. – *Anal. chim. Acta*, **25** : 513. 1961.
905. KŘÍŽ, M. – *Sklář Keram.*, **6** : 140. 1956.
905a. KRLEŽA, F. – *Glasn. hem. tehnol.*, *B i H*, 13–14, 15. 1964–1965.
906. KROUPA, M. – *Hutn. Listy*, **13** : 922. 1958.
907. KUEBLER, W., W.J. SHANEMAN, and J. GALLAGHER. – *Chemist Analyst*, **18** : 6. 1929.
908. KULČICKÝ, I. and F. ŠVÁCHA. – *Colln. Czech. Chem. Commun.*, **23** : 1582. 1958; *Chemické Listy*, **52** : 340. 1958.
909. KURJAKOVIĆ-BOGANOVIĆ, M. and R. PLEPELIC. – *Kemija Ind.*, **11** : 700. 1962.
910. LACROIX, S. – *Anal. chim. Acta*, **1** : 260. 1947.
911. LANDI, M.F. and A. BATTAGLIA. – *Metallurgia ital.*, **59** : 650. 1967.
912. LANDI, M.F. and L. BRAICOVICH. – *Metallurgia ital.*, **54** : 389. 1962.
913. LANGE, B. *Kolorimetrische Analyse*, p. 93. 1952.
914. LANGMYHR, F.J. and H. KRISTIANSEN. – *Anal. chim. Acta*, **20** : 524. 1959.
915. LANGMYHR, F.J. and A.R. STORM. – *Acta chem. scand.*, **15** : 1461. 1961.
916. LANG, R. and J. REIFER. – *Z. analyt. Chem.*, **93** : 161. 1933.
917. LARSSON, A. – *Yernkontorets ann.*, **137**, No.6 : 172. 1953.
918. LASZLOVSZKY, J. – *Mikrochim. Acta*, No.3 : 441. 1962.
919. LEIMBACH, G. – *Ber. dt. chem. Ges.*, **25** : 3161. 1922.
920. LEO, R. and G. KÖNIG. – *Metallurgie u. GiessTech.*, No.3 : 26. 1955.
921. LEPPER, H.A. *Official Methods of Analysis of the Association of Official Agricultural Chemists*, p. 97. 1950.
922. LEWIS, L.L., M.J. NARDOZZI, and L.M. MELNICK. – *Analyt. Chem.*, **33** : 1351. 1961.
923. LEWIS, J.H. and C.A. TAYLOR. – *J. Appl. Chem.*, **8** : 223. 1958.
923a. LIEBMAN, A.M. – *Analyt. Chem.*, **29** : 899. 1957.
924. LIGHTOWLERS, E.C. – *Analyt. Chem.*, **34** : 1398. 1962.
925. LILIE, H. – *Chem. Tech., Berl.*, **9** : 364. 1957.
926. LILIE, H. and H. ROSIN. – *Z. analyt. Chem.*, **160** : 261. 1958.
927. LINNELL, R.H. and T.H. RAAB. – *Analyt. Chem.*, **33** : 154. 1961.

928. LITEANU, C. and M. COSMA. — *Studia Univ. Babeş Bolyai, Chem.,* No. 2 : 63. 1959.
929. LITEANU, C. and I. CRIŞAN. — *Studii Cerc. Chim. Cluj,* 12 : 261. 1961.
930. LITEANU, C., I. CRIŞAN, and C. CALU. — *Revta Chim.,* 10 : 351. 1959.
931. LITEANU, C., I. CRIŞAN, and C. CALU. — *Studia Univ. Babeş Bolyai, Chem.,* No. 2 : 105. 1959.
932. LITEANU, C., I. LUCÁCS, and C. STRUSIEVICI. — *Anal. chim. Acta,* 24 : 200. 1961.
933. LITTLE, I. — *Austral. J. Soil Res.,* 2 : 76. 1964.
934. LONGUYON, I.G. — *Ber. dt. keram. Ges.,* 35 : 155. 1958.
935. LOUNAMAA, N. — *Spectrochim. Acta,* 7 : 356. 1956.
936. LOVASI, J. and L. ZOMBORY. — *Microchem. J.,* 11 : 277. 1966.
937. LUFF, G. — *Chemikerzeitung,* 46 : 366. 1922.
938. LUKE, C.L. — *Analyt. Chem.,* 24 : 1122. 1952.
939. LUKE, C.L. and K.C. BRAUN. — *Analyt. Chem.,* 24 : 1120. 1952.
940. LUNDELL, G.E.F. and H.B. KNOWLES. — *Ind. Engng. Chem.,* 14 : 1136. 1922.
941. LUNDELL, G.E.F. and H.B. KNOWLES. — *J. Am. Chem. Soc.,* 45 : 676. 1923.
942. LUNDELL, G.E.F. and H.B. KNOWLES. — *J. Res. Natn. Bur. Stand.,* 3 : 86, 91. 1929.
943. LUSKER, J.A. and F. SEBBA. — *J. Appl. Chem.,* 15 : 577. 1965.
944. LYDERSEN, D. — *Z. analyt. Chem.,* 139 : 401. 1953.
945. LYLE, S.J. and D.L. SOUTHERN. — *Talanta,* 11 : 1239. 1964.
946. MAGEE, R.J. and I.A.P. SCOTT. — *Talanta,* 3 : 131. 1959.
947. MAJUMDAR, A.K. and K. CHATTERIEE. — *Mikrochim. Acta,* No. 4 : 663. 1967.
948. MAJUMDAR, A.K. and C.P. SAVARIAR. — *Z. analyt. Chem.* 174 : 197, 269. 1960.
949. MAJUMDAR, A.K. and B. SEN. — *Anal. chim. Acta,* 8 : 378, 384. 1953.
950. MALAT, M. — *Anal. chim. Acta,* 25 : 289. 1961.
950a. MALINEK, M. and M. SOUDNÝ. — *Hutn. Listy,* 16 : 358. 1961.
951. MALISSA, H. and H. KOTZIAN. — *Anal. chim. Acta,* 26 : 128. 1962.
951a. MANCHEN, W. — *Neue Hütte,* 1 : 163. 1956.
952. MARCZENKO, Z. — *Chemia analit.,* 4 : 437. 1959.
953. MARCZENKO, Z. — *Ibid.,* 9 : 1093. 1957.
954. MARCZENKO, Z. and M. MOISKI. — *Chemia analit.,* 12 : 1155. 1967.
955. MARCZENKO, Z. and A. STEPIEN. — *Chemia analit.,* 5 : 247. 1960.
956. MAREC, D.J., E.D. SALESIN, and K.L. GORDON. — *Talanta,* 8 : 293. 1961.
957. MARGERUM, D.W., W. SPRAIN, and C.V. BANKS. — *Analyt. Chem.* 25 : 249. 1953.
958. MATHUR, N.K. and S.P. BHARGAVA. — *Indian J. Appl. Chem.,* 1 : 138. 1963.
959. MATYJA, R. — *Chemia analit.,* 8 : 533. 1963.
960. MAXWELL, J.A. and R.P. GRAHAM. — *Chem. Rev.,* 46 : 482. 1950.
961. MAYR, C. and A. GEBAUER. — *Z. analyt. Chem.,* 113 : 200. 1938.

962. McLEAN, E.O., M.R. HEDDLESON, and R.T. BARTLETT. – *Proc. Soil Sci. Soc. Am.,* **22** : 382. 1960.

963. MEHLIG, J.P. and C.J. DERNBACH. – *Chim. analyt.,* **32** : 80. 1943.

964. MENDELOWITZ, A. – *Anal. chim. Acta,* **14** : 235. 1956.

965. MENETRIER, M. – *Mém. scient. Revue Métall.,* **64** : 241. 1967.

966. MEYER, S. and D.G. KECH. – *Mikrochim. Acta,* No. 5 : 720. 1959.

967. MIDDLETON, K.R. – *Analyst, Lond.,* **89** : 421. 1964.

968. MIEHR, W., P. KOCH, and J. KRATZERT. – *Angew. Chem.,* **43** : 250. 1930.

969. MIKULA, J.J. and M. CODELL. – *Anal. chim. Acta,* **9** : 467. 1953.

970. MILLER, C.C. and R.A. CHALMERS. – *Analyst, Lond.,* **78** : 686. 1953.

971. MILLER, L.B. – *Soil Sci.,* **26** : 435. 1928.

972. MILLNER, T. – *Z. analyt. Chem.,* **113** : 83. 1938.

973. MILLNER, T. and K. HORKAY. – *Acta chim. hung.,* **33** : 201. 1961.

974. MILLNER, T. and F. KUNOS. – *Z. analyt. Chem.,* **113** : 102. 1938.

975. MILNER, G.W.C. and J.L. WOODHEAD. – *Anal. chim. Acta,* **12** : 127. 1955.

976. MILNER, G.W.C. and J.L. WOODHEAD. – *Analyst, Lond.,* **79** : 363. 1954.

977. MILNER, G.W.C. and J.L. WOODHEAD. – A.E.R.E. Rep., No. 1400. 1954.

978. MILNER, O.I. and L. GORDON. – *Talanta.* – **4** : 115. 1960.

979. MILNER, F.J., R.P. DEGRAZIO, C.R. FORREY, and T.C. JONES. – *Anal. chim. Acta,* **22** : 214. 1960.

980. MISCICKA, M. and J. BOBER. – *Szklo Ceram.,* **11** : 60. 1960.

981. MISRA, M.K. and R.K. NANÁ. – *J. Indian Chem. Soc.,* **42** : 267. 1965.

982. MOELLER, T. – *Analyt. Chem.,* **22** : 686. 1950.

983. MOELLER, T. – *Ind. Engng. Chem. Analyt. Edn.,* **15** : 346. 1943.

984. MOELLER, T. and A.J. CHONEN. – *J. Am. Chem. Soc.,* **72** : 3546. 1950.

985. MOELLER, T. and SHU-KUNG-CHU. – *J. Inorg. Nucl. Chem.,* **28** : 153. 1966.

986. MOHR, E. – *Chem. Techn.,* **11** : 598. 1959.

986a. MOLKON, P.H.H. and W.A.K. VAN NESTE. – *Anal. chim. Acta,* **39** : 267. 1967.

987. MORIC, G.P. and T.R. SWEET. – *Analyt. Chem.,* **37** : 1552. 1965.

988. MORIC, G.P. and T.R. SWEET. – *Anal. chim. Acta,* **34** : 314. 1966.

989. MOSER, L. – *Monats.,* **53** : 39. 1929.

990. MOSHIER, R.W. and J.E. SCHWARBERG. – *Talanta,* **13** : 445. 1966.

991. MOTOJIMA, K. – *J. Chem. Soc., Japan, Pure Chemistry Section,* **76** : 903. 1955.

992. MOTOJIMA, K. and H. HASHITANI. – *Bull. Chem. Soc. Japan,* **29** : 458. 1956.

993. MOTOJIMA, K., H. HASHITANI, and T. INAHASHI. – *Analyt. Chem.,* **34** : 571. 1962.

994. MOTOJIMA, K. and N. ISHIWATARI. – *J. Nucl. Sci. Technol.,* **2** : 13. 1965.

995. MOYER, H.V. and W.J. REMINGTON. – *Ind. Engng. Chem., Analyt. Edn.,* **10** : 212. 1941.

996. MUKHEDKAR, A.J., S.B. KULKARNI, and K.P. CHAPHALKAR. – *J. Univ. Poona, Sci. Technol.* No. 32 : 47. 1966.

996a. MUKHERJE A.K. and A.K. DEY. – *Z. analyt. Chem.*, **152** : 424. 1956.

997. MUNSHI, K.N. and A.K. DEY. – *J. prakt. Chem.*, **18** : 233. 1962.

998. MURAWLEFF, L. and O. KRASSNOWSKI. – *Z. analyt. Chem.*, **69** : 389. 1926.

999. MURTHY, A.S., T.P. SARMA, and B.S.V. RAGHAVA RAO. – *Z. analyt. Chem.*, **145** : 418. 1955.

1000. MUSIL, J. – *Hutn. Listy*, **23** : 649. 1968.

1001. MUSSAKIN, A.P. – *Z. analyt. Chem.*, **105** : 351. 1936.

1002. MUTAGUTI, M. – *Rep. Cast. Res. Lab. Waseda Univ.*, **No. 6** : 71. 1955.

1003. MYERS, V.C., J.W. MULL, and D.B. MORRISON. – *J. Biol. Chem.*, **78** : 595. 1928.

1004. NAKAGAWA, J., T. TANAKA, and S. HONDA. – *Bull. Govt. – Industr. Res. Inst., Osaka*, **12**, No. 2 : 199. 1961.

1005. NANDA, R.K. and S. ADITYA. – *Indian Chem. Soc.*, **40** : 660. 1963.

1006. NAREBSKI, W. – *Bull. Acad. pol. Sci. Sér. Sci. géol. géogr.*, **10** : 185. 1962.

1007. NAREBSKI, W. – *Kwart. geol.*, **6** : 1. 1962.

1008. NASÄNEN, R. and J. VEIVO. – *Suomen kemistilehti*, **29** : B213. 1956.

1009. NEEB, K.H. – *Z. analyt. Chem.*, **194** : 255. 1963.

1010. NEEB, K.H. – *Z. analyt. Chem.*, **221** : 200. 1966.

1011. NELSON, F., K.M. RUSCH, and K.A. KRAUS. – *J. Am. Chem. Soc.*, **82** : 339. 1960.

1012. NICOLAS, J., P. DOUILLET, and M. QUINTIN. – *Bull. Grpe. fr. Argiles*, **19**, No.1 : 5. 1967.

1013. NIESSNER, M. – *Z. analyt. Chem.*, **76** : 135. 1928.

1014. NIEUWENBURG, C.J. and G. UNTENBROCK. – *Anal. chim. Acta*, **2** : 88. 1948.

1015. NIGAUD, L. – *Chim. analyt.*, **47** : 180. 1965.

1016. NOLL, C.A. and L.J. STEFANELLI. – *Analyt. Chem.*, **35** : 1914. 1963.

1017. NOVÁK, K. and V. MIKA. – *Chemický Prům.*, **13** : 360. 1963.

1018. NOVOTNÝ, M. – *Sklář Keram.*, **16**, No. 7 : 1. 1966.

1018a. NOWICKA-JANKOWSKA, T. and J. MINCZEWSKI. – *Chemia analit.*, **10** : 129. 1965.

1019. NUDPARNI, M.N., M.S. VARDE, and V.T. ATHAWABA. – *Anal. chim. Acta*, **16** : 421. 1957.

1020. NYDAHL, F. – *Talanta*, **4** : 141. 1960.

1021. OCKENDEN, H.M. and J.K. FOREMAN. – *Analyst, Lond.*, **82** : 592. 1957.

1022. ODEBLAD, E. and S. ODEBLAD. – *Anal. chim. Acta*, **15** : 114. 1956.

1023. OEHLMANN, F. – *Chem. Tech., Berl.*, **8** : 544. 1956.

1024. OELSCHLÄGER, W.Z. – *Analyt. Chem.*, **154** : 321,329. 1957.

1025. OHNESORGE, W.E. and A.L. BURLINGAME. – *Analyt. Chem.*, **34** : 1086. 1962.

1026. OKADA, M. – *Rep. Govt. Chem. Ind. Res. Inst., Tokyo*, **58** : 7. 1963.

1027. OKA, J. and A. MURATA. – *J. Chem. Soc., Japan*, **69** : 179. 1948.

1028. OKURA, T., K. GOTO, and T. JOTUYANAGI. – *Analyt. Chem.*, **34** : 582. 1967.

1029. OLSEN, A.L., E.A. GEE, and V. McLENDON. – *Ind. Engng. Chem.*, **18** : 60. 1926.

1030. ONICIU, L. and R. HALAS. – *Studia Univ. Babęs Bolyai, Chem.*, **7**, No. 2 : 7. 1962.

1031. ONICIU, L. and E. SCHMIDT. – *Studii Cerc. Acad RPR*, **11** : 363. 1963.

1032. ONICIU, L., E. SCHMIDT, and I. CADARIU. – *Rev. roumaine Chim.*, **9** : 849. 1964.

1033. OSBORN, C.H. and A. JEWSBURY. – *Anal. chim. Acta*, **3** : 108. 1949.

1034. OSTERTAG, H. and J. CAPPELIEZ. – *Compt. rend.*, **246** : 1550. 1958.

1035. OSTROWSKI, S. – *Zesz. nauk. Politech. gdańsk.*, No. 7 : 59. 1957.

1036. OTOMO, M. – *Bull. Chem. Soc. Japan*, **36** : 809. 1963.

1037. OWEN, A.G. and W.J. PRICE. – *Analyst, Lond.*, **85** : 221. 1960.

1038. OWENS, E.G. and J.H. JOE. – *Analyt. Chem.*, **31** : 384. 1959.

1039. OWENS, E.G. and J.H. JOE. – *Talanta*, **8** : 505. 1961.

1040. PAGE, J.A., D.H. SIMPSON, and R.P. GRAHAM. – *Anal. chim. Acta*, **16** : 199. 1957.

1041. PAKALNS, P. – *Anal. chim. Acta*, **32** : 57. 1965.

1042. PALMER, S.M. and G.T. REYNOLDS. – *Z. analyt. Chem.*, **216** : 202. 1966.

1043. PANDE, C.S. and T.S. SRIVASTRAWA. – *Z. analyt. Chem.*, **173** : 195. 1960.

1044. PARISSAKIS, G. and P.B. ISSOPONLOS. – *Mikrochim. Acta*, No. 1 : 28. 1965.

1045. PARKER, C.A. and A.P. GODDARD. – *Anal. chim. Acta*, **4** : 517. 1950.

1046. PARKER, R.I. – *Metallurgia*, **55** : 103. 1957.

1047. PARKS, T.D. and L. LYKKEN. – *Analyt. Chem.*, **20** : 1102. 1948.

1048. PATROVSKÝ, V. – *Chemické Listy*, **47** : 676. 1953.

1049. PATROVSKÝ, V. and M. HUKA. – *Colln. Czech. Chem. Commun.*, **21** : 1599. 1956.

1050. PATTNAIK, R.K. and S. PANI. – *J. Indian Chem. Soc.*, **38** : 349. 1961.

1051. PAYNE, S.T. – *Light Metals*, **17** : 195. 1954.

1052. PELCH, M. and L. ENGLISCH. – *Soil Sci.*, **57** : 167. 1944.

1053. PELLOWE, E.F. and F.R.F. HARDY. – *Analyst, Lond.*, **79** : 225. 1954.

1054. PENDER, H.W. – *Analyt. Chem.*, **31** : 1107. 1959.

1055. PENTSCHEFF, N.P. and B. EVTIMOVA. – *Doklady Bulgarska Akademiya na Naukite*, **18** : 1127. 1965.

1056. PERKINS, M. and G.F. REYNOLDS. – *Anal. chim. Acta*, **18** : 616, 625. 1958.

1057. PERKINS, M. and G.F. REYNOLDS. – *Anal. chim. Acta*, **19** : 625. 1958.

1058. PHILIP, J.C. and A. BRANDLEY. – *J. Chem. Soc.*, **103** : 795. 1913.

1059. PHILLIPS, J.P., J.F. DEYE, and T. LEACH. – *Anal. chim. Acta*, **23** : 131. 1960.

1060. PHILLIPS, J.P. and L.L. MERRIT. – *J. Am. Chem. Soc.*, **71** : 3984. 1949.

1061. PIGOTT, E.C. – *J. Soc. Chem. Ind., Lond.*, **58** : 139. 1939.

1062. PILGRIM, W.E. and W.R. FORD. – *Analyt. Chem.*, **35** : 1735. 1963.

1063.   PILZ, W. – *Math.-naturwiss. Kl.*, Sec. 11b, **162** : 47. 1953; *Monatsch. Chem.*, **84** : 471. 1953.

1064.   PITSCH, R. and P. LUDWIG. – *Mikrochim. Acta*, No. 6 : 1115. 1964.

1065.   PODDAR, S.N., N.R. SENGUPTA, and J.K. ADHYA. – *Sci. Cult.*, **29** : 258. 1963.

1066.   POETHKE, W. and C. JACKEL. – *Pharmazie*, **19** : 203. 1964.

1067.   POHL, F.A. – *Z. analyt. Chem.*, **142** : 19. 1954.

1068.   POHL, H. – *Aluminium (BRD)*, **38** : 162. 1962.

1069.   POHL, H. – *Z. analyt. Chem.*, **133** : 322. 1951.

1070.   POKRAS, L. and P.M. BERNAYS. – *J. Am. Chem. Soc.*, **73** : 7. 1951.

1071.   POLLAK, F.F. and B.S. PELLOWE. – *Metallurgia*, **16** : 281. 1950.

1072.   POLLOCK, E.N. – *Energia nucl.*, **10** : 496. 1963.

1073.   PODOBNIK, B., M. DULAR, and J. KOROŠIN. – *Mikrochim. Acta*, No. 4 : 713. 1966.

1074.   POOLE, P. and H.D.J. SEGROVE. – *Glass. Technol.*, **39** : 205, 7. 1955.

1075.   PORTA, A. – *Metallurgia*, **50** : 325. 1958.

1076.   POSSIDONI, A.J.F. – *An. Asoc. quim. argent.*, **57** : 96. 1963.

1077.   POVONDRA, P. and F. ELIAŠ. – *Hutn. Listy*, **17** : 665. 1962.

1078.   PRAJZLER, J. – *Colln. Czech. Chem. Commun.*, **3** : 407. 1931.

1079.   PŘIBIL, R., J. CIHALIK, J. DOLEŽAL, V. SIMON, and J. ZYKA. – *Čslka. Farm.*, **2** : 223. 1953.

1080.   PŘIBIL, R., Z. KOUDELA, and B. MATYSKA. – *Chemické Listy*, **44** : 222. 1950.

1081.   PŘIBIL, R., Z. KOUDELA, and B. MATYSKA. – *Colln. Czech. Chem. Commun.*, **16** : 80. 1951.

1082.   PŘIBIL, R. and V. VESELÝ. – *Chemist Analyst*, **54** : 46. 1965.

1083.   PŘIBIL, R. and V. VESELÝ. – *Hutn. Listy*, **18** : 512. **1963**.

1084.   PŘIBIL, R. and V. VESELÝ. – *Talanta*, **9** : 23. 1962.

1085.   PŘIBIL, R. and V. VESELÝ. – *Talanta*, **10** : 233, 383, 1287. 1963.

1086.   PRICE, J.B. and S.T. PAYNE. – *Analyst, Lond.*, **74** : 641. 1949.

1087.   PRITCHARD, D.T. – *Anal. chim. Acta*, **32** : 184. 1965.

1088.   PRITCHARD, D.T. – *Analyst, Lond.*, **92** : 103. 1964.

1089.   PROCIV, D. – *Colln. Czech. Chem. Commun.*, **1** : 95. 1929.

1090.   Product. Group U.K. Atomic Energy Author. Rept., No. 27 (S). 1962.

1091.   Product. Group U.K. Atomic Energy Author. Rept., No. 29S (S). 1962.

1092.   Product. Group U.K. Atomic Energy Author. Rept., No. 390(S). 1964.

1093.   PRUSSIN, S.G., J.A. HARRIS, and J.M. HOLLANDER. – *Analyt. Chem.*, **37** : 1127. 1965.

1094.   PUNGOR, E. and E.E. ZAPP. – *Magy. Kém. Foly.*, **65** : 436. 1959.

1095.   PUNGOR, E. and E.E. ZAPP. – *Z. analyt. Chem.*, **171** : 161. 1965; **197** : 404. 1963.

1096.   QUITTNER, P., A. SIMONITS, and A. ELEK. – *Talanta*, **14** : 417. 1967.

1097.   RADMACHER, W. and W. SCHMITZ. – *Brennst.-Chem.*, **38** : 225. 1957.

1098.   RAINE, P.A. – *Analyst, Lond.*, **74** : 364. 1949.

1099. RAMAKRISHNA, T.V., P.W. WEST, and J.W. ROBINSON. – *Anal. chim. Acta*, 39 : 81. 1967.

1100. RAMAN, N.V. and M.A.R. VACUDEVA. – *J. Scient. Ind. Res.*, BC18, No.12, B537. 1959.

1101. RAO, C.B. and V. VENKATESWARLU. – *Z. analyt. Chem.*, 178 : 277. 1961.

1102. RAY, H.N., S.S. BISWAS, and S. RAY. – *Z. analyt. Chem.*, 228 : 114. 1967.

1103. RAY, P. – *Z. analyt. Chem.*, 86 : 13. 1931.

1104. RAY, P. and A.K. CHATTOPADHYA. – *Z. anorg. allg. Chem.*, 169 : 99. 1928.

1105. REDDI, M.L.N. and U.V. SESHAIAH. – *Indian J. Chem.*, 2 : 34. 1964.

1106. REES, W.T. – *Analyst, Lond.*, 87 : 202. 1962.

1107. REMY, H. *Lehrbuch der anorganischen Chemie*, Vol.1. Leipzig. 1957.

1108. REMY, H. and A. KUHLMAN. – *Z. analyt. Chem.*, 65 : 161, 166. 1924–1925.

1109. REUTEL, C. – *Metall Erz.*, 38 : 170. 1941.

1110. REYNOLDS, C.F. – *Z. analyt. Chem.*, 173 : 24. 1960.

1111. REYNOLDS, G.F. and T.J. WEBBER. – *Anal. chim. Acta*, 19 : 293. 1958.

1112. RHODES, D.F. and W.E. MOTT. – *Analyt. Chem.*, 84 : 1507. 1962.

1113. RIBEIRO, C.M.E. – *Técnica, Barcelona*, 29, No. 370 : 533. 1967.

1114. RICHTER, F. – *Z. analyt. Chem.*, 126 : 426. 1944.

1115. RICHTER, F. – *Z. analyt. Chem.*, 127 : 113. 1944.

1116. RINALDI, F. and P. AGUZZI. – *Metallurgia ital.*, 59 : 655; 1967.

1117. RINCK, E. and P. FESCHOTTE. – *Compt. rend.*, 240 : 1618. 1955.

1118. RINGBOM, A. *Complexation in Analytical Chemistry*. New York, Interscience. 1963.

1118a. RINGBOM, A. and B. WILKMAN. – *Acta chem. scand.*, 3 : 22. 1949; *C.A.*, 43 : 7861. 1949.

1119. ROLFE, A.C., F.R. RUSSEL, and N.T. WILKINSON. – *J. Appl. Chem.*, 1 : 170. 1951.

1120. ROLLER, P.S. – *J. Am. Chem. Soc.*, 55 : 2437. 1933.

1121. ROONEY, R.C. – *Analyst, Lond.*, 83 : 546. 1958.

1122. ROSSOTE, R. – *Chim. analyt.*, 38 : 250. 1956.

1123. ROYEN, H.J. and H. GREEWE. – *Arch. EisenhüttWes.*, 4 : 17. 1930.

1124. ROYEN, H.J. and H. GREEWE. – *Ibid.*, 7 : 517. 1934.

1125. RUBINS, E.J. and G.R. HAGSTROM. – *J. Agric. Fd. Chem.*, 7 : 722. 1959.

1126. RUME, V. – *Chemický Prům.*, 5 : 480. 1955.

1127. RUNGE, E.F. and F.R. BRYAN. – *Appl. Spectrosc.*, 13 : 116. 1959.

1128. RYBA, O., J. CIFKA, D. JEŽKOVÁ, M. MALÁT, and V. SUK. – *Colln. Czech. Chem. Commun.*, 23 : 71. 1958.

1129. SAGRERA, J.L. – *Revta Ciencia apl.*, 21 : 97. 1967.

1130. SAJO, I. – *Acta chim. hung.*, 6 : 333, 251. 1955.

1131. SAJO, I. – *Kohász. Lap.*, 9 : 445. 1954.

1132. SAJO, I. – *Magy. Kém. Foly.*, 59 : 319. 1953.

1133. SAJO, I. – *Magy. Kém. Foly.*, 60 : 269. 1954.

1134. SAJO, I. – *Magy. Kém. Foly.*, 62 : 56. 1956.

1135. SAJO, I. – *Talanta*, 10 : 493. 1963.

1136. SAJO, I., G. POSGAY, and M. HORVATH. − *Femip. Kut. Intez. Közl.*, **6** : 307. 1966.
1137. SAJO, I. and B. SIPOS. − *Z. analyt. Chem.*, **222** : 23. 1966.
1138. SALSKA, S. and S. HELD. − *Chemia analit.*, **3** : 543. 1958.
1139. SAMUELSON, O., L. LUNDEN, and K. SCHRAMM. − *Z. analit. Chem.*, **140** : 330. 1953.
1140. SAMUELSON, O. and B. SJÖBERG. − *Anal. chim. Acta*, **14** : 121. 1956.
1141. SARUDI, I. − *Z. analyt. Chem.*, **163** : 34. 1958.
1142. SASTRI, C.L., G. SRIRAMULA, and R.Bh.S. RAGHAVA. − *J. Scient. Ind. Res.*, **14**, No.4 : B171. 1955.
1143. SAVELLI, C. − *Metallurgia ital.*, **59** : 671. 1967.
1144. SAYLOR, J.H. and J.W. LEDBETTER. − *Anal. chim. Acta*, **30** : 427. 1964.
1145. SAYLOR, J.H. and J.W. LEDBETTER. − *Anal. chim. Acta*, **32** : 398. 1965.
1146. SCHERRER, J.A. and W.H. SMITH. − *J. Res. Natn. Bur. Stand.*, **21** : 105. 1938.
1147. SCHIRM, E. − *Chemikerzeitung*, **33** : 877. 1909.
1148. SCHIRM, E. − *Ibid.*, **35** : 897. 1911.
1149. SCHMIDT, W., K. KONOPICKY, and J. KOSTYRA. − *Z. analyt. Chem.*, **206** : 174. 1964.
1150. SCHMITZ, B. − *Dt. ApothZtg.*, **95** : 637. 1955.
1151. SCHMITZ, W. − *Bergb.-Rdsch.*, **11** : 661. 1959.
1152. SCHOME, S.C. − *Analyst, Lond.*, **75** : 27. 1950.
1153. SCHWARBERG, J.E., R.W. MOSHIER, and J.H. WALSCH. − *Talanta*, **11** : 1213. 1964.
1154. SCHWARZENBACH, G., R. GUT, and G. ANDERREG. − *Helv. chim. Acta*, **37** : 937. 1954.
1155. SCHWARZ-BERGKAMPF, E. − *Z. analyt. Chem.*, **83** : 345. 1931.
1156. SCOOLES, P.H. and D.V. SMITH. − *Analyst, Lond.*, **83** : 615. 1958.
1157. SEIDEL, W. and W. FISCHER. − *Z. anorg. allg. Chem.*, **247** : 367. 1941.
1158. SEN, A.B. and V.B. CHAHAN. − *Z. analyt. Chem.*, **195** : 255. 1963.
1159. SEN, A.B. and S.N. KAPOOR. − *J. prakt. Chem.*, **22** : 314. 1963.
1160. SEUTHE, A. − *Stahl u. Eisen*, **64** : 493. 1944.
1161. SHIOU CHUAN-SUN. − *Analyt. Chem.*, **31** : 1322. 1957.
1162. SHORT, H.G. − *Analyst, Lond.*, **75** : 420. 1950.
1163. SHULL, K.E. − *J. Am. Wat. Wks. Ass.*, **52** : 779. 1960.
1164. SINGH, D.R. − *J. Scient. Res. Banaras Hindu Univ.*, **10** : 152. 1959−1960.
1165. SINGH, D.R. and G.C. SAXENA. − *Indian J. Chem.*, **2** : 251. 1964.
1166. SINGH, K., B. SAHOO, and D. PATNAIK. − *Proc. Indian Acad. Sci.*, **A50** : 129. 1959.
1167. SINGHAL, G.K. and K.N. TANDON. − *Talanta*, **15** : 707. 1968.
1168. SINGLETON, W. and R.C. CHIRNSIDE. − *J. Soc. Glass. Technol.*, **12**, No.45 : 18. 1928.
1169. SIR, Z. and R. PŘIBIL. − *Colln. Czech. Chem. Commun.*, **20** : 871. 1955.
1170. SKŘIVÁNEK, V. and P. KLEIN. − *Rudy*, **11**, No.3 : 89. 1963.

1171. SKŘIVÁNEK, V. and P. KLEIN. – *Z. analyt. Chem.*, **184** : 360. 1961.

1172. SLADE, R.E. – *Z. Elektrochem.*, **17** : 265. 1911.

1173. SLÁMA, J. – *Hutn. Listy*, **19** : 51. 1964.

1174. SLAVIN, W. and D.C. MANNING. – *Analyt. Chem.*, **35** : 253. 1963.

1175. SMALLS, A.A. – *Analyst, Lond.*, **72** : 14. 1947.

1176. SMITH, G.F. and F.W. CAGLE. – *Analyt. Chem.*, **20** : 574. 1948.

1177. SMITH, G.S. – *Analyst, Lond.*, **64** : 577. 1939.

1178. SMITH, H.F. and R.A. ROYER. – *Analyt. Chem.*, **35** : 1098. 1963.

1179. SMITH, W.H., E.E. SAGER, and I.J. SIEWERS. – *Analyt. Chem.*, **21** : 1334. 1949.

1180. SOLAJA, P. – *Chemikerzeitung*, **49** : 337. 1925.

1181. SOLAJA, P. – *Z. analyt. Chem.*, **80** : 334. 1930.

1182. SOMMER, L. and H. NOVOTNA. – *Talanta*, **14** : 457. 1967.

1183. SOSIN, Z. and I. STRZESZEWSKA. – *Chemia analit.*, **9** : 425. 1964.

1184. SPAUSZUS, S. and W. MÜLLER. – *Kohász. Lap.*, **95** : 439. 1962.

1185. SPAUSZUS, S. and C. SCHWARZ. – *Neue Hütte*, **7** : 180. 1962.

1186. SPECKER, H. and H. HARTKAMP. – *Z. analyt. Chem.*, **140** : 353, 1437. 1953.

1187. SPECKER, H., M. KUCHTNER, and H. HARTKAMP. – *Analyt. Chem.*, **142** : 166. 1954.

1188. SPRAIN, W. and C.V. BANKS. – *Anal. chim. Acta*, **6** : 363. 1952.

1189. SRIVASTAVA, S.C., S.N. SINHA, and A.K. DEY. – *J. prakt. Chem.*, **20** : 70. 1963.

1190. SRIVASTAVA, S.N. and MANOHAR. – *J. Indian Chem. Soc.*, **37** : 299. 1960.

1191. STACY, B.D. – *Biochemie*, **56** : 47. 1954.

1192. STAMMLER, M. and A.D. PEGNITZ. – *Metall.*, **13** : 103. 1959.

1192a. STANESCU, G. – *Comun. inst. cercetari chim.*, **1** : 304. 1957–1962.

1193. STARY, J. and J. SMIZANSKA. – *Anal. chim. Acta*, **29** : 545. 1963.

1194. STEELE, M.C. – *Am. Foundryman*, **25** : 56. 1954.

1195. STEELE, M.C. and L.J. ENGLAND. – *Anal. chim. Acta*, **16** : 148. 1957.

1196. STEELE, S.D. and L. RUSSEL. – *Iron Steel, Lond.*, **16** : 182. 1942.

1197. STEFAN, A. and E. TURKIEWICZ. – *Rudy Metale Nieżel.*, **12** : 76. 1967.

1198. STEINBACH, J.F. and H. FREISER. – *Analyt. Chem.*, **26** : 375. 1954.

1199. STENGER, V.A., W.R. KRAMER, and A.W. BESHGETOOR. – *Ind. Engng. Chem.*, **44** : 797. 1952.

1200. STOCK, A. – *Ber. dt. chem. Ges.*, **33** : 548. 1900.

1201. STONE, K.G. and L. FRIEDMAN. – *J. Am. Chem. Soc.*, **69** : 209. 1947.

1202. STAFFORD, N. and P.F. WYATT. – *Analyst, Lond.*, **68** : 319. 1943.

1203. STAFFORD, N. and P.F. WYATT. – *Analyst, Lond.*, **72** : 54. 1947.

1204. STRELOW, F.W.E. – *Analyt. Chem.*, **31** : 1974. 1959.

1205. STRELOW, F.W.E. – *Ibid.*, **33** : 542. 1961.

1206. STRELOW, F.W.E. – *Ibid.*, **35** : 1279. 1963.

1207. STRELOW, F.W.E. – *Jl. S. Afr. Chem. Inst.*, **16**, No. 2 : 38. 1963.

1208.   STUMPF, K.E. – *Z. analyt. Chem.*, **138** : 30. 1953.
1209.   SUDÔ, E. – *J. Chem. Soc. Japan*, **72** : 718. 1951.
1210.   SUDÔ, E. – *Sci. Rep. Res. Insts. Tohoku Univ.*, **A 8** : 375. 1956.
1211.   SÚK, V. and M. MALÁT. – *Chemist Analyst*, **45** : 30. 1956.
1212.   SÚK, V. and V. MIKĚTUKOVÁ. – *Colln. Czech. Chem. Commun.*, **24** : 3629. 1959.
1213.   SWANK, U.W. and A.A. MELLON. – *Ind. Engng. Chem., Analyt. Edn.*, **9** : 406. 1937.
1214.   SZABO, Z. and M. BECK. – *Acta chim. hung.*, **7** : 211. 1954.
1215.   SZARVAS, P., I. KORONDÁN, and I. RAISZ. – *Magy. Kém. Lap.*, **22** : 149. 1967.
1216.   SZÜCS, A.I. and O.N. KLUG. – *Chemia analit.*, **12** : 939. 1967.
1217.   TAHLER, H. and F.H. MÜHLBERGER. – *Z. analyt. Chem.*, **144** : 241. 1955.
1218.   TAIMNI, I.K. and S.N. TANDON. – *Anal. chim. Acta*, **21** : 502. 1959.
1219.   TAKAO, Z. and S. MIYOSHI. – *Tetsu-to-Hagané Overseas*, **1**, No. 2 : 28. 1961.
1220.   TAKASIMA, J. – *J. Chem. Soc. Japan, Pure Chemistry Section*, **80** : 619. 1959.
1221.   TALESNICK, I. and J.A. PAGE. – *Talanta*, **10** : 1055. 1963.
1222.   TANAKA, H. – *J. Soc. Chem. Ind. Japan, Suppl.*, **33** : 489 B. 1930.
1223.   TAYLOR, M.P. – *Analyst, Lond.*, **80** : 153. 1955.
1224.   TEICHER, H. and L. GORDON. – *Analyt. Chem.*, **23** : 930. 1951.
1225.   TEICHER, H. and L. GORDON. – *Anal. chim. Acta*, **9** : 507. 1953.
1226.   TEITELBAUM, M. – *Z. analyt. Chem.*, **82** : 366. 1930.
1227.   THEIS, M.L. – *Z. analyt. Chem.*, **144** : 105. 1955.
1228.   THRUN, W.E. – *Analyt. Chem.*, **20** : 1117. 1948.
1229.   THRUN, W.E. – *Ind. Engng. Chem., Analyt. Edn.*, **2** : 8. 1930.
1230.   TODEASA, A., D. CIOLAN, A. KOVACS, and C. TURCANU. – *Revta Chim.*, **9** : 577. 1958.
1231.   TREADWELL, F.P. *Kurzes Lehrbuch der analytischen Chemie*, Vol. 2. 1930.
1232.   TREADWELL, F.P. and BERNASCONI. – *Helv. chim. Acta*, **13** : 500. 1930.
1233.   TULLO, W., W.J. STRINGER, and G.A.F. HARRISON. – *Analyst, Lond.*, **74** : 296. 1949.
1234.   TURNER, S.E. – *Analyt. Chem.*, **28** : 457. 1956.
1235.   UNDERHILL, E.P. and F.I. PETERMAN. – *Am. J. Physiol.*, **90** : 1. 1929.
1236.   UNDERWOOD, E.E. and A.L. UNDERWOOD. – *Talanta*, **3** : 249. 1960.
1237.   United Kingdom Atomic Energy Authority, Industrial Group IWO–AM/S-159. 1959.
1238.   URUBAY, S., J. KORKISCH, and G.E. JANAUER. – *Talanta*, **10** : 673. 1963.
1239.   VARMA, A. – *Bull. Chem. Soc. Japan*, **35** : 1444. 1962.
1240.   VASSILIADIS, C., C.T. KAWASSIADES, T.P. HADJIOANNON, and G.COLO-VOS. – *Anal. chim. Acta*, **36** : 115. 1966.
1240a.   VECSERNYES, L. – *Magy. Kém. Foly.*, **72** : 372. 1966.
1241.   VETEJŠKA, R. and J. MAZÁČEK. – *Colln. Czech. Chem. Commun.*, **25** : 2245. 1960.

1242. VILLA, L. and S. MAGARIAN. – *Bull. Ass. fr. Chim. Ind. Cuir.*, **20**, No. 1 : 1. 1958.

1243. VOINOVITCH, I.A. – *Chemia analit.*, **7** : 511. 1962.

1244. VOINOVITCH, I.A. and A. LEFRANC-KOUBA. – *Chim. analyt.*, **42** : 543. 1960.

1245. WACYKIEWICZ, K. – *Prace Inst. Ministerstwa Hutnictwa*, **7** : 35. 1955.

1246. WAINER, E. – *J. Chem. Educ.*, **11** : 526. 1934.

1247. WALRAF, M. – *Zem.-Kalk-Gips*, **9**, No. 5 :186. 1956.

1248. WÄNNINEN, E. and A. RINGBOM. – *Anal. chim. Acta*, **12** : 308. 1955.

1249. WEHBER, P. – *Z. analyt. Chem.*, **158** : 321. 1957.

1250. WEIDMANN, H.Z. – *Metall.*, **44** : 565. 1953.

1251. WEINLAND, E.F. and F. ENSGROBER. – *Z. anorg. allg. Chem.*, **84** : 340. 1914.

1252. WEISSLER, A. and C.E. WHITE. – *Ind. Engng. Chem., Analyt. Edn.*, **18** : 530. 1946.

1253. WELLS, J.E. and D.P. HUNTER. – *Analyst, Lond.*, **73** : 67. 1948.

1254. WENGER, P., E. ABRAMSON, and Z. BASSO. – *Helv. chim. Acta*, **29** : 49. 1946.

1255. WENTURELLO, G. – *Annali Chim.*, No. 4 : 46. 1956.

1256. WERNER, O. – *Metall.*, **3** : 146. 1949.

1257. WERNER, O. – *Z. Ver. dt. Chem., Suppl.*, **48** : 92. 1944.

1258. WERZ, W. and A. NEUBERGER. – *Arch. EisenhüttWes.*, **26** : 205. 1955.

1259. WETLESEN, C.V. – *Anal. chim. Acta*, **24** : 294. 1961.

1260. WETLESEN, C.U. and S.H. OMANG. – *Anal. chim. Acta*, **24** : 294. 1961.

1261. WHITE, C.E. and C.S. LOWE. – *Ind. Engng. Chem., Analyt. Edn.*, **12** : 229. 1940.

1262. WHITE, C.E., H.C.E. McFARLONE, J. FOGT, and R. FUCHS. – *Analyt. Chem.*, **39** : 367. 1967.

1263. WHITE, J.C. *Oak Ridge National Laboratory, Paper Presented at Pittsburgh Conference on Analytical Chemistry and Applied Spectroscopy.* Pittsburgh. 1957.

1264. WIBERLEY, S.E. and L.G. BASSET. – *Analyt. Chem.*, **21** : 609. 1949.

1265. WIEBEL, M. – *Z. analyt. Chem.*, **184** : 322. 1961.

1266. WIELE, H. – *Angew. Chem.*, **67** : 126. 1955.

1267. WILKINS, D.H. – *Anal. chim. Acta*, **18** : 372. 1958.

1268. WILKINS, D.H. – *Ibid.*, **23** : 309. 1960.

1269. WILL, F. – *Analyt. Chem.*, **33** : 1360. 1961.

1270. WILLARD, H.H. – *Anal. chim. Acta*, **22** : 1372. 1950.

1271. WILLARD, H.H. and J.A. DEAN. – *Analyt. Chem.*, **22** : 1264. 1950.

1272. WILLARD, H.H. and C.A. HORTON. – *Analyt. Chem.*, **24** : 862. 1952.

1273. WILLARD, H.H. and N.K. TANG. – *Ind. Engng. Chem., Analyt. Edn.*, **9** : 357. 1937; *J. Am. Chem. Soc.*, **59** : 1190. 1937.

1274. WILLIS, J.B. – *Nature*, **207** : 715. 1965.

1275. WILSON, A.D. – *Analyst, Lond.*, **88** : 18. 1963.
1276. WILSON, A.D. and G.A. SERGEANT. – *Analyst, Lond.*, **88** : 109. 1963.
1277. WILSON, H.N. – *Anal. chim. Acta*, **1** : 330. 1947.
1278. WINTER, O.B., W.E. THRUN, and O.D. BIRD. – *J. Am. Chem. Soc.*, **51** : 2721. 1929.
1279. WOOD, J.K. – *J. Chem. Soc.*, **93** : 423. 1908.
1280. WOODWARD, C. and H. FREISER. – *Talanta*, **15** : 321. 1968.
1281. WRANGELL, M.V. and E. KOCH. – *Landw. Jbr.*, **63** : 682. 1926.
1282. YOE, J.H. and W.L. HILL. – *J. Am. Chem. Soc.*, **49** : 2395. 1927.
1283. YOE, J.H. and W.L. HILL. – *Ibid.*, **51** : 2721. 1929.
1284. YOE, J.H. and W.L. HILL. – *Ibid.*, **50** : 748. 1928.
1285. YOKOYAMA, Y. – *Sci. Rep. Res. Insts. Tôhoku Univ.*, **A13** : 8. 1961.
1286. YOUNGDAHL, C.A. and F.E. DE BOER. – *Nature*, **184**, No. 4679 : 54. 1959.
1287. YUAN, T.L. and J.G.A. FISKELL. – *J. Agric. Fd. Chem.*, **7** : 115. 1959.
1288. ZALESSKY, Z. and I.A. VOINOVITCH. – *Indian Ceram.*, No. 532 : 287. 1961.
1289. ZIBULSKY, H., M.F. SLOWINSKI, and J.A. WHITE. – *Analyt. Chem.*, **31** : 280. 1959.
1290. ZIEGLER, M. – *Z. analyt. Chem.*, **180** : 348. 1961.
1291. ZIEGLER, M. and O. GLEMSER. – *Z. analyt. Chem.*, **157** : 19. 1957.
1292. ZULKOWSKY, F. – *Am. Chem. Pharm.*, **202** : 200. 1880.

SUBJECT INDEX